*INTERNATIONAL SERIES OF MONOGRAPHS ON
PURE AND APPLIED BIOLOGY*

Division: **BIOCHEMISTRY**

GENERAL EDITORS:
R. K. CALLOW, F.R.S., P. N. CAMPBELL, S. P. DATTA, L. L. ENGEL

VOLUME 2

THE CHROMATOGRAPHY OF STEROIDS

DEPARTMENT OF BIOCHEMISTRY
JOHNSTON LABORATORIES,
UNIVERSITY OF LIVERPOOL.

OTHER TITLES IN THE SERIES ON PURE AND APPLIED BIOLOGY

BIOCHEMISTRY DIVISION

Vol. 1. PITT-RIVERS AND TATA—*The Thyroid Hormones*
Vol. 3. ENGEL—*Physical Properties of Steroid Hormones*

BOTANY DIVISION

Vol. 1. BOR—*Grasses of Burma, Ceylon, India and Pakistan*
Vol. 2. TURRILL (Ed.)—*Vistas in Botany*
Vol. 3. SCHULTES—*Native Orchids of Trinidad and Tobago*
Vol. 4. COOKE—*Cork and the Cork Tree*

MODERN TRENDS IN PHYSIOLOGICAL SCIENCES DIVISION

Vol. 1. FLORKIN—*Unity and Diversity in Biochemistry*
Vol. 2. BRACHET—*The Biochemistry of Development*
Vol. 3. GEREBTZOFF—*Cholinesterases*
Vol. 4. BROUHA—*Physiology in Industry*
Vol. 5. BACQ AND ALEXANDER—*Fundamentals of Radiobiology*
Vol. 6. FLORKIN (Ed.)—*Aspects of the Origin of Life*
Vol. 7. HOLLAENDER (Ed.)—*Radiation Protection and Recovery*
Vol. 8. KAYSER—*The Physiology of Natural Hibernation*
Vol. 9. FRANÇON—*Progress in Microscopy*
Vol. 10. CHARLIER—*Coronary Vasodilators*
Vol. 11. GROSS—*Oncogenic Viruses*
Vol. 12. MERCER—*Keratin and Keratinization*
Vol. 13. HEATH—*Organophosphorus Poisons*
Vol. 14. CHANTRENNE—*The Biosynthesis of Proteins*
Vol. 15. RIVERA—*Cilia, Ciliated Epithelium and Ciliary Activity*
Vol. 16. ENSELME—*Unsaturated Fatty Acids in Atherosclerosis*

PLANT PHYSIOLOGY DIVISION

Vol. 1. SUTCLIFFE—*Mineral Salts Absorption in Plants*
Vol. 2. SIEGEL—*The Plant Cell Wall*

ZOOLOGY DIVISION

Vol. 1. RAVEN—*An Outline of Developmental Physiology*
Vol. 2. RAVEN—*Morphogenesis: The Analysis of Molluscan Development*
Vol. 3. SAVORY—*Instinctive Living*
Vol. 4. KERKUT—*Implications of Evolution*
Vol. 5. TARTAR—*Biology of Stentor*
Vol. 6. JENKIN—*Animal Hormones*
Vol. 7. CORLISS—*The Ciliated Protozoa*
Vol. 8. GEORGE—*The Brain as a Computer*
Vol. 9. ARTHUR—*Ticks and Disease*
Vol. 10. RAVEN—*Oogenesis*
Vol. 11. MANN—*Leeches (Hirudinea)*

BUSH, IAN ELCOCK
THE CHROMATOGRAPHY OF STEROID
000247551

HCL QP752.S7.B97

BC

STAFF LIBRARY

DUE

LIVERPOOL UNIVERSITY LIBRARY
WITHDRAWN
FROM
STOCK

THE CHROMATOGRAPHY
OF STEROIDS

BY

I. E. BUSH

M.A., PH.D., M.B., B.CH. (*Cantab.*)

BOWMAN PROFESSOR OF PHYSIOLOGY

THE UNIVERSITY OF BIRMINGHAM

DEPARTMENT OF BIOCHEMISTRY
JOHNSTON LABORATORIES,
UNIVERSITY OF LIVERPOOL.

PERGAMON PRESS

OXFORD · LONDON · NEW YORK · PARIS

1961

PERGAMON PRESS LTD.
Headington Hill Hall, Oxford
4 & 5 Fitzroy Square, London W.1

PERGAMON PRESS INC.
122 East 55th Street, New York 22, N.Y.
Statler Center 640, 900 Wilshire Boulevard
Los Angeles 17, California

PERGAMON PRESS S.A.R.L.
24 Rue des Ecoles, Paris V^e

PERGAMON PRESS G.m.b.H.
Kaiserstrasse 75, Frankfurt am Main

Copyright
©
1961
Pergamon Press Ltd.

Library of Congress Card Number 61-9097

Set in Imprint 11 on 12 pt. and printed in Great Britain by
J. W. ARROWSMITH LTD., BRISTOL

TO MY FATHER
who, after all, started me off

ACKNOWLEDGEMENTS

I am grateful to the Editors of the *Biochemical Journal* for permission to publish Figs. 3.2, 5.3 and 6.11: to the Editors of *Helvetica Chimica Acta* and Dr. R. Neher for permission to publish Figs. 3.22 and 3.23: to the Editors of the *Journal of Endocrinology* for permission to publish Figs. 3.3 and 4.3: and to Messrs. J. & A. Churchill, Ltd., London, and Professor C. J. O. R. Morris for permission to publish Fig. 3.8a.

I am also grateful to the following for many useful discussions, and for their generosity in providing unpublished material in advance or for drawing my attention to special features of their publications. Any errors of interpretation are, however, my own responsibility.

Dr. W. Acklin, Dr. H. Dowlatabadi, Dr. L. L. Engel, Dr. A. Izzo, Dr. A. T. James, Dr. E. H. Keutmann, Dr. W. Klyne, Dr. M. Lewbart, Dr. S. Lieberman, Dr. A. G. Long, Professor G. F. Marrian, F.R.S., Dr. R. Neher, Dr. R. Peterson, Professor U. Prelog, Professor T. Reichstein, Dr. L. Reineke, Dr. J. J. Schneider, Dr. & Mrs. J. Tait, F.R.S., Dr. P. Vestergaard, Dr. L. I. Woolf, and Dr. A. Zaffaroni.

CONTENTS

	PAGE
Preface	xi
Glossaries	xiv
1. Symbols and terms	xiv
2. Description of solvent systems	xvii
3. The structure and nomenclature of the steroids	xvii
I. The Basic Theory of Chromatography and Some General Principles	1
1. Introductory	1
2. Derivation of R_F and R_M values	6
3. Solvent–solute interactions	11
4. Chemical structure and chromatographic behaviour	19
5. Partition coefficients and solvent systems	27
6. The effect of temperature	31
7. Formation of several species of solute	31
II. Chromatographic Separation of Steroids	33
Section 1: Descriptive	33
1. Introduction	33
2. Special features of the chromatography of steroids	34
3. Adsorption methods	36
4. Partition chromatography	43
(*a*) General survey and history	43
(*b*) Sources of error in determining R values	68
(*c*) Description of results: relations between the different R functions	74
(*d*) Separation and resolving power	80
Section 2: Quantitative Treatment of Steroid Behaviour on Chromatograms	83
1. ΔR_{Mg} values in typical solvent systems	85
2. ΔR_{Mg} values in atypical solvent systems	96
3. ΔR_{Mg} and R_{Mr} values	102
4. Solvent–solute interactions and ΔR_{Mg} values	118

	PAGE
III. TECHNIQUES AND APPARATUS	133
1. Transfer operations and volumetric error	133
(a) Transfer to adsorption columns	134
(b) Transfer to partition columns	136
(c) Transfer of material to paper chromatograms	138
2. Column chromatography	144
(a) Preparation of adsorption columns	144
(b) Preparation of partition columns	145
(c) Design and operation of columns	148
(d) Multiple-column chromatography	157
3. Paper chromatography	158
(a) Preparation of sheets and strips	158
(b) Impregnation of paper with stationary phases	162
(c) No-touch technique for preparing paper chromatograms	168
(d) Operation of chromatography tanks	170
(e) Ways of running paper chromatograms	173
(f) Temperature control	174
(g) General features of methods of detection	176
(h) Special features of methods of detection	180
(i) The use of dyes and markers	180
(j) Elution of material from paper	182
(k) Preparative paper chromatography	186
4. Preparation of extracts suitable for chromatography	191
(a) General	191
(b) Animal tissues	192
(c) Blood and plasma	193
(d) Sweat, faeces, urine	196
IV. QUANTITATIVE CHROMATOGRAPHY: COLORIMETRIC AND RADIOISOTOPIC TECHNIQUES	201
1. Introduction	201
2. General procedure	203
3. Chromatographic selectivity	208
4. Modifying steps	214
5. Chromatographic technique in quantitative methods	218
6. Direct scanning of paper chromatograms	222
(a) Treatment of paper strips with reagents	224
(b) Solvents and reagents suitable for scanning methods	226
(c) Treatment of strips after the reagent has been applied	229
(d) Optical requirements	234
(e) Geometrical and planimetric methods	238
(f) Electronic recording apparatus	241
(g) Fully automatic apparatus	243

CONTENTS

	PAGE
7. Logistics of quantitative estimations using chromatography	246
8. Use of radioactive isotopes in quantitative chromatography	247
9. Complication and simplicity	253

V. STRUCTURAL ANALYSIS AND IDENTIFICATION OF STEROIDS BY
   CHROMATOGRAPHY — 256
   1. The basis of the classical method — 256
   2. The basis of chromatographic identification — 258
   3. General procedure for structure determination — 262
   4. Examples of structure determination — 264
      (a) Cortisol — 264
      (b) 2α-Me-P^4-6β, 11β, 17α, 20ξ, 21-ol-3-one — 271
      (c) 9α-F-αP-3α, 11β, 17α, 21-ol-20-one — 275
      (d) 18-Hydroxyöestrone — 276
   5. Systematic investigation of unknown steroids — 280
   6. General properties of simple functional groups — 283
      (a) Changes by chemical modification $-\Delta R_{Mr}$ values — 283
      (b) Solvent changes $-\Delta R_{Ms}$ values — 288
   7. Special properties of simple and complex functional groups — 290
      (a) ΔR_{Mr} on modification — 290
      (b) ΔR_{Ms} on change of solvent system — 294
   8. Spectroscopic properties, colour and fluorescence reactions — 297
   9. Future developments — 300
   10. General procedure using ΔR_{Ms} and ΔR_{Mg} values — 304

VI. SOME TYPICAL ANALYTICAL PROBLEMS OF STEROID BIOCHEMISTRY — 311
   1. Introduction — 311
   2. Cortisol in human blood — 311
   3. Oestrogens in human urine — 312
   4. Progesterone in blood or urine — 316
   5. Aldosterone in urine — 317
      (a) Direct fluorimetry on paper — 319
      (b) Fluorimetry in potassium-t-butoxide — 321
      (c) Assay of tritium-labelled acetoxyl groups — 321
   6. Testosterone in blood or urine — 324
   7. Difficult separations of non-polar steroids — 327
   8. Conjugates of urinary steroids — 328
   9. Fractionation and estimation of urinary steroids — 336
      (a) Fraction 1 — 339
      (b) Fraction 2 — 339
      (c) Fraction 3 — 340
      (d) More complete fractionation of urine — 341

	PAGE
(e) Δ^4-3-Ketones	343
(f) Pregnane-3α, 20α-diol	344
(g) Reducing metabolites in the presence of steroid analogues	344
(h) Non-alcoholic steroid ketones	345
10. Conclusion	346

APPENDICES 347
  I. PURIFICATION OF REAGENTS AND MATERIALS 347
   1. Solvents 347
   2. Adsorbents 352
   3. Kieselguhr ("Celite") 353
   4. Cellulose 354
   5. Glassware 355
   6. Reagents 356
  II. MICROCHEMICAL REACTIONS FOR STEROIDS 358
  III. METHODS OF DETECTION ON PAPER CHROMATOGRAMS 372
   1. Selective Absorption of Light 372
   2. Colour and Fluorescence Reactions 373
   3. Detection of Radioactive Material 382
  IV. CALCULATIONS WITH R_F AND R_M VALUES 383
   How to Use the Tables 383

REFERENCES 389
NAME INDEX 409
SUBJECT INDEX 420

PREFACE

THERE are now many books on chromatography and no excuse is needed for most of them, dealing as they do with a technique that has revolutionized biochemistry. On the other hand, scientists are in danger of being overwhelmed by the modern plethora of papers and books on increasingly specialized and esoteric subjects, so that I feel slightly guilty at having accepted the invitation to write the following chapters which are devoted largely to one rather special field of application of chromatographic methods. My main excuse for doing so is that I have been struck for a long time by two features of this field which are probably both largely due to a similar underlying cause: first the small extent to which most practitioners of chromatography in the steroid field have drawn upon the general theory and background of partition chromatography which was built up with other families of organic compounds; and second the negligible extent to which those interested in the general theory of chromatographic behaviour have realized that in many ways the steroids represent an ideal group to study. As a physiologist fumbling my way into the steroid field and forced to work out my own methods it took me six months to realize the force of the first aspect; having broken the back of the technical problem however, the second aspect was lost to view among the many interesting applications that presented themselves and because of which, indeed, the original work had been done. This is probably true of many biochemists; chromatographic methods for particular classes of compound had to be worked out primarily in order to tackle problems which had to be solved. Few had the leisure to enquire into methodology any deeper than was necessary to obtain or copy a workable method.

Therefore, while I hope that this book may be of interest and practical use to those who actually use chromatographic methods in their work with steroids, its principal aim is to try and remedy the situation described above. Thus the steroid specialist may at first be surprised by references to work on fatty acids, flavonoids, sugars and amino acids, which often demonstrate general principles underlying technical details of work with steroids far better than the techniques for steroids themselves. Again, the best example of a general rule must in some cases be drawn from gas–liquid systems although the latter cannot as yet be used for work with the majority of steroids.[1]

In order to preserve a reasonably continuous argument a large number of topics have been relegated to the Appendices. A further problem was the enormous extent of the literature and the vast number of different compounds to be considered. If all the compounds, solvent systems, conditions of operation and R_F values were to be tabulated in the usual way over 100,000 items would have had to be listed. Furthermore, such a list would be of limited use because many of the data, although useful in themselves are not sufficiently accurate to be compared with data from other laboratories. I have attempted to solve this problem by giving a general table of values from which the R_M and hence the R_F value of almost any steroid can be calculated for the aqueous methanol systems (Appendix IV). This together with the information given in Chapter II should enable anyone to work out the approximate R_F value of a new steroid in other types of solvent systems with reasonable accuracy when the R_F values of a few standard steroids under local conditions are known. For similar reasons I have not attempted to cover the whole field descriptively but have drawn examples mainly from the fields I know best in order to illustrate general principles. It is not unfair to observe, however, that these fields do in fact provide the most complete set of results at present on which to base a general theory. Those who have attempted to assemble ΔR_M values from the literature will realize how often an apparently voluminous set of R_F values may yield very few reliable ΔR_M values, because one or two key compounds have not been run in the solvent system under consideration for reasons of practical convenience.

It is easy to become too scholastic over theories of methodology but it is equally easy to become the slave of empirical details because of the absence of a useful general theory. Those looking for cut-and-dried recipes in this book will be disappointed, but I hope that they will be able to overcome any resistance they may have to a somewhat theoretical treatment and be stimulated to apply some of its ideas in practice.

I would like to take this opportunity of expressing my gratitude to those who have supported my own work in this field, particularly to Professor E. D. (now Lord) Adrian, O.M., F.R.S., Sir Charles Harington, F.R.S., Sir George Pickering, F.R.S., F.R.C.P., the Medical Research Council, the Commonwealth Fund and St. Mary's Hospital Medical School. I also

---

[1] See p. 328.

benefited greatly from the hospitality of Dr. L. T. Samuels (Salt Lake City, Utah) and Dr. L. L. Engel (Boston) and from the early encouragement and advice of Professor C. W. Shoppee, F.R.S., Dr. C. S. Hanes, F.R.S., and the late and unique Dr. Konrad Dobriner. Perhaps my greatest debts however are to Dr. P. R. Lewis (Cambridge) who so tactfully guided my early footsteps and made me appreciate the value and ubiquity of theory: and to my early supervisor and present editor, Dr. R. K. Callow, F.R.S., for his invariably timely and objective intolerance.

GLOSSARIES

1. Symbols and Terms

Some symbols and terms which are commonly used in discussing chromatography are given below, but many of the more fanciful neologisms and some well-established terms have been omitted.

Adsorption—The excessive concentration of a solute at an interface. In chromatography the interface is most commonly that between a liquid or a gas and a finely divided solid.

Ascending chromatography—Commonly refers to one of the two principal ways of running paper chromatograms, namely, that in which the paper dips from above into the mobile solvent phase which ascends through the paper by capillary forces.

Descending chromatography—Commonly refers to the other principal method of running chromatograms in which the paper hangs from a trough containing the mobile phase which passes down the paper by reason of both capillary and gravitational forces.

Development—Commonly used in adsorption chromatography to refer to the percolation through a column of a solvent which produces movement of the solutes adsorbed on the column, as distinct from the original solvent with which the solutes were transferred to the column. It is sometimes used in connexion with partition chromatography simply to mean the running of the chromatogram, but it should be avoided because of its other uses such as to mean the treatment of paper chromatograms with reagents giving colour reactions.

Displacement—A change in distribution of a solute between two phases which is produced by the addition of another solute. Displacement in adsorption chromatography is due to competition between the displaced and the displacing solutes for adsorptive points on the solid stationary phase. Displacement conditions are recognized by the emergence of the solute suddenly from the column as a relatively sharp front followed by a constant concentration of the solute in the effluent until the displacer emerges. The same mechanism however is partially responsible for some types of chromatography in which this characteristic of displacement chromatography is not observed.

Effluent—The liquid or gas emerging from a chromatographic system. With paper chromatograms this is often referred to as the "run-off".

Eluate—The solution of a substance which is obtained by allowing a suitable solvent to percolate through a porous medium which originally contained the solute. This is commonly used in connexion with fragments cut from paper chromatograms, but is also used to refer to fractions of effluents and not infrequently as a substitute for effluent itself.

Eluent—The solvent used to obtain an eluate. It thus applies to chromatographic solvents proper, as well as to solvents used to extract fragments of paper, etc.

Elution—The removal of a solute from a porous medium by allowing a suitable solvent to percolate through it. This and the verb "to elute" are used for the process of removing solutes from a chromatographic column in the course of chromatography proper, as well as for the extraction of small segments of a chromatogram.

Gradient elution—A method of running chromatograms in which the concentration of one or more components of the solvent system is changed gradually rather than in discontinuous steps.

Interstitial volume—The volume of fluid or gas in the interstices of a chromatographic system. This usually means the volume of the mobile phase within the system, but is sometimes used to mean the total volume of the column minus the volume of the supporting material: in this case it equals the sum of the volumes of the stationary and mobile phases if a partition system is under consideration. The latter use is avoided in this book.

Ion-exchange—A process whereby ions of the same charge sign replace one another in a given phase. In chromatography, the term usually refers to systems in which the stationary phase is made up of an ionic polymer. This can either be a synthetic resin or a mineral material such as the zeolites and specially treated alumina.

Mobile phase—The moving phase of a chromatographic system, gas or liquid.

Stationary phase—The stationary phase of a chromatographic system. This is made up of the surface of an adsorbent or the liquid or gel held by the supporting material of a partition system. It does not refer to the supporting material itself unless this is known to play an important part in the distribution of the substances under consideration, e.g. by means of adsorption or ion exchange. In many cases, it is referred to without any definite knowledge of its composition.

Partition—The distribution of a solute between two phases which behave as if they were both fluids. It includes both liquid–liquid and liquid–gas systems.

Polar—In organic chemistry this usually means a substance made up of molecules with a large dipole moment. This is not always synonymous with its use in discussions of chromatographic behaviour, in which a polar substance means one which has a distribution coefficient in favour of a polar phase. Since the affinity of substances for polar solvents depends both upon their dipole moments and their molar volumes (among other steric factors) it is clear that polarity in the strict sense will not always be synonymous with this particular sort of solubility property. For instance, benzene will be described as a more "polar" solvent than toluene in chromatographic systems although only the latter has a dipole moment.

R value—A general term to mean any parameter describing the mobility of substances in a chromatographic system in terms of some ratio of mobilities, e.g. R_x usually means the ratio of the mobility of the substance under consideration to that of a reference substance x. Such values are directly proportional to the R_F value of the substance.

R_F *value*—The ratio of the velocity of the substance under consideration to the velocity of the mobile phase in a chromatographic system. It was originally introduced for paper chromatograms and means "R relative to the solvent front".

R_M *value*—This is defined by the equation $R_M = \log(1/R_F - 1)$ and is not referred to as if in the class of R values (q.v.) in this book.

ΔR_M—Literally means a difference between two R_M values. Originally used largely for what are called ΔR_{Mg} values in this book (q.v.).

ΔR_{Mg}—A ΔR_M value due to the simple substitution of another group for a hydrogen atom.

ΔR_{Mr}—A ΔR_M value due to a change of molecular structure other than the simple substitution of a hydrogen atom, e.g. two or more substitutions of hydrogen atoms, epimerization of a substituent, complex degradations, or the result of treatment with a reagent whose action one cannot, or does not wish to, specify, etc. ΔR_{Mg} is thus a special case of ΔR_{Mr}.

ΔR_{Ms}—(a) The difference between the R_M value of a substance in two different solvent systems; or (b) The difference in the ΔR_{Mg} value of a substituent or part of a molecule in two different solvent systems.

(See Chapter 2, Section (2) for further discussion of ΔR_M values.)

Retention volume—The volume of solvent which emerges from a chromatographic system between the introduction of a solute at the start of the system and its emergence at the end of the system. A more useful variable is the "true retention volume" which is the above minus the volume of mobile phase contained in the column. The former phrase is nowadays quite commonly taken automatically to mean the latter but will not be so used in this book.

Reversed-phase—The commonest solvent systems for partition chromatography involve a stationary "more polar" and a mobile "less polar" phase. "Reversed-phase" describes the less common type of system in which the less polar phase forms the stationary phase. In a few cases this is inferred from several properties of the system rather than from a direct knowledge of the phase composition.

"Straight"—This will be used in this book when necessary as the antonym of "reversed-phase".

2. Description of Solvent Systems

The composition of solvent systems is abbreviated in the text where possible as follows:

L	light petroleum
D	decahydronaphthalene
X	xylene
T	toluene
B	benzene
F	formamide
PG	propylene glycol
XYxy/	x vol of X+y vol of Y where X and Y are the less polar constituents of the solvent systems.
/x	x% methanol in water by vol where the total volume equals that of the less polar constituents given in front of the oblique stroke.

/Ax	x% acetic acid in water by vol, the total volume being equal to that of the less polar solvent before the oblique stroke. For example, LT 21/85 means the system made up of light petroleum–toluene–methanol–water (66·6: 33·3: 85: 15 by vol.) (*Note* that the figures before the oblique stroke indicate a ratio and those after it a percentage.)
Group R	Reversed-phase systems suitable for extremely hydrophobic steroids, e.g. sterols and esters of non-polar steroids.
Group A	Any solvent system giving low or moderate R_F values with non-polar steroids, e.g. androstenedione, androsterone, etc.
Group B	Any solvent system giving low or moderate R_F values with moderately polar steroids, e.g. 11-oxygenated 17-oxo-steroids, monoacetates of adrenocortical steroids.
Group C	Any solvent system giving low or moderate R_F values with polar steroids, e.g. cortisol, many cardiac aglycones, etc.
Group D	Any solvent system giving low or moderate R_F values with very polar steroids, e.g. cortol, the less polar cardiac glycosides, etc.

(It is not at present profitable to make further classifications.)

3. The Structure and Nomenclature of the Steroids

An excellent account of this will be found in Klyne (1957) but nomenclature still remains a rather thorny topic. In order to save space and facilitate the reading of the tables, an abbreviated form of nomenclature has been adopted in this book which is based upon the official system based on the Ciba Conference of 1950, but reverts to certain features of the older system. The saving of space will be seen to be considerable, especially with the more complex compounds, and in the author's experience anyone with a working knowledge of conventional nomenclature finds no difficulty in reading the abbreviated form almost at first sight. The other main advantages lie in the adoption of the older order of substituents, which enables a table of compounds to be scanned rapidly for closely related substances; and in the increased information value gained by shedding inessential prefixes and suffixes. The latter cannot be justified on grounds of euphony, and merely add a bogus formal rectitude to names which are of technical rather than literary origin; their information value is nearly always nil.

The abbreviated form of systematic nomenclature used here is as follows:

Parent Compounds

Substitute the capital letters for the standard full names given alongside:

Androstane	A	Cardanolide	D (Digitalis)
Pregnane	P	Bufanolide	B
Cholane	G (Gallensäure)	Spirostan	S
Cholestane	C	Furostan	F
	Oestra–1,3,5(10)–triene	O	

Configuration:

An unnumbered Greek prefix (α, β, or ξ) to the capital denoting the parent refers always and only to configuration at C-5 (e.g. αP for 5α-Pregnane).

20ξ refers to unknown configuration at C-20 in derivatives of C, D, B, S, and F when it precedes the capital.

a or *b* as prefix to the capital indicates the configuration of C-22 in derivatives of S or F.

Assymmetric substituents are indicated in the usual way.

Inversions of configuration at known positions in the parent compounds will be indicated in the usual way by a number, and "epi" as prefix to the capital.

Double bonds:

By analogy with the old nomenclature these will be given as numbers post-superscript to the capital of the parent (i.e. as in the old nomenclature except that the numbers follow the parent capital and the "Δ" is omitted).

Elimination of Carbon Atoms, Ring Enlargement, etc.:

The conventional numbers and prefixes will be used as prefixes to the parent capital (i.e. nor, homo, seco, bisnor).

Substituents:

These will be abbreviated as given below, prefixed by number and Greek letter as in the standard form, but numerating particles will be omitted (e.g. -3α, 17α, 20α-ol and not -3α, 17α, 20α-triol). Halogen and hydrocarbon substituents will precede the capital of the parent in that order; other substituents will follow the capital in alphabetical order.

Ester groups will be numbered and follow the full name of the unesterified parent in brackets (e.g. βP-3α, 11β-ol-(3-OAc) and not βP-3α-OAc-11β-ol).

Methyl	Me	Fluoro	F
Methylene	me	Chloro	Cl
Ethyl	Et	Bromo	Br
Vinyl	Vi	Iodo	I
n–Propyl	Pr	Hydroxyl	ol
Acetate	OAc	Ketone	one
n–Propionate	OPr	Aldehyde	al
etc.		Carboxylic acid	yl
		Epoxide	ox
		etc.	

Examples

Pregnanediol 3α,20α-Dihydroxy-5β-pregnane	βP-3α,20α-ol
Progesterone Pregn-4-ene-3,20-dione	P^4-3,20-one
Cortisone 21-acetate 21-Acetoxy-17α-hydroxypregn-4-ene-3,11,20-trione	P^4-17α,21-ol-3,11,20-one (-21-OAc)
Triamcinolone 9α-Fluoro-11β,16α,17α,21-tetrahydroxypregna-1,4-diene-3,20-dione	9α-F-$P^{1,4}$-11β,16α,17α,21-ol-3,20-one
Cholic acid 3α,7α,12α-Trihydroxy-5β-cholane-24-carboxylic acid	βG-3α,7α,12α-ol-24-yl
Digoxigenin 3α,12β,14-Trihydroxycard-20(22)-enolide	$βD^{20(22)}$-3α,12β,14-ol
Marrianolic acid 3-Hydroxy-16,17-secooestra-1,3,5(10)-triene-16,17-dioic acid	16,17-seco-0-3-ol-16,17-yl

Trivial Names:

In addition to those accepted by the International Union of Pure and Applied Chemistry, the following will be used since they are in common use and largely unambiguous:

Dehydroepiandrosterone	A^5-3β-ol-17-one
Tetrahydrocortisone	βP-3α,17α,21-ol-11,20-one
Tetrahydrocortisol	βP-3α,11β,17α,21-ol-20-one
Allotetrahydrocortisol	αP-3α,11β,17α,21-ol-20-one
Dihydrocortisol	βP-11β,17α,21-ol-3,20-one

(and similarly for the analogous derivatives of corticosterone)

Adrenosterone	A^4-3,11,17-one
Pregnanediol	βP-3α,20α-ol
Cortisol	P^4-11β,17α,21-ol-3,20-one
Cholesterol	C^5-3β-ol
Androstenedione	A^4-3,17-one

They will be compounded with prefixed or suffixed substituents where this seems convenient and unambiguous, e.g. 11β-hydroxytestosterone.

CHAPTER 1

THE BASIC THEORY OF CHROMATOGRAPHY AND SOME GENERAL PRINCIPLES

1. Introductory

THE CHIEF purpose of chromatography is to achieve the separation of closely similar substances. Most separations depend upon the distribution of substances between two phases, which are chosen so as to obtain the maximum difference between the distribution of the desired substance and that of unwanted material called "impurities". Precipitation and crystallization are special cases in which the substance itself becomes a separate phase during the process. The distinctive feature of chromatography is the amplification of the distribution process by dividing the phases into very small elements fixed in space, so that each element of one phase (the moving phase) comes in contact with the elements of the other (the stationary phase) in a fixed order. The continuously successive contacts allow solutes to distribute between the elements of the two phases and to reach equilibrium in a small length of the system. This is achieved by supporting one phase as a finely divided powder in a tube or by using a porous material such as paper, and allowing the other phase to percolate through the first by a pressure difference or by capillary attraction.

It is unnecessary to give here a full account of the theory of chromatography; this is to be found in Martin and Synge (1941), Glueckauf (1949a,b, 1955, 1958), Cassidy (1957), and Martin (1949), among others. However, the main features of this theory must be reviewed, since the steroids form a special case in which it has been realized too seldom that apparently esoteric details of technique are usually the direct consequence of the general theory. Such seemingly specialized techniques are merely a result of certain properties of steroids which are less common or less accentuated with other classes of substances (Bush, 1954b).

The central variable in chromatographic theory, whether of adsorption or partition type, is the distribution coefficient of a substance in the two phases of the system under examination. In adsorption systems this coefficient varies considerably with the concentration of substance being distributed and with the temperature; at any given temperature this can be represented by the function of the quantity of substance in one phase and the total amount in the system, or "adsorption isotherm", which in this case will be more or less strongly curved (Fig. 1.1a). An important feature of adsorption systems is that such isotherms are usually greatly

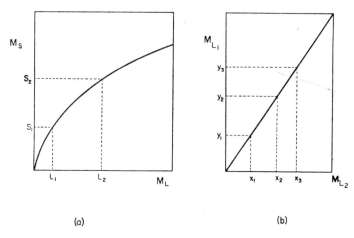

Fig. 1.1. Distribution isotherms. (a) *Curved:* Typical of adsorption systems (solid–liquid, $\alpha_s = M_s/M_L$) with all types of solute, and of some liquid–liquid systems, particularly with ionizing or associating solutes.

$$\alpha_{s,1} = S_1/L_1 \quad \alpha_{s,2} = S_2/L_2$$

(b) *Linear:* Typical of partition systems with all solutes and suitable phases. Seen occasionally with certain adsorption systems and certain solutes (Green and Kay, 1952).

$$\alpha_{L1} = y_1/x_1 = \alpha_{L2} = y_2/x_2 = \text{, etc.}$$

M_x = mass of solute contained in unit weight of phase x.

altered by the presence of other substances in the system; this is a consequence of the limited surface of the solid phase and the predominant role of unimolecular layers in adsorption (Langmuir, 1916). This effect, known as displacement, can occur with substances related or unrelated to the substance under examination. In contrast, the isotherms for most partition (liquid–liquid or gas–liquid) systems are approximately linear over a fairly large range of concentrations and independent of moderate concentrations of other substances (Fig. 1.1b). This is the most important and distinguishing feature of partition systems and the reason for their enormous success.

Many earlier observations were made of the movement of dyes and other materials in cloth, filter paper, or blotting paper (Goppelsroeder, 1909) but were essentially observations of miniature adsorption chromatograms; only when Martin and Synge (1941) demonstrated the importance of the partition principle (i.e. distribution between two liquid phases with approximately linear isotherms) was it possible to extend the technique to its present range.

The term "partition chromatogram" will be used in this book to mean any system in which substances behave "as if" by distribution between two bulk phases, i.e. with linear distribution isotherms, and in which there is presumptive evidence of the existence of two bulk phases; and it will be assumed that the other consequences of an ideal partition system follow from this (Bush, 1954a). It should be emphasized however that in the vast majority of these systems there is no doubt that two bulk phases exist in fact and that more or less simple liquid–liquid distribution occurs. Objections to this idea have been discussed by Martin (1950) and it is worth while summarizing his arguments here since they often seem to be forgotten. They are mainly relevant to cellulose chromatography, but are readily extended to cover any technique where the supporting material swells to form a gel with the solvent used for the stationary phase.

Fully hydrated cellulose contains about 6% by weight of firmly bound water which is taken up with a large heat of sorption, and a further 16% which is taken up with a small heat of sorption. Diffusion in the firmly bound water (the first 6% taken up by dry cellulose) is slow and has a high temperature coefficient, and this water has a high density. The remaining 16% appears to be but little different from ordinary liquid water, having an apparent specific gravity close to 1·0, and showing much higher diffusion constants both for water and other molecules. This loosely bound fraction of the water of hydrated cellulose is in the form of an expanded gel whose properties as a solvent are close to those of water, or, as Martin points out, even more closely resembled by a concentrated solution of a polysaccharide. This water is unevenly distributed, being mainly associated with the amorphous rather than the crystalline regions of the cellulose fibres. The former only make up about 40% of the total fibre in native cellulose, and this proportion varies with different types of paper.

This concept of hydrated cellulose (Hermans, 1949) is both in accord with X-ray and other studies of cellulose structure, and also explains such things as the successful use of "miscible" or "one-phase" solvent systems; the polysaccharide gel can form a separate water-rich phase by a process analogous to that of salting out by strong electrolytes. Similar concepts give adequate explanations of the behaviour of hydrated silica gel, or of a rubberized or acetylated paper loaded with hydrophobic solvents. In the case of the kieselguhrs (e.g. Celites) however, the stationary phase is held on the supporting particles in a different way; in this case it is contained between the microscopic spicules of the individual diatoms by surface tension (Fig. 1.2).

While the liquid–liquid distribution certainly dominates most successful solvent systems designed as partition systems, this is not to say that complications due to adsorption and other factors do not occur. Martin has suggested that the low R_F values of basic amino acids in chromatograms based on cellulose or starch, which are less than expected on the basis of partition coefficients in the pure solvents of a given system, may be due to ion-exchange with carboxyl groups or to association with the partially dissolved polysaccharide molecules of the stationary phase. Adsorption, probably on the crystalline regions of cellulose, is a more serious complication in that the distribution isotherms are often sufficiently curved to cause tailing. In the case of organic phosphates interaction with metallic impurities causes serious anomalies such as double spots and other artifacts (Hanes & Isherwood, 1949), and acetylated amino acids tailed badly on silica gel columns unless adsorption was prevented by small amounts of alcohols in the solvent systems (Martin, 1949). Steroids are adsorbed very strongly from water or hydrocarbon solvents by paper and much larger amounts of alcohols must be used in such solvent systems to achieve true partition systems (Bush, 1952).

If a technique is achieved whereby ideal partition conditions obtain for a class of substances then it can be shown that a given substance moves through the system at a rate determined by its partition coefficient between the phases of the system, the rate of movement of the mobile phase, and the relative cross-sectional areas of the two phases. The original treatment of Martin and Synge (1941), based on the counter-current theory of fractional distillation, used the rate of fall of the mobile phase in the part of the column *above* the stationary phase as a variable. In order to deal with paper chromatograms, Consden, Gordon and Martin (1944) used instead the rate of movement of the solvent front; the same treatment can be used for columns and is in some ways simpler, so that it will be used here. In columns the velocity of the mobile phase can be measured directly by means of a substance having a very high partition coefficient in favour of the mobile phase (Butt, Morris and Morris 1951) or by calculation

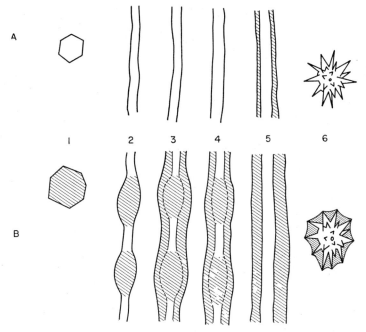

Fig. 1.2. Different types of stationary phase (diagrammatic).
A. Dry state before uptake of solvent.
B. After adsorption or inhibition of stationary phase: shaded area shows gel and/or liquid phase:
1. Silica gel: direct addition of solvent.
2. Cellulose: equilibration with aqueous solvent via vapour phase.
3. Cellulose: impregnation with formamide and good swelling.
4. Cellulose: impregnation with propylene glycol and moderate swelling.
5. Rubber: equilibration via vapour phase, non-polar solvent taken up.
6. Kieselguhr: direct addition of solvent, polar (normal) or non-polar (siliconed).

Unshaded areas in B are crystalline and/or unswollen areas of the cellulose fibre.

when the cross-sectional areas of the phases are known. Movement of the mobile phase will therefore be considered here in terms of its passage through the interstices of the system proper for both columns and paper. Consequently R_F values will be used in the treatment of the great majority of systems.

The most important assumption underlying the theoretical description of the chromatographic process is that equilibrium between the two phases is obtained after each infinitesimal movement of the mobile phase (i.e. the mobile phase is in equilibrium with the stationary phase in each "theoretical plate" by the time it emerges from it). This assumption, drawn from the theory of distillation columns, is based on the "separating-funnel concept" of counter-current processes. It is directly applicable to the units, or "plates", of a Craig machine (Craig and Craig, 1956) but only an approximation for chromatographic systems (Glueckauf, 1955; see p. 80). The effective length in which equilibrium occurs depends upon the diffusion constants of the solutes in both phases, the dimensions of the finely divided elements of the phases, the velocity of the mobile phase, and the temperature. This length is known as the height of an effective theoretical plate (Martin and Synge, 1941). The diffusion coefficient for many solutes varies as approximately the cube root of the molecular weight and has a temperature coefficient of about 1·25 per 10°C (Höber, 1945; p. 16). Ideal conditions will obviously be obtained most easily with elevated temperatures, solutes of low molecular weight, solvents of low viscosity, a finely divided supporting material for the stationary phase, and a low velocity of the mobile phase.

As ideal conditions are approached the limit to the efficiency of the technique is set by the diffusion of solute longitudinally in the system. Far more important limitations in practice are set by inhomogeneities in the phases, lack of uniform velocity of the mobile phase in different parts of the system (e.g. "edging" or "coning"), the spread of material over a finite length of the system during its initial introduction before the run is begun, and excessive concentration of solutes, especially salts, acids, or bases (Martin and Synge, 1941).

2. Derivation of R_F and R_M Values

Martin and Synge (1941) showed that the movement of a solute in a partition system was such as to form a zone in which the concentration/length curve approximated to the curve of normal error as the number of infinitesimal elements through which the zone had passed tended to infinity. The velocity of the position of maximum concentration of a solute, when expressed as a ratio of the velocity of the mobile phase (i.e. R_F value) is given by the equation

$$R_F = \frac{A_M}{A_M + \alpha_S A_S} \qquad (1.1)$$

BASIC THEORY OF CHROMATOGRAPHY

where A_M and A_S are the cross-sectional areas of the mobile (M) and stationary (S) phases, and α_S is the partition coefficient of the solute expressed as concentration in S/concentration in M. This equation is readily arranged to derive a simple and important expression for α_S as

$$\alpha_S = \frac{A_M}{A_S}\left(\frac{1}{R_F} - 1\right) \qquad (1.2)$$

For a given technique A_M/A_S is a constant, so that the ratio of the partition coefficients of two solutes in the system will be directly proportional to the ratio of the expressions in the bracket of equation 1.2 for the two compounds. This function of the R_F value of a solute owes its importance to its direct proportionality to the partition coefficient.

The logarithm of this function is itself of importance because of an expression derived by Martin (1949) relating α_S to chemical structure; an enquiry which led to a novel and extremely significant conclusion. In an ideal solution the chemical potential of a solute A in a phase S is related to its mole fraction in this phase, N_A^S, by the equation

$$\mu_A^S = \mu_A^{S_0} + RT \ln N_A^S \qquad (1.3)$$

where $\mu_A^{S_0}$ is its chemical potential in a standard state. For two phases S and M in equilibrium the chemical potentials of all components are equal so that

$$\mu_A^M - \mu_A^S = 0 \qquad (1.4)$$

and hence

$$\mu_A^M - \mu_A^S = \mu_A^{M_0} - \mu_A^{S_0} + RT \ln N_A^M - RT \ln N_A^S = 0 \qquad (1.5)$$

Calling the difference $(\mu_A^{S_0} - \mu_A^{M_0})$ of the standard chemical potentials in the two phases $\Delta\mu_A$

$$\Delta\mu_A = RT \ln N_A^M - RT \ln N_A^S \qquad (1.6)$$

and hence

$$\Delta\mu_A = RT \ln \left(\frac{N_A^M}{N_A^S}\right) = RT \ln \alpha_A \qquad (1.7)$$

where α_A is in terms of mole fractions; and

$$\ln \alpha_A = \frac{\Delta\mu_A}{RT} \qquad (1.8)$$

where $\Delta\mu_A$ is the free energy needed to move one mole of A from the phase M to phase S.

If it is now assumed that $\Delta\mu_A$ is approximately the sum of the free energies needed to move each constituent group of A from S to M (the crux of Martin's argument) then the addition of a group X to the substance A to form a new substance B should result in the relation

$$\Delta\mu_B = \Delta\mu_A + \Delta\mu_X \tag{1.9}$$

and from equation 1.8 the partition coefficient of the substance B will be given by

$$\ln \alpha_B = \frac{\Delta\mu_A}{RT} + \frac{\Delta\mu_X}{RT} \tag{1.9a}$$

whence (from equation 1.8)

$$\ln \alpha_B - \ln \alpha_A = \frac{\Delta\mu_X}{RT} \tag{1.10}$$

and

$$\ln \frac{\alpha_B}{\alpha_A} = \frac{\Delta\mu_X}{RT} \tag{1.11}$$

Thus the relation between the partition coefficients of two substances differing only by the addition of a group X is determined solely by the solvent employed and the nature of the group X. A misprint in Martin (1949) has unfortunately led to an erroneous version of this equation being given in at least one exposition of this argument.

It will be seen now why the expression in the brackets of equation 1.2 is so useful for the comparison of chromatographic mobilities of different substances. Remembering that A_M/A_S is constant for a given system and putting

$$R_M = \log\left(\frac{1}{R_F} - 1\right)$$

(Bate-Smith and Westall, 1950), we have from equation 1.2

$$\log \alpha = k \cdot R_M \tag{1.12}$$

and hence from equation 1.10 that, after rearrangement

$$2 \cdot 303 k (R_{M_B} - R_{M_A}) = \frac{\Delta\mu_X}{RT} \tag{1.13}$$

BASIC THEORY OF CHROMATOGRAPHY

We can then say that the difference in $R_M(\Delta_n^X R_M)$ caused by introducing n groups X into a substance is given by the equation

$$\Delta_n^X R_M = 2{\cdot}303nk\frac{\Delta\mu_X}{RT} \qquad (1.14)$$

Equation 1.13 implies that the introduction of a group X into *any molecule* causes the same change in R_M value *in a given solvent system*. Equation 1.14 suggests that the introduction of a number of similar groups into one type of molecule gives rise to a change in R_M value directly proportional to the number of such groups. It is commonly supposed intuitively that differences between molecules have effects proportional to the rest of the molecule; e.g. that an additional hydroxyl group in a small molecule will cause a larger change in partition coefficient than the addition of a similar group in a large molecule. Again it is often supposed that two derivatives made by adding a group of high molecular weight will be more difficult to separate than the parent substances; Martin's treatment showed that this was not necessarily so (see Chapter 2).

Martin (1949) himself was careful to point out the limitations of the application of this argument, and we shall see that the steroids provide many interesting examples both of obedience to the rule, and exceptions which are only to be expected if these limitations are kept in mind.

At first sight, however, the steroids appear to disobey these rules almost completely, a fact which is fortunate for the purely practical need to separate the many isomers found in the steroid family. Thus secondary hydroxyl groups in different positions on the steroid nucleus cause very different changes in R_F value (and hence in R_M): differences which are usually but not always the same for several different solvent systems. Savard (1953) first pointed out that the different effects of many epimeric hydroxyl groups in steroids were strictly correlated with their conformations, equatorial hydroxyls always causing a larger change in R_F value than axial hydroxyls (however see p. 91).

It will be seen in Chapter 2, however, that Martin's treatment is not entirely inapplicable. For any given type of solvent system the effect of a substituent in a *given position and orientation* is highly characteristic, in terms of ΔR_M values, when compared with other positions and orientations, and is almost independent of the type of steroid into which such a group is introduced. In other words Martin's treatment is obeyed well by the steroids (ΔR_M values will be considered in detail in Chapter 2) if the word "group" is taken to mean not just primary or secondary hydroxyl, carbonyl, or carboxyl, etc., but also defined by position and orientation. It may be objected that such a restriction on the meaning of "group" so reduces the generality of the theory as to make it useless, but in my view this is not so since this restriction makes the theory workable over a wide range of steroids and of great practical value in the everyday use of chromatography in the elucidation of the structures of unknown compounds.

The theoretical justification for this restriction is also considerable and will be considered in more detail in the next section. At the moment it can be summarized as follows. Martin himself suggested that the main reason for disagreement with the simple treatment above would lie in stereochemical factors unless obvious electrical or bonding interactions between groups occurred. It is reasonable to suppose that such steric factors are more pronounced in the steroids than in almost any other family of organic crystalloids, since the nucleus consists of four rigidly joined rings in which there is no possibility of major conformational inversions and relatively little possibility of minor conformational "bending" (e.g. "chair–boat" transformations). Furthermore the extended ring system provides a large area of hydrocarbon skeleton for interactions with solvent molecules by Van der Waal's (London) forces; because of the relative rigidity of this skeleton and the very short range of these forces, such interactions will be far more sensitive to steric factors than those of molecules in which free rotation and other conformational changes of low energy difference can occur. It will be seen later that this effect can be used to advantage in the selection of solvent systems: some separations that are poor or impossible with aliphatic solvents are easily achieved, for instance, by using an alicyclic or aromatic substitute without changing the other components of the system.

The same theoretical treatment of the chromatographic process can be used for adsorption chromatography whenever the solute concentrations are low enough to be in the linear or nearly linear part of the adsorption isotherm (Martin, 1949). The effect of substituents on distribution coefficients is, however, much more complicated, and the theory becomes very complicated when curved isotherms and displacement have to be considered. Le Rosen, Monaghan, Rivet and Smith (1951) derived an interesting but rather complicated theory of the behaviour of solutes in adsorption systems by assuming that four types of interaction were predominant, namely electron donation and acceptance, hydrogen bonding, and dispersion forces. Constants were set up by taking various adsorbents and solutes as standards and using them to calculate the necessary constants for other adsorbents. Quite good predictions of R_F values were obtained except where salt formation or other special interactions were expected. This theory has not been applied to the steroids and it is doubtful if it would be adequate in view of the importance of steric factors which were not taken into account by Le Rosen's equation. Indeed, the theory broke down quite badly with relatively simple branched and cyclic ketones (Smith and Le Rosen, 1951). Since these theoretical aspects are of less interest in steroid chromatography in its practical uses they will not be dealt with systematically. Some of the more important practical consequences will be discussed in later chapters.

3. Solvent–Solute Interactions

Classical Studies of Solubility

The essence of Martin's theory of partition coefficients is the idea that to a first approximation the chemical potential of a dissolved substance is an additive function of constituent parts of its molecules. The further application of this theory to partition chromatograms by Bate-Smith and Westall (1950), using the R_M function, is a direct consequence of this basic idea. It is worthwhile, therefore, seeing how far this idea is substantiated by straightforward studies of dilute solutions; it will reinforce the validity of the application of such arguments to partition chromatography if it appears that similar additive rules have worked for simple solutions.

Most theoretical treatments of solubility start from some form of Van Laar's theory (1894 *et seq.*) the main feature of which was the consideration of the interactions of a solute molecule with its adjacent solvent molecules on the assumption that solute and solvent molecules were of similar size, and that interactions with molecules outside the first co-ordination shell were negligible. This theory considered the possible configurations of solute and solute molecules and added up the potential energies of these configurations. Interactions beyond the first co-ordination shell may in fact make up 20% of the total energy of solute–solvent interactions, and the theory has also been modified to allow for differences in the size of solvent and solute molecules (Rushbrooke, 1938; Kirkwood, 1939). The theory was extended by many workers and is discussed fully by Hildebrand and Scott (1950) and Moelwyn-Hughes (1957).

One extension of this sort of treatment was to break up the molecular interactions into the interactions between their separate parts (Butler, Ramchandani and Thomson, 1935; Butler and Ramchandani, 1935; Butler and Harrower, 1937; Butler, 1937). These workers in fact made the first thorough experiments showing that the excess free energy of a variety of homologous series dissolved in water was an approximately additive function of the number of methylene groups in the molecules of the solute. Although their treatment has been surpassed in detail and precision by that of Pierotti, Deal and Derr (1959) the results of Butler and his collaborators bear such a close relation to Martin's theory and to many of the findings on partition chromatograms that they are worth considering at some length.

Using the vapour pressure method, Butler et al. (1935) obtained the free energy of hydration of homologous series of ethers, alcohols, ketones and primary amines. The increase in free energy with the addition of each methylene group was approximately constant, and was the same for all four series of compounds. The mean value of ΔF_{CH_2} was $+165$ cal/mole and varied between about 100 and 240 cal/mole for straight chain homologues. The actual free energy of hydration of a solute of any given carbon number depended upon the nature of the polar group. Butler and Ramchandani (1935) then used the structure for water proposed by Bernal and Fowler (1933) and their figure of 6000 cal/mole for the interaction energy between two water molecules to derive approximate figures for the energy of the interactions between water and the constituent groups of the solutes they had studied. Using a treatment similar to Van Laar's they assumed that the excess free energy of the solute was the sum of the number of water–water interactions broken to form a cavity for the solute minus the sum of the (water)–(solute group) interactions by which they were replaced. Using the symbol $\gamma_{x-x'}$ to mean the energy of the interaction between any two groups x and x' in the solute or the solvent:

$$F = n/2 \cdot \gamma_{w-w} - \sum \gamma_{s-w} \qquad (1)$$

where w stands for water, and s for solute molecules.

The main assumption was that, on Bernal and Fowler's picture of the structure of water, a terminal methyl group was surrounded by three water molecules, a methylene group by two, and a terminal polar group by three: and that for every pair of water molecules bounding the cavity holding the solute molecule one water–water bond was broken.

This equation is expanded for different solutes as follows:

$$F_{C_2H_6} = 3\gamma_{w-w} - 6\gamma_{c-w} \qquad (2)$$

for ethane where c stands for a methyl group;

$$F_{EtOH} = 4\gamma_{w-w} - 5\gamma_{c-w} - 3\gamma_{p-w} \qquad (3)$$

for ethanol where p stands for the polar group, in this case a hydroxyl group. The equations for butanol and diethyl ether were:

$$F_{\text{BuOH}} = 6\gamma_{\text{w-w}} - 9\gamma_{\text{c-w}} - 3\gamma_{p\text{-w}} \qquad (4)$$

and

$$F_{\text{EtOEt}} = 6\gamma_{\text{w-w}} - 10\gamma_{\text{c-w}} - 2\gamma_{p\text{-w}} \qquad (5)$$

Knowing the mean figure for ΔF per methylene group, the difference in F for the hydrocarbon and the homologous primary alcohol, and using the figure 6000 cal/mole for $\gamma_{\text{w-w}}$, they then derived the interaction energies for the different groups from these equations as

$\gamma_{\text{c-w}}$ = 2900 cal/mole

$\gamma_{\text{OH-w}}$ = 5000 ,, ,,

$\gamma_{\text{Oether-w}}$ = 4700 ,, ,,

$\gamma_{\text{Oketone-w}}$ = 4500 ,, ,,

$\gamma_{\text{NH}_2\text{-w}}$ = 4900 ,, ,,

Butler et al. (1935) pointed out that the additive property of these interactions was strictly dependent upon steric factors and gave as examples the apparent reduction of the OH−W interactions in ethylene glycol and glycerol. If the deficits were averaged over the hydroxyl groups in these solutes the average $\gamma_{\text{OH-w}}$ values were 4400 for ethylene and 4000 for glycerol. They also showed that while the addition of a methylene group gave a ΔF of 165 cal/mole, the addition of a methyl group to the α-carbon atom (converting a primary to a secondary alcohol) gave a ΔF of 265 cal/mole. There was thus a ΔF of approximately 100 cal/mole between all pairs of primary and secondary alcohols with the same number of carbon atoms. These estimates have been revised by later work (see Barrer, 1957).

This simple treatment was greatly extended and amplified by Pierotti et al. (1959) who took into account a much larger number of interactions and also added terms allowing for the effects of steric "dilution" of certain interactions by enlargement of the hydrocarbon parts of the solute molecule (cf. Kirkwood, 1939). They stress however that they use the same assumptions as Butler et al. (1935), namely that the excess free energy is the resultant of the sum of the energies of the interactions between the separate parts of the solvent and solute molecules. Their basic equation as set out for methylene homologues R_1X_1 dissolved in methylene homologues R_2X_2 is

$$\widehat{F}_1^E/2{\cdot}3RT = \log\gamma_1^0 = A_{12} + B_2n_1/n_2 + C_1/n_1 + D(n_1-n_2)^2 + F_2/n_2$$

where

$\overline{F}_1$ = partial molal free energy of solute 1 at infinite dilution in excess of that predicted by Raoult's Law
γ_1^0 = activity coefficient of solute 1 at infinite dilution in 2
n_1, n_2 = number of carbon atoms in hydrocarbon parts of 1 and 2
A_{12} = coefficient dependent on the nature of functional groups X_1 and X_2
B_2 = coefficient dependent upon nature of X_2
C_1 = coefficient dependent upon nature of X_1
D = coefficient independent of X_1 and X_2
F_2 = coefficient dependent upon nature of X_2.

The first A-term represents the energy of the X_1-X_1, X_1-X_2, and X_2-X_2 interactions in the limiting case of zero carbon number. The B-term allows for the increase in the number of X_2-X_2 interactions that must be broken as n_1 increases and the increase in the number of R_1-X_2 interactions. The C-term allows for the "dilution" of X_1-X_1 interactions as n_1 increases; the F-term similarly allows for the same effect of n_2 on X_2-X_2 interactions. The D-term corrects for the effect of n_1 and n_2 on the hydrocarbon–hydrocarbon interactions, namely R_1-R_1, R_1-R_2 and R_2-R_2.

It is readily appreciated that this type of equation represents a considerable sophistication of earlier work while still being based firmly on the additive principle. For different series of homologues the terms related to the various coefficients are alterable to take into account special steric factors: for example the C-term is expanded to cover the general case of branched and straight chain alcohols to the form $C(1/n_1-1)-(1/n'_1-1)- -(1/n''_1-1)$ where the n's give the carbon number of each hydrocarbon chain, in each case counting from and including the carbon atom linking two branches. For secondary alcohols the last term, and for primary alcohols the last two terms, drop out; and in the last case the original form of the equation results. The same expansion is used for other polar groups substituted as primary, secondary or tertiary functions.

The treatment of Peirotti *et al.* (1959) was applied to the prediction of activity coefficients of an enormous number of solvent–solute combinations at three or four temperatures and is quite extraordinarily successful when the predicted values are compared with experimental results. Most of the predicted values of γ^0 are within 1 or 2% of the observed values over a range of activity coefficients between 1·5 and 10^4. These authors also used their figures to calculate distribution coefficients of dilute solutes between two phases; at infinite dilution these are simply obtained by calculating the ratios of the activity coefficients in the two solvents when these are relatively immiscible.

It seems then that the results of orthodox thermodynamic studies of simple solutions have found an extremely satisfactory mechanistic explanation in terms of some sort of additive treatment of the interactions between the parts of solute and solvent molecules. Furthermore the most recent work has gone on from this to a direct application of the results to the prediction of partition coefficients between two liquid phases. If, then, partition chromatography is really based on partition between two liquid phases Martin's approach follows immediately.

There still seems to be a tendency to reject the simplicity of the mechanism of partition chromatography suggested by Martin and Synge (1941) and Martin (1950) and to appeal to various mechanisms involving adsorption and ion-exchange for the explanation of many phenomena in chromatographic systems, especially on cellulose or starch-supporting media. While such mechanisms undoubtedly play a part in determining the R_F values of certain classes of compound, e.g. basic amino acids (Martin, 1950) they should not be considered as anything other than exceptions to the general rule that partition between two liquid phases is the main mechanism involved in most types of so-called partition chromatography. In the case of the steroids, where steric factors are extremely important and require a slight modification of the approach, it is particularly tempting to appeal to complicated interactions with cellulose in order to explain the mobilities of steroids on paper chromatograms; in certain cases it will be seen that this is reasonable (p. 67), but again this appears to be the exception rather than the rule. Later on, an attempt will be made to give a mechanistic explanation of some of the subtle effects seen with closely related steroids in terms of solute–solvent interactions. Since it may seem that some of these are rather more specific than could be expected for interactions with a liquid phase, it is useful to consider the results of orthodox studies of activity coefficients in order to show that such subtle interactions are in fact well known to occur in liquid phases.

Complex Interactions in Liquid Phases

Even relatively sophisticated treatments of solubility such as that of Pierotti *et al.* (1959) only make a statistical allowance for steric factors. Thus the C and F terms of their equation, which express the increasing interference with the mutual interaction of functional groups as the hydrocarbon radical increases in size, assume that these interactions and their interference are randomly oriented. Even in the simple case of branched carbon chains (as with secondary and tertiary alcohols) it was necessary to expand the C term, and the interactions with each branch were then considered as random. In general this sort of treatment, and other statistical treatments of molecular interactions, work very well for molecules which are simple and flexible. They tend to break down as soon as very polar molecules, or assymmetrical molecules with conformational rigidity, are considered. It is of great interest that such deviations from simple theories are very common with molecules containing cyclic structures. Thus cyclohexane–benzene mixtures show pronounced deviations which were not accounted for by equations which do make some allowance for molecular shape and non-random interactions (Scatchard, Wood and Mochel, 1939). Similarly, Pierotti, *et al.* (1959) found it best to consider aromatic rings as single functional groups rather than as the sum of their elements. This worked well for n-alkylbenzenes but broke down badly with dicyclohexylbenzene; furthermore, differences were observed between anthracene and phenanthrene which could only be ascribed to steric factors not taken into account by their equations.

It appears then that in pure solvents, and in the absence of the supporting materials found in partition chromatography, interactions are observed between non-polar molecules which are highly sensitive to molecular shapes and sizes and which are not accounted for by assuming random orientations for them. It is reasonable to expect that such interactions will produce effects in partition chromatograms and that in general there is no need to appeal to special features, such as the nature of the supporting materials, in order to account for them. It is particularly noteworthy that such interactions become very significant when the solute molecules reach the size and complexity of three fused rings. The steroids, which contain four fused rings, will be expected therefore to show partition coefficients which are highly sensitive to steric factors.

Solvent Structure and Interactions with Solutes

The physical properties of liquids show that their molecules are very close to one another. Although they are not rigidly held in the type of lattice found in crystalline solids, the molecules of liquids are probably not very much further apart than in the solid state, and over small elements of space they are more or less oriented as in the solid crystal. They thus have a considerable amount of "short-range" order. This order is usually very much greater for solvents with very polar molecules. In non-polar solvents such as the paraffins it is slight but becomes much larger again when rigid asymmetrical molecules are involved. The molecules of acyclic hydrocarbons tend to adopt a linear shape in the liquid state and such solvents may be pictured as relatively disordered assemblages of flexible chains and ovoids. Large or asymmetrical molecules such as decalin, tetralin, and xylene behave more like elastic slabs; to be closely packed they must tend to arrange themselves in quasi-crystalline arrays and such solvents may be pictured as something like a pile of dominoes or draughtsmen shaken about in a box. To a smaller degree benzene and cyclohexane must behave in the same way. The solvent–solvent interactions of non-polar liquids, however, are still weak and order decreases rapidly with increasing temperature.

With polar solvents the interactions are, in general, strong and the molecules highly ordered. In the case of water at room temperatures a highly ordered tetrahedral structure exists which is very little different from that of ice over small elements of space (Bernal and Fowler, 1933; Hunter, 1954). With acetic acid dimerization occurs and the dimers are probably fairly rigid slabs in shape because of stabilization by resonance and the steric limitations of the hydrogen bonds linking the monomers (Hunter, 1954). Although the coordination numbers are altered from integral quantities by steric factors in some molecules it can be taken as an approximation that one water molecule can form hydrogen bonds with four others; a primary polar group in an organic molecule with three water molecules or two alcohol molecules; and an ether-linked oxygen atom with two water molecules (Butler and Ramchandani, 1935; Hildebrand and Scott, 1950, p. 172; Hunter, 1954). The structure of other solvents commonly used in steroid chromatography such as the glycols, formamide, and aqueous methanol is not known in detail but can be inferred from the interaction energies given by Butler *et al.* (1935) and consideration of the probable coordination numbers. The interaction energy between primary alcoholic hydroxyl groups and water is considerable, namely about 4700 cal/mole compared with 6000 cal/mole for the water–water interaction, and spectroscopic evidence shows that alcohols of low molecular weight are strongly associated solvents. The H-bond between primary alcohols has an energy slightly greater than the same bond in liquid water. The glycols can be considered as water-like solvents in which the total interaction energy is reduced by steric reduction in coordination number. Methanol–water mixtures on the other hand can also be thought of as water-like, but in this case the methyl radical not only reduces the coordination number of the hydroxyl group of the methanol molecule itself, but also that of neighbouring water molecules. With propylene glycol the two hydroxyl groups are not equivalent: the secondary hydroxyl group is hindered both by the terminal methyl group and by the primary hydroxyl group. This solvent can be considered as water-like but with rather weaker interaction energy than ethylene glycol (See p. 94), and with greater possibilities of steric effects than with ethylene glycol or water-methanol.

These polar solvents can all be considered as approximating to a disrupted and weakened water lattice. In the water–methanol mixtures the methanol molecules break up the water lattice; in the glycols the water-like hydroxyl groups are themselves attached to the disrupting hydrocarbon radical. It will not be surprising to find that these solvents all show very similar interactions with steroids when these are evaluated in terms of R_M values (see Chapter 2). Water itself is not a good solvent for the majority of steroids because its coordination, and hence its interaction energy, is too great: non-polar steroids cannot compete with the tendency of the water lattice to re-form and are, as it were, "squeezed" out of solution.

Theories of solubility tend to vary between the extremes of a purely statistical approach and a highly mechanistic one. Similarly, the structure of liquids was originally considered to be close to that of a compressed gas, but with the advent of X-ray and Raman spectra there came a tendency to regard them as more like slightly disturbed crystal lattices. The modern picture is somewhere between the two extremes in each case and it is recognized that particular situations may require an approach which emphasizes one or the other aspect according to the nature of the interacting molecules. A thorough account of these concepts and some of the dangers of adopting one or other extreme position will be found in Hildebrand and Scott (1950, pp. 186, 194) and Moelwyn-Hughes (1957, p. 676).

The results of orthodox studies suggest that we can expect that the application of Martin's approach to the behaviour of steroids in partition chromatographic systems should be reasonably successful if steric factors are taken into account. ΔR_M values should be approximately constant for given solvent systems, but it will not be surprising if the steric complexity of the steroids leads to rather greater variations of these values than would be expected with simpler compounds. In particular the existence of subtle effects not obeying the simple additive rules should not be attributed to anything other than liquid–solute interactions unless there is specific evidence to the contrary. There is plenty of evidence for these with molecules far simpler than steroids, and in the case of the triterpenes and steroids there is already evidence for the existence of long-range steric effects on reaction rates in solution which will be expected to produce significant effects on partition coefficients in certain solvent systems (Barton, 1955). In passing on to the question of the chromatographic behaviour of various solutes, therefore, it will be taken as understood that this behaviour is a simple consequence of the results of orthodox studies of solubility and partition coefficients. Chromatograms are merely highly sensitive indicators of the latter which can be measured by the appropriate function of mobility.

4. Chemical Structure and Chromatographic Behaviour

The R_M Treatment of Simple Compounds

One of the main purposes of this book is to show that the behaviour of steroids in chromatographic systems is a consequence of general laws and, moreover, that it can be treated quantitatively with reasonable accuracy. This is an extension of earlier descriptive rules and gives stronger support to the use of chromatographic methods in identifying steroids; it also allows us to formulate some general rules of procedure for the latter task (Chapter 5), and also for designing methods of quantitative estimation based on chromatographic methods (Chapter 4). Before considering the steroids themselves it is therefore necessary to examine the behaviour of simpler series of compounds in the light of the R_M method of approach (Bate-Smith and Westall, 1950). This has already been done for various moderately complicated solutes but first let us consider the behaviour of some simple homologous series, some of which have not been subjected to this treatment. For this purpose R_M values have been calculated from papers in the literature using the curve plotted in Fig. 2.6 (p. 76): since the original authors did not often apply this method, which requires a high standard of technique for really accurate results (Bate-Smith and Westall, 1950), it is all the more remarkable how good an agreement is obtained with the R_M rule.

Figure 1.3 shows the R_M values for three series of aliphatic homologues

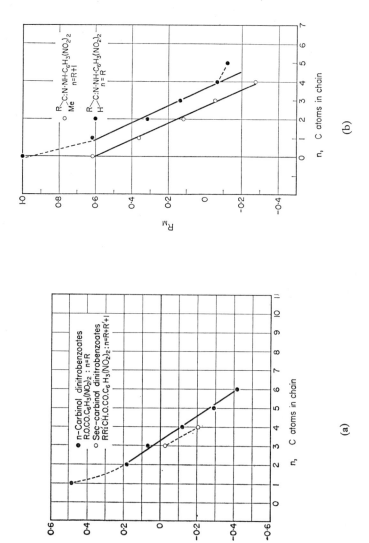

FIG. 1.3. R_M values of homologous series. (a) Carbinol dinitrobenzoates: Solvent system heptane–methanol, 1:1. ΔR_M/methylene group = -0.25. (Data of Meigh, 1952). (b) Carbonyl dinitrophenylhydrazones: Solvent system heptane–methanol, 1:1. ΔR_M/methylene group = -0.23 (Meigh, 1952).

in the system heptane–methanol on paper (Meigh, 1952). It is seen that a nearly linear relation is obtained and that for all three series the linear part of the curve has a slope corresponding to a ΔR_{Mg} of about $-0\cdot23$ per methylene group. Figure 1.4 shows a similar plot calculated from the

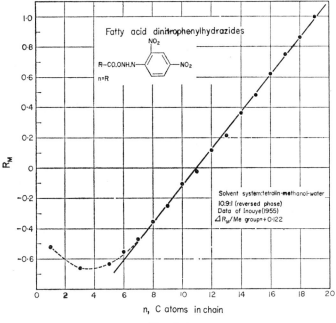

FIG. 1.4.

R_F values given by Inouye, Noda and Hirayama (1955) (Inouye and Noda (1955)) for fatty acid dinitrophenylhydrazides in the reversed-phase system tetralin—90% methanol. For the range $R = C_5 - C_{21}$ a linear relation is obtained with a ΔR_{Mg} per methylene group of approximately $+0\cdot12$. It will be noticed that serious deviations from theory occur within the range $R = C_1 - C_{3 \text{ or } 4}$. It may be suggested that this is due to the conformational rigidity and size of the tetralin molecule; only when the aliphatic chain of these homologues becomes considerably larger than the solvent molecules will random interactions predominate over oriented interactions, which would be expected with the dinitrophenyl radical (cf. Briegleb and Kambeitz, 1934) to increase affinity for the stationary phase and cause a positive deviation of R_M from that predicted, as observed (cf. Butler, 1937).

Figure 1.5 illustrates another point. Some available R_F values for fatty acid hydroxamates (Fink and Fink, 1949; Thompson, 1951) have been used to obtain plots of R_M against methylene content in three different solvent systems on paper. It is seen that the slopes (i.e. ΔR_M per methylene group) differ. Similar differences have been found with amines on gas–liquid

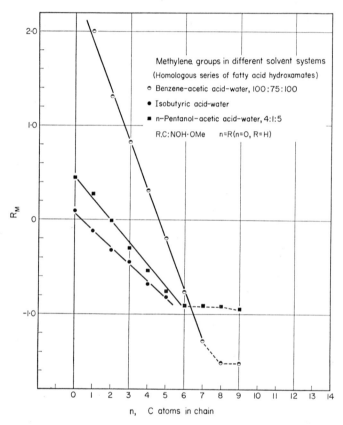

FIG. 1.5. R_M values of homologous series. Characteristic changes with change of solvent system. From data of Fink and Fink (1949) and Thompson (1951).

○ ΔR_M/methylene group = −0·55
■ ΔR_M/methylene group = −0·21
● ΔR_M/methylene group = −0·18

columns using different types of stationary phase (James, 1952), and much larger differences with carboxylic acids with different numbers of carboxyl groups on changing from acid to neutral or basic solvent systems (Reichl, 1955, 1956). It is seen that the ratio of ΔR_{Mg} values in two different solvents, if chosen appropriately, is often characteristic of the type of added group. In principle, it should be possible to identify the nature of the several groups giving rise to the R_M value of a substance by measuring its value in a suitable set of solvent systems (ΔR_{Ms} values, p. 102) (see also Van Duin, 1953a or b). Since all groups could be "investigated" in principle by a wide enough range of systems the identification of compounds purely on the basis of mobilities in chromatographic systems is possible in principle. In practice, it will be seen that it is easier with the steroids to use several chemical ways of producing changes of R_M (ΔR_{Mr} values, p. 102) but several types of solvent system remain to be explored which may be useful in giving significant ΔR_{Ms} values with steroids, and some are already promising (p. 107).

In Fig. 1.6 the figures given by I. Smith (1958) have been transformed into R_M values for four sets of rather complex homologous series. Only in the case of the alkyl cysteines is a serious deviation from theory observed;

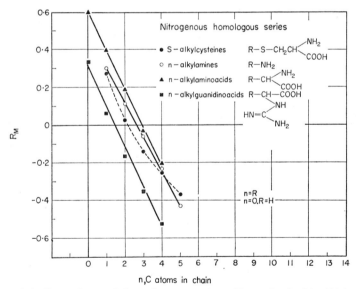

FIG. 1.6. R_M values of homologous series. From I. Smith (1958). Solvent system n-butanol–acetic acid–water (see text, p. 23).

it is possible that the observed deviation is due to the very large contribution of dispersion forces expected from the sulphur atom: the latter is increasingly hindered by enlargement of the alkyl side chain.

Bate-Smith and Westall (1950) first emphasized the value of using R_M units, and only the size of the molecules they studied places their work so late in the argument. They showed that a wide variety of natural polyphenolic compounds of the flavone type obeyed Martin's theory very closely indeed. For our purpose it is interesting to observe that these compounds were mainly C_{15} compounds with two fused rings and a third ring attached by a single bond permitting fairly free rotation (Fig. 1.7).

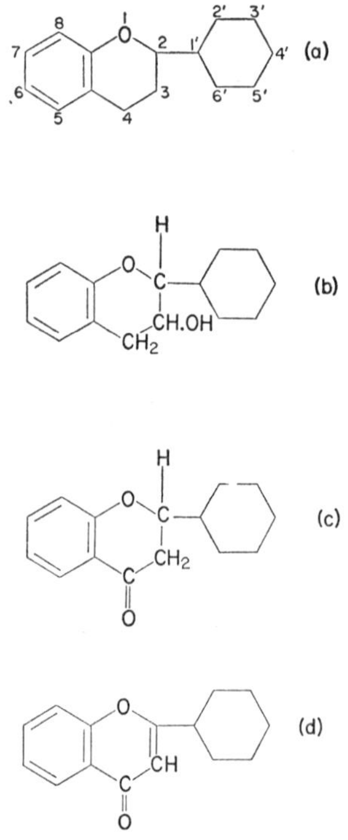

FIG. 1.7. Flavonoids: (a) skeleton; (b) flavanols; (c) flavonones; (d) flavones.

Excellent agreement with Martin's theory was observed for substituents (—OH, —OMe) at positions where little or no steric hindrance was involved (e.g. 4, 6, 7, 8). Also, because these rings are aromatic, the substituents were in the plane of the rings, i.e. strictly equatorial. The only deviation with these groups was with 1,2-dihydroxy-compounds; ΔR_M for a *meta-* or *para-*hydroxyl was nearly twice that of an *ortho-*hydroxyl, an effect similar to that seen with certain steroid *cis-*diols (p. 119).

When substituents in the pyran ring were examined (Bradfield and Bate-Smith, 1950), however, it was found that epimeric hydroxyl groups had significantly different ΔR_M values. Thus D-catechin had an R_M value of -0.50, while L-epicatechin had one of -0.27. It will be remembered that Pierotti *et al.* found the largest deviations from predicted activity coefficients with dicyclohexylbenzene, and it is interesting to find that in chromatographic systems, as with solubility measurements, disagreement with simple theories of behaviour begins to appear when two or three fused rings occur in the solute, especially when one or more of these rings are saturated so that epimeric substituents show conformational differences (see review by Barton and Cookson, 1956).

With most steroids at least four fused alicyclic rings are involved (Fig. 1.8), so that it is not surprising that far more of their substituents show ΔR_M values which are strongly dependent upon steric factors. It will be shown in Chapter 2, however, that if allowance is made for these, relatively constant R_M values are found for given substituents.

Van Duin (1952, 1953) made a study similar to that of Bate-Smith and Westall (1950) but used Celite or silica-gel columns. He showed that the true retention volume (see Butt, Morris, Morris and Williams, 1951, who also introduced this parameter) was analogous to the function $(1/R_F - 1)$ for paper chromatograms and that, therefore, differences in $\log V_r$ should obey the same additive rules as R_M values. He was able to show good agreement with this rule for the dinitrophenylhydrazones and dinitrophenylhydrazides of several homologous series of carbonyl compounds and esters. This study is valuable in that the Celite columns undoubtedly provide two phases exactly as postulated by the original theory of Martin and Synge (1941); there is no doubt here that the rule observed is a direct consequence of the behaviour of liquid–liquid systems. He also pointed out the consequences of this obedience to an additive rule, namely that, as Martin (1949) had suggested, small differences of molecular structure were equally effective in producing separations in large and small molecules. In contrast, others had found that with adsorption methods separations became more and more difficult with homologous series of compounds as the length of the alkyl chain was increased above six carbon atoms (e.g. Gordon, Wopat, Burnham and Jones, 1951).

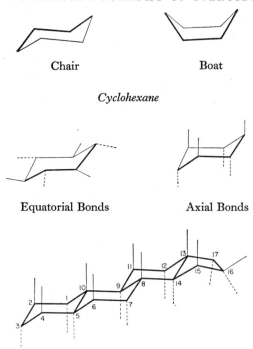

FIG. 1.8. Conformations of Steroids.

α-Bonds ---- β-Bonds ———

The equatorial bonds have been omitted from the steroid skeleton for clarity: the axial bonds are vertical.

Note the alternation of axial and equatorial correspondence with α and β orientations. The 16 position gives intermediate conformations for both α and β substituents. The 15 position is abnormal due to the 5-membered D-ring but the 15α orientation behaves as if equatorial to ring C (after Klyne, 1957).

Other successful treatments using the R_M function have been carried out with glucose polymers (Jeanes, Wise and Dimler, 1951), with alkyl amines and diamines (Bremner and Kenton, 1951) and with the influence of the water content of the mobile phase on the mobilities of sugars (Isherwood and Jermyn, 1951). In the intermediate range of R_F values approximately linear relationships will also be obtained by using R_F values if these only cover a small range (see Fig. 2.6). Nicholas and Rimington (1951) used this type of relation to predict the number of carboxyl groups in porphyrins, Prochazka (1954) similarly with steroid acids, and Grundy, Simpson and Tait (1952) a rather similar approach to compare neutral steroids. The latter authors were careful to point out that their equation was only applicable to single solvent systems and only gave qualitative results in terms of the probable order of R_F values. The use of R_F values in this way must be considered only as a rough qualitative procedure; much more valuable information can be gained from the use of R_M values, which alone allow one to compare the differences between compounds of widely different R_F values, and the differences between solvent systems, on a quantitative basis.

5. Partition Coefficients and Solvent Systems

The previous argument has assumed that the partition coefficient of a solute is unaffected by its concentration, and the approximate truth of this assumption is undoubtedly the basis of the success of partition methods. It is only approximately true, however, and although the curvature of liquid–liquid distribution isotherms is very much less than most isotherms for adsorption systems, the reasons for such departure from linearity need discussion since they probably lie at the root of the question of what makes a good solvent system for partition chromatography.

The first thing to recognize is that differences of solubility are due to deviations of solutions from ideality, that is from obedience to Raoult's law. In general, such deviations can be positive or negative and vary with the mole fraction so that they pass through a maximum or minimum. The necessity for this is easily seen from the fact that at infinite mole fraction (one component only) and zero mole fraction the activity of a substance must be unity and zero respectively (Fig. 1.9, points O and X). If two solvents are sufficiently different to be partially immiscible, a third substance which is only very slightly soluble in one of the pair is likely to be much more soluble in the other member, and their deviations from ideality in the two phases are likely to be in opposite directions giving two curves of the type shown in Fig. 1.10. If we consider the equilibrium of the solute and the two solvents it is clear that the partition coefficient of the solute is obtained by taking the ratios of the mole fractions at different positions of equal activities such as points on y, in Fig. 1.10. Only over the small range of dilute conditions is this ratio independent of concentration.

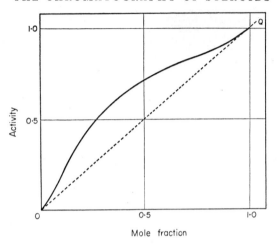

Fig. 1.9. Activity-concentration curves.

It is clear from Fig. 1.10 that a distribution isotherm for a solute in a liquid–liquid system will only approach linearity when the envelope enclosed by the two curves between O and X is kept to a minimum, and

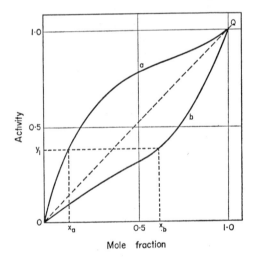

Fig. 1.10. Activities in two immiscible solvents.
Partition coefficient α_b for total quantity of solute $(x_a + x_b)$ is given by x_b/x_a. It is seen that x_b/x_a varies with $(x_a + x_b)$. (After Hildebrand and Scott: diagrammatic).

for any system the closest approach to linearity will be obtained at low concentrations of solute. In general, the more different the two solvent phases, the bigger the difference in the solubility of a substance in the two phases; and the greater the area of the envelope in Fig. 1.10, the more likely it is that the deviations from ideality will be in the opposite direction and the smaller the range of concentration in which the distribution isotherm will be linear. This is probably one of the reasons* for the empirical rule recognized very early in the history of paper chromatography that a solvent pair was unlikely to be really useful unless the members of the pair were miscible to the extent of about 10% or more; unless this is so, the absolute solubilities of the solutes are unlikely to be good in both phases. It also explains why the partition coefficient of a substance in two liquid phases is not in general the ratio of its *solubilities* in the separate phases;† this is true only for the condition of complete saturation. The extremely curved distribution isotherms for the anions of organic acids in most immiscible pairs of solvents and association of the acids in the non-polar phase, is the reason for the use of "swamping acids" or buffers in solvent systems for these compounds (Synge, 1946) (see p. 331). The alternative is to use a basic system with a sufficiently high water content of the mobile phase (Lugg and Overell, 1947, 1948; Isherwood, 1954). It is also clear that in any system the deviation from linear distribution isotherms will be greatest with compounds whose partition coefficients differ most from unity in either direction.

At first sight this factor seems to be too small to be a potential cause of serious streaking on chromatograms since individual solutes are usually at concentrations in the range 0·001–0·01 molar. However, it may be appreciable with very immiscible solvents and solutes of low absolute solubility. Thus a deviation of 5% from linearity of the activity–concentration curves of a solute in opposite directions in the two phases should give very approximately a 10% change in α_S and hence a 1-cm change on a 40-cm run in the position of a zone normally running with an R_F of 0·5.

Sjövall (1954) using the system heptane—70% acetic acid found that deoxycholic acid tailed badly when more than 5 μg was run on filter paper; this was not due to lack of sensitivity of the colour reagent since the results

---

* The other was the need to reduce ΔR_{Mg} values to a convenient value to enable a large number of substances to be run on one sheet of paper.

† N.B. Still less the ratio of solubilities in the two main *components* of the phases before mutual saturation.

were checked with radioactive material. He found that this was prevented by adding quite small amounts of isopropyl ether to the systems and suggested that the phenomenon was due to the formation of a choleic acid which was dissociated by small concentrations of the ether. Whatever the type of interaction involved it seems probable that the ether was effectively increasing the miscibility of the two phases and overcoming an effect of the above type.

The serious streaking of steroids on paper chromatograms run with water as stationary phase (Bush, 1952) is probably a combination of this effect with actual adsorption to the paper. The good results with systems based on paper impregnated with propylene glycol, despite its very low miscibility with the usual hydrocarbon mobile phases, are probably due to the very much larger volume of the stationary phase in these systems; the displacement effects seen with the less polar steroids in ligroin or heptane-propylene glycol (Savard, 1953; Rubin, Dorfman and Pincus, 1953a, 1954; Rubin, Rosenkrantz, Dorfman and Pincus, 1953b; Axelrod, 1953) are however possibly due to this effect, the two phases in these cases being almost completely immiscible. Apart from thermodynamic considerations, other factors enhance the importance of using solvent systems with a fair degree of intermiscibility of the two major components, and in which the solutes under examination have as high a solubility as possible. Among these are the fact that such systems will be far less likely to be upset by adsorption of solutes to the supporting material, by complex formation between solutes and by minor degrees of overloading. Mixtures with very considerable intermiscibility, on the other hand, are much more sensitive to changes of temperature and may give inconveniently low ΔR_{Mg} values (see p. 125) (e.g. collidine–water with 48% water in the mobile phase (L. I. Woolf, 1958)).

Another factor which enters into steroid chromatography is due to the use of stationary phases with low or zero water content on paper or cellulose. The common solvent systems used for amino acids and like compounds have a high water content with the result that the amorphous regions of the cellulose support are almost completely swollen. In these systems the concentrated polysaccharide solution which forms the stationary phase (Martin, 1950) is an expanded gel in which the polysaccharide chains are probably relatively unconstrained so that the degrees of freedom of interactions between them and solutes of small molecular weight are numerous. In systems with low water content, however, the degree of swelling is likely to be considerably reduced, with the result that the gel is probably a highly ordered structure; large asymmetrical molecules in this situation may interact with the polysaccharide chains in orientations which are sterically limited (see however p. 129).

Kress and Bialkowski (1931) state that the order of swelling capacity correlates well with the dielectric constant of the liquids that have been studied and is: formamide ⩾ water > ethylene glycol > methanol > furfuryl alcohol > ethanol > furfural > propanol > butanol > oils. Aqueous solutions of acids, bases, and salts may swell cellulose much more than water alone, but the effects are complicated. The porosity paradoxically *increases* as paper is hydrated but it seems probable that the effective volume of the stationary phase increases more than the volume of the interstices. This effect, however, seems to be small with most papers and the volatile solvent systems used for steroids. With impregnated paper systems, however, the technique of impregnation can cause very considerable changes in R_F values, by raising the (cross-sectional) ratio, stationary phase area : mobile phase area (Reineke, 1956; Neher, 1958a).

6. The Effect of Temperature

In general, elevated temperatures are useful for raising the capacity of the system and obtaining closer correspondence with ideal conditions at all solute concentrations. They also reduce any adsorption to the supporting material. The general effect of a rise in temperature is usually to increase the miscibility of the solvents so that the two phases become more alike (see also p. 125). In addition, the specificity of short-range interactions which are sterically dependent will be reduced by a rise in temperature so that most separations are reduced at high temperatures. This effect has been noticed with steroids in the Craig counter-current machine (Carstensen, personal communication) and sharper separations have been obtained by working at 0–2 °C. Reduced temperatures are unlikely to be useful in the partition chromatography of the majority of steroids, however, because of the limitations imposed by low solubilities and the tendency to adsorption.

7. Formation of Several Species of Solute

As with all solubility phenomena, chromatographic behaviour depends upon both solvent and solute behaving as if they were each one species. In fact numerous species of transitory solvent–solute "complexes" are postulated to explain solubility phenomena, but the essential feature of these is that the velocity of their formation and dissociation is very large and governed by close approximation to ideal behaviour. If a solute forms two species which have widely different partition coefficients between the two phases of a system and if the equilibrium between the two species, or the partition coefficient of one or other, deviates considerably from ideal behaviour then the zone occupied by that solute will be enlarged over that predicted by theory. Similar effects will be produced by any interaction which achieves equilibrium much more slowly than the main ones responsible for the partition coefficient of the solute under examination. The two main types of such interactions which interfere with partition chromatography are adsorption (including ion-exchange and compound formation, as with metallic impurities and carboxyl groups in paper) and the dissociation of ionizing solutes (Synge, 1946; Hanes and Isherwood, 1949; Isherwood and Hanes, 1953, Lugg and Overell, 1947, 1948; Munier, 1952). The usual way of overcoming the latter is to use conditions designed to ensure that the solute is entirely in one form or another. Thus, acids are run as their salts in a system containing an excess of a suitable base, which may be buffered (McFarren, 1951; McFarren and Mills, 1952) or in hydroxide form to obtain a high pH; or alternatively the system contains a large excess of a "swamping" acid chosen so as to maintain a pH well below the pK of the acid solutes to be separated. These techniques are only relevant to the steroid conjugates, bile acids and etianic acids, and, to a much smaller extent, oestrogens (Engel, Slaunwhite, Carter and Nathanson 1950).

The effects seen with complicated mixtures are often extremely variable and not easily understood. A considerable amount of adjustment may be needed before good results are obtained with ionizing substances in complicated mixtures. In n-butanol–ammonia, for instance, steroid sulphates and glucuronosides are both ionized and run without tailing. In acetic acid–t-butanol systems, however, the sulphates are still ionized and streak, although not enough to impair most separations. Only in HCl-containing systems are the sulphates present largely in undissociated form giving symmetrical zones without streaking (see p. 289).

CHAPTER 2

CHROMATOGRAPHIC SEPARATION OF STEROIDS

SECTION I: DESCRIPTIVE

1. INTRODUCTION

IN THIS chapter we shall be considering chromatography proper; that is to say the various chromatographic methods that have been found useful for separating steroids, as distinct from the various ancillary techniques required for detecting or estimating them. With few exceptions the discussion will be confined to an account of the behaviour of steroids in chromatographic systems without reference to the technical details required to obtain such behaviour. These technical details would confuse the argument and have been collected for convenience in the next chapter.

The main purpose of this chapter is to give a practical illustration of the theoretical treatment briefly outlined in Chapter 1 and to lay the basis for the arguments of Chapters 4 and 5. Practical problems, including most of those concerned with the selection of the most suitable solvent systems for various separations, and methods of detection and estimation, will be relegated largely to Chapters 3, 4 and 6. One further explanation is needed of the rather one-sided approach adopted for this chapter; adsorption methods will receive relatively little attention. This is not to imply that such methods have not been useful, or indeed that they will not continue to be useful; it is because the factors determining adsorption isotherms are in general more complicated than those determining partition isotherms and that they are still only understood in rather empirical terms (p. 10). In addition, adsorption methods have been very thoroughly reviewed in recent years (Lederer and Lederer, 1957; Neher, 1958a, 1959a), while it seems to the author that although very useful reviews of the literature of partition methods for steroids have been published, they have not attempted systematically to relate the behaviour of steroids in such systems to the general theory of chromatography.

The practical usefulness of partition chromatography in steroid problems of all kinds has tended to obscure both the relevance of results with other classes of substances, and also the great interest of the steroids as models for testing the theory of chromatography. This interest is largely due to two features of the steroids. First, their linked polycyclic structure provides a considerable degree of rigidity to their hydrocarbon skeletons, and this in turn means that the shapes, or "conformations", of any one steroid are fairly limited in number. In conjunction with information from X-ray crystallography and studies of dipole moments (Kumler, 1945), conformational analysis (Barton, 1950) provides for nearly every steroid an idea of its shape which is far more definite than with most other classes of organic substances because it is less subject to variations due to bending and twisting movements of its constituent groups. Secondly, the large size of the steroid nucleus and its more or less extended shape means that many steroids have numerous substituents attached to the nucleus which are nevertheless far enough apart in the molecule to be considered as independent groups (cf. however, Barton, 1955). This is, of course, the ideal requirement for the application of the theoretical approach outlined in the previous chapter.

2. Special Features of the Chromatography of Steroids

The steroids are large, more or less slab-shaped, molecules with molecular weights ranging from about 270 to 1000. They are extraordinarily numerous and have a wide range of properties. They include for instance the extremely hydrophobic esters of the sterols, the androgens, the phenolic oestrogens, the adrenal cortical hormones and their metabolites, the sapogenins and cardiac aglycones, and also the water-soluble conjugated steroids such as the glycosides, bile salts, and glucuronosides (Fig. 2.1). Apart from the difficulty of finding solvent systems giving favourable partition coefficients for the more hydrophobic steroids, most steroids have rather low solubilities in convenient solvents and are strongly adsorbed unless special measures are taken to prevent this. They are known to associate with proteins and probably do so with other substances of high molecular weight. An additional difficulty is that their stability is very variable. For instance, most oestrogens and bile acids are fairly stable but the adrenocortical hormones are rather unstable and many steroids undergo rearrangements on activated alumina (Reichstein and Shoppee, 1949; Neher, 1958a, p. 28) or charcoal (Meyer, 1953; Levy and Kushinsky, 1954). The majority of steroids are neutral substances, thus excluding the

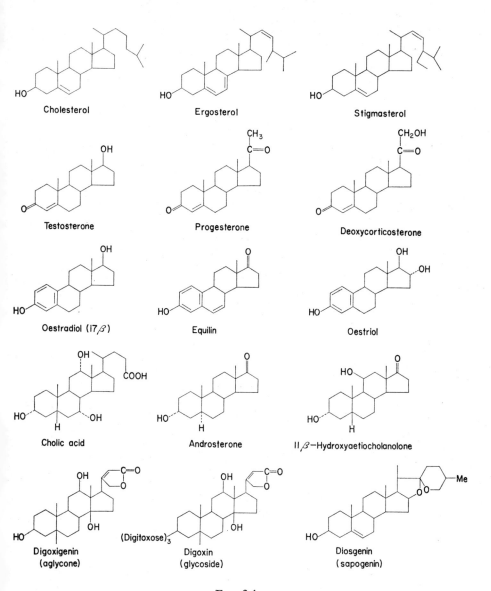

Fig 2.1.

use of changes of pH as a means of improving separations; acidic and basic solvents are, however, of use in the separation of the oestrogens, bile acids and their amino-acid conjugates, the etianic acids, and the acidic conjugates of urine.

An extremely important feature of this field arises from the mere number of compounds to be dealt with. Nature fortunately limits us among the naturally occurring steroids to a relatively small number of isomeric possibilities in each of the principal families which make up the steroid group as a whole. However, the number of compounds is still enormous, with the result that identification by simple chromatographic properties alone is rarely feasible outside of certain situations where previous work has shown that a particular steroid or type of steroid is predominant. Steroid chromatography is therefore based upon a very wide range of solvent systems covering a large range of solute polarity, a large number of methods of detecting different types of substituent, and above all upon the use of microchemical degradations and the formation of simple derivatives giving characteristic changes in chromatographic and other properties (Chapters 2 and 5).

3. Adsorption Methods

Two main types of use have been made of adsorption systems. In the first a complicated extract is run on a large column with small step-wise changes in the eluting solvent and the effluent is collected in suitably sized fractions. The fractions are examined by the evaporation of solvent and weighing (Reichstein and Shoppee, 1949), or, in addition, by other methods such as infrared spectroscopy (Dobriner, Lieberman and Rhoads, 1948; Lieberman, Dobriner, Hill, Fieser and Rhoads, 1948; Lieberman and Dobriner, 1948). This method is flexible and the eluting solvents can be changed according to a set rule, or for instance when the weight or ultraviolet absorption of eluted material has been at a constant or decreasing small value for two or three successive fractions (Reichstein and Shoppee, 1949). This method is usually used for preparative and isolation work rather than for estimation. The other main type of method has been with smaller columns and is mainly used for estimating small quantities of a fairly small number of compounds. Here a regular sequence of solvents is used, the adsorbent and other conditions need very careful standardization, and gradient elution or other special methods may be needed for the best results. Such methods have been used for separating and estimating urinary 17-oxosteroids (Callow, 1939; Callow and Callow, 1939; Dingemanse and Huis in't Veld, 1946; Dingemanse, Huis in't Veld and de Laat, 1946; Henry and Thevenet, 1953; Dingemanse, Huis in't Veld and Hartogh-Katz, 1948, 1952; Lakshmanan and Lieberman, 1954), adrenal steroids and their metabolites (Pincus and Romanoff, 1950; Romanoff, Wolf, Constandse and Pincus, 1953; Heftmann and Johnson, 1954) and oestrogens and pregnanediol (Stimmel, 1946; Stimmel, Randolph and Conn, 1952; Brown, 1955). A third type of method is really a crude version of the second general method. In this case a relatively small column is used to separate groups of compounds into a relatively small number of fractions; these fractions are then further separated by a more selective technique or used for estimating one limited type of compound by means of a more or less specific absorptiometric method. This method is very useful in certain fields and especially useful for separating a rather crude extract of organic material into fractions which are clean enough for partition chromatography on paper or small columns (Romanoff et al., 1953; Savard, 1954; Nelson and Samuels, 1952; Bush and Sandberg, 1953; Ayres, Garrod, Simpson and Tait, 1957a).

Alumina, magnesium and other silicates, and silica gel have been used most frequently. Certain changes in relative mobility of steroids have been found with different adsorbents but these are relatively few compared with other fields (Neher, 1958a, p. 31; Lieberman, Fukushima and Dobriner, 1950).

In the steroid field the main factor determining adsorption has been the number and type of polar groups in the molecule, the configuration of ring junctions, and, to a lesser extent, the number of carbon atoms in the molecule. As with other classes of substance the polar groups dominate the picture and the order of elution of steroids from an alumina column bears a close resemblance to the order found with partition systems in which the polar phase is stationary. Hydroxyl groups retard steroids more than ketone groups with the exception of the 11β-hydroxysteroids in which the reverse is the case. This is understandable in that the considerable steric hindrance of this hydroxyl group makes its approach to a rigid solid surface difficult.

With other classes of substance it has been found that the relative effects of double bonds and hydroxyl groups are considerably altered by changing adsorbents. Thus sucrose has a high affinity for hydroxyl groups; and magnesium hydroxide has a lower affinity for hydroxyl groups but a higher affinity for double bonds (Strain, 1948). Such effects might also be useful with steroids and sterols and some important inversions of usual relative mobilities have been reported by Neher (1958a, p. 31) with magnesium silicate.

Lakshmanan and Lieberman (1954) made a careful study of the effects of different degrees of hydration of alumina. They devised a neat method of controlling this by exposing the alumina to air over salt solutions of known vapour pressure. The best separations of urinary 17-oxosteroids were obtained with alumina containing 7–8% water. Other methods such as grinding the alumina with water to a measured increase in weight or with a measured volume of water have been used (Kellie and Wade, 1957) but would seem to be intrinsically more subject to error due to inhomogeneity or difficulties due to mixing in a damp atmosphere. However, Kellie and Wade preferred it and found that the best separations required a water content of 4–5%. Reichstein and Shoppee (1949) found that small amounts of alkali in alumina gave the best separations but had to be neutralized for use with sensitive substances such as ketols, lactones, and some esters. Despite the enormous number of reactions undergone by steroids on certain types of alumina, and the paramount importance of using the acetates of steroid ketols in their classical work on adrenal gland extracts, Reichstein's group have more recently found that free α-ketols can be separated on alumina of moderate activity without mishap as long as they are only exposed to the adsorbent for a few hours (Prof. Reichstein, personal communication).

Reichstein introduced the now standard method of using adsorption columns, namely the elution of material from the column in liquid fractions (Durchlauf method; "liquid" or "flowing" chromatogram) rather than the earlier method of extruding the adsorbent and cutting it into sections for elution. His sequence of solvents (Fig. 2.2) is similar to others described by Trappe (1940a, b) and Strain (1942). Acetone is not advisable with alumina because it readily undergoes condensation reactions to oily materials on active or alkaline grades. Another danger to be remembered is that organic peroxides are strongly adsorbed by alumina (p. 349); if solvents are not peroxide-free, high concentrations may accumulate in the upper zones of the column (Dasler and Bauer, 1946) and cause unwanted side-reactions with slow-moving or stationary steroids in the early stages of a chromatogram.

An enormous number of degradations and isomerizations of steroids on alumina has been described. How many of these are due to simple reactions involving only steroid and adsorbent is unknown; many of them may have involved peroxides or other impurities in the solvents, or impurities in the mixture put on the column. Many of these reactions are known to be due to alkali in the alumina (Reichstein and Shoppee, 1949). A very useful reaction is the hydrolysis of steroid 3-formates by alumina (Norymberski and Sermin, 1953). A very full review is given by Lederer and Lederer (1957, p. 62).

FIG. 2.2. Solvent sequence for alumina columns.

Steiger and Reichstein (1938)
  Pentane   Benzene
  Pentane–benzene 4:1   Ether
  Pentane–benzene 1:1   Acetone
Trappe (1940)
  Light petroleum   Chloroform
  Cyclohexane   Ether
  Carbon tetrachloride   Ethyl acetate
  Trichloroethylene   Acetone
  Toluene   n-Propanol
  Benzene   Ethanol
  Methylene chloride   Methanol

The Sloan-Kettering group (Dobriner, Lieberman and Rhoads, 1948; Lieberman *et al.*, 1948) used alumina in their earlier work on urinary steroids but have largely changed to the use of silica gel and magnesium silicate mixed with Celite, (Dobriner, *et al.*, 1948; Lieberman, *et al.*, 1950). The separations are said by some to be less sharp than with alumina, but the results with the more polar C_{21} steroids are more satisfactory. The Celite adsorbs some non-steroidal material but otherwise its main function is to improve the running properties of the column. Silica gel has been found useful for relatively crude separations of urinary and other steroids into broad groups as a preliminary to more refined separation on paper chromatograms (Romanoff *et al.*, 1953; Levy *et al.*, 1954; Neher, 1958a, p. 24; Savard, 1954; Bush and Sandberg, 1953; Ayres *et al.*, 1957a). Magnesium silicate is probably preferable for all easily hydrolysed esters and has been used specifically for acetyl-glycosides by Aebi and Reichstein (1950) and by Lichti, Tamm and Reichstein (1956). Meyer (1949), however, observed hydrolysis of the 16α-acetoxyl groups of sapogenins on Florisil.

Adsorbents such as silica gel or silicic acid are not entirely free from trouble with side-reactions. Isomerization of steroids (Soloway, Considine, Fukushima and Gallagher, 1954) and of terpenes (Arbouzov and Isaeva,

1953) has been described. Some of these isomerizations could also be produced simply by heating, which introduces the question of whether this was the basis of the phenomenon on silica gel. Dry silica gel adsorbs water and some polar solvents with a very considerable evolution of heat; changes in composition of the eluting solvent need to be gradual to avoid this and the effect is increased by any water in the solvent. The same danger with peroxides is to be expected with silica gel as with alumina (e.g. Peterson, 1957; and see Appendix I).

Silica gel, and possibly other adsorbents, can swell appreciably not only with water but also with alcohols. If this is done purposely a column results which operates as if it were a partition column, a method specifically introduced by Kritchevsky *et al.*, (Kritchevsky, 1954; Katzenellenbogen, Kritchevsky and Dobriner, 1952). However, it seems to the author that many methods based on silica gel may involve this process unwittingly when large proportions of polar solvents are added to a non-polar one to effect elution—as distinct from the small proportions used for displacement in concentrations ranging from 0·1% to 5%. It depends upon all the factors of column dimensions, rate of solvent flow, and the nature of the solvents used, whether significant troubles arise in this situation, but it is conceivable that some steroids will be partly caught by the advancing front of the "partition zone" and that this fraction will be strongly retarded as the partition mechanism come into play, thus giving rise to "splitting" or "tailing" of the steroids running in the critical zone. As a general rule the sequence of solvents used with silica gel when an adsorption mechanism is desired should be designed so as to avoid this possibility. Thus solvent mixtures involving a large proportion of a very polar solvent such as methanol or ethanol, in a very non-polar solvent such as light petroleum or benzene, should be avoided in favour of a smaller proportion of polar solvent in a more polar major component such as ethyl acetate, or a mixture of solvents, neither of which is very polar, such as benzene—chloroform.

Some situations are known in which a system apparently based upon adsorption behaves very much like a partition system with particular classes of solutes. That is to say little tailing is seen even with relatively large amounts of solute, suggesting that the distribution isotherm is linear over a large range of solute concentration as in liquid–liquid systems. This seems most common with silicic acid (Green and Kay, 1952) and has been used as an argument against considering partition systems to be essentially different from adsorption systems (see p. 15). In most cases, however, the main problem with adsorption systems is that the distribution isotherm for most solutes is strongly curved resulting in "tailing" because of the greater retardation of zones of low solute concentration. This nuisance can be reduced to workable proportions by the method of gradient elution (Alm, Williams and Tiselius, 1952) which has been applied with great success to the separation of certain steroids (Lakshmanan and Lieberman, 1954). The gradient needed for good separation is adjusted empirically to secure the best compromise between one giving rapid elution but poor separation and one giving excellent separation but in the form of low broad peaks taking a long time for complete elution. The best results are obtained with gradients "concave upwards" resulting from the addition of the second solvent; the aim is to push the slower moving compounds forwards and the slower the compounds the greater the concentration of displacing or polar solvent that is needed to secure a given mobility.

The mechanism of adsorption systems has received little attention in the steroid field; practical results were relatively easily achieved and were of an importance that pushed methodological considerations into the background.

Strain (1942) among others has already pointed out that adsorption involves three essential components—solute, adsorbent, and bulk phase (solvent or gas)—none of which can be neglected in any treatment of the phenomenon. In most systems the solute is probably adsorbed mainly in monomolecular layers and from the sensitivity of adsorption isotherms to geometrical changes of the solute it is probable that a large area of contact is involved between the solute and the adsorbent. The energy of this association probably varies with different types of "active point" on the adsorbent, and numerous complications ensue if certain parts of the solute are attached more firmly than others. Thus it is conceivable that the quasi-partition behaviour of some adsorption systems (Green and Kay, 1952) is because only a limited part of the solute is very firmly attached to the adsorbent; for certain solutes it will be possible for adsorption to occur with most of the molecule unattached—as it were floating in the bulk phase like a weed attached to a stone on a river bed—so that many more molecules can be packed into a unimolecular layer than if they were associated with the adsorbent over a large area of the molecule. In such a case the interactions of the rest of the molecule with the bulk phase will often play the major part in determining the sensitivity of the system to structural changes in the solute rather than those which affect the attached group itself. It has been seen (p. 16) with liquid–liquid systems that such interactions may in certain cases be almost as sensitive to geometrical changes in the solute as any of those found with liquid–solid systems and which have usually been attributed to the adsorbent rather than the bulk phase. On the other hand, it remains true that numerous cases exist in which the nature of the solute, e.g. the carotenes (Martin, 1950) is such that it is unlikely that the sensitivity of solute–solid interactions to the structure of the solute is ever likely to be approached by solute–liquid interactions.

It will be seen later that the main disadvantage of chromatographic mobility as a property for identifying steroids (or any other substance) *vis-à-vis* the mixed melting point is that mobility in partition systems is uninfluenced by moderate amounts of impurities. The "mixed chromatogram" (Burton, Zaffaroni and Keutmann, 1951a, b), therefore, while very useful for checking mobilities on over-run paper chromatograms, cannot be considered as analogous to the mixed melting point (Bush, 1954a) when done with partition systems. The phenomenon of displacement in adsorption systems however involves intermolecular forces similar to those found in organic crystals. Mixed adsorption chromatograms done *under conditions where displacement is known to occur* (e.g. Hamilton and Holman, 1952) might reasonably be held to have a validity comparable to that of the mixed melting point in the proof of chemical identity. Constant specific activity of a labelled steroid in fractions collected from a column run under displacement conditions with a known amount of the suspected reference steroid as carrier would be a very stringent proof of the identity of the labelled compound and its carrier. Similar techniques with partition systems (Udenfriend, 1950; Tait and Tait, 1959) or with counter-current liquid–liquid systems (Baggett and Engel, 1957) provide excellent checks on identity but both suffer from the fundamental logical shortcomings described in Chapter 5 (p. 258) when compared with mixed melting points. To my knowledge this technique has not been used with steroids and it should be emphasized that the conditions commonly used for elution chromatography of steroid mixtures will not be suitable for displacement chromatography.

The great advantage of adsorption chromatography over partition chromatography is that for separations on the preparative scale smaller columns can be used, and because of the solid nature of the stationary phase, such columns are far easier to prepare and use. Thus they need less careful temperature control (unless rigid elution schemes are used) and are not subject to the instability of liquid–liquid systems due to detergent substances in the extract, salts, or hygroscopic substances.

On the other hand artifacts due to a variety of causes may occur and overlapping bands can only be avoided by prolonged elution sequences or by the use of gradient elution with its own complications. On the microgram or even smaller scale it appears to be definitely inferior to the partition method or else to offer no advantages, but in some cases this is a matter of opinion. One problem only partly solved by partition methods (p. 101) is the separation of the 5-epimeric saturated 3-oxosteroids: these epimers are resolved relatively easily on adsorption columns (Lieberman et al., 1948). Steroids differing only by one unconjugated double bond, or in features of the hydrocarbon skeleton (e.g. 14α-(H) and 14β-(H) epimers; Schmidlin, Anner, Billeter, Heusler, Ueberwasser, Wieland and Wettstein, 1957) can be separated by partition systems in certain cases (p. 101), but complete separation of such compounds is often difficult on adsorption columns as well (Lieberman and Dobriner, 1948). With both methods the epoxides have been preferred for achieving complete separation (Seebeck and Reichstein, 1943; Rubin et al., 1953). Their stability again makes adsorption columns preferable on the industrial scale where possible.

Group separations of steroid sulphates and glucuronosides have been achieved on alumina (Barlow, 1957; Crépy, Malassis, Meslin and Jayle, 1957b). Individual separations of some 17-oxosteroid glucuronosides have been obtained by Edwards and Kellie (1956).

4. Partition Chromatography

General Survey and History

The nature of these systems has already been touched upon but certain features are in no danger of suffering from over-emphasis. The most important one is that they behave as if they involved distribution of solutes between two bulk liquid phases, this being shown by apparently linear distribution isotherms over a moderately large range of concentrations and by the absence of proper displacement effects—that is to say displacement by like substances in *small* concentrations. In many cases, of which paper impregnated with solvent, and loaded Celite, are the clearest examples, it is known that two liquid phases exist and that liquid–liquid distribution, or "partition", accounts completely and exactly for the chromatographic processes observed (Martin, 1950; Tait and Tait, 1960). That the nature of the stationary phase is complicated and only approximately understood in the case of paper (Martin, 1950, Cassidy, 1957), starch (Stein and Moore, 1948) and silica gel (Martin and Synge, 1941); that certain separations and certain systems are complicated by adsorption (often to the benefit of the experimenter); and that certain examples of true displacement do occur (Martin and Synge, 1941; Silberman and Thorp, 1953; Savard, 1953, 1954) in no way detracts from the necessity of treating partition systems on the basis of the theory of distribution between two liquid phases.

In Chapter 1 it was shown how, to a first approximation, the partition coefficient of a substance was determined by the free energy of moving its constituent groups from one phase to the other. It was also shown that the function $R_M = \log(1/R_F - 1)$ was directly proportional to this free energy and that, hence, changes in the constituents of a substance should lead to linearly proportional changes in its R_M value in a particular solvent system (Bate-Smith and Westall, 1950). It will be shown in this chapter that with the restrictions placed upon the definition of "group" in Chapter 1, this supposition is largely correct for the steroids and that, in this respect, they resemble the other classes of substances dealt with in the previous chapter. In other words the steroids, despite their complexity, behave very much as one would expect the sum of their constituent groups to behave if certain stereochemical effects and interactions between closely spaced groups are taken into account. First of all, however, we must discuss the subject in a descriptive way, if only to point out some of the difficulties of the theoretical treatment.

Partition chromatography of steroids is now a simple and routine procedure, although it still requires a rather closer attention to technical detail than similar procedures for amino acids, sugars, and purines, among others. These technical difficulties are due to the fact that favourable partition coefficients are only obtained for the majority of steroids with solvent systems that are rather volatile, and therefore rather more sensitive to changes of temperature and to failure to achieve complete equilibrium in the system. Indeed the *average* standard of technique used in many other fields is below that needed for steroid work. This is simply because the solvents commonly used in other fields are less sensitive to deviations from ideal conditions and good practical results are more easily obtained. It should be remembered, however, that several workers have already emphasized that really consistent and reproducible results with these easier solvent systems are only achieved when the technique is of a high standard (e.g. Bate-Smith and Westall, 1950).

The main types of method have been classified before (Bush, 1954b; Heftmann, 1955; Neher, 1958a) and an up-to-date summary is given in Table 2.1.

Table 2.1. Examples of Partition Chromatography Applied to the Steroids

imp. = impregnated-paper method (p. 64).
± imp. = orthodox or impregnated-paper method can be used (p. 167)
* = low ΔR_{Mg} values are expected due to an unusually large intermiscibility of the phases being probable.
( ) = a large or even major part of the component in brackets is probably in the other phase.

The classification by type is according to the definition given on p. 83 but has not been proved in all cases. Alkaline or acid systems in which it is known or probable that the ΔR_{Mg} of non-ionizing groups is "typical" are classed with "typical" systems (p. 214). It is clear that the ΔR_{Mg} values of ionizing groups will be greatly different in such systems (see p. 289).

The classification by solubility groups A, B, C, D and R is largely on the basis of neutral steroids; it is not, therefore, to be taken as a recommendation for neutral steroids in the case of alkaline or acid systems except in the case where the main steroid example indicates that such systems have in fact been used for neutral steroids (see Glossary II).

Certain papers giving exceptionally useful details of points arising from the source of the steroids involved are indicated by putting the source in brackets after the example.

Type of system	Solubility group	Stationary phase	Mobile phase	Supporting material	Main example	Non-steroid examples	References
Reversed phase (typical)	Group R Very non-polar, e.g. sterols and their esters	Hydrocarbons, Paraffin, Carbon tetra-chloride	Aq. alcohols	Paper (imp.)	Non-polar steroids	—	Sykora and Prochazka (1953)
		Vaseline	Aq. alcohols	Acetylated paper	—	Non-polar dinitrophenyl-hydrazones	Kostir and Slavik (1950)
		Vaseline	Aq. alcohols	Paper (imp.)	—	DDT derivatives	Winteringham et al. (1950)
		Paraffin oil	Aq. alcohols	Paper (imp.)	—	—	Quaife, Geyer and Bolliger (1956)
		Kerosene	Aq. alcohols	Paper (imp.)	Vitamin D group	—	Kodicek and Ashby (1954)
		Kerosene	Aq. alcohols	Paper (imp.)	—	Paint-resins, terpenes, etc.	Mills and Werner (1952, 1955)
		Cyclohexane	Aq. alcohols	Siliconed paper	Sterols	—	R. P. Martin (1957) Kucera, Prochazka and Veres (1957) Kritchevsky and Tiselius (1951)
		Carbon tetra-chloride	Aq. alcohols	Aluminium soap on paper	Sterols	—	Kiss and Szell (1956)
		Kerosene	Aq. alcohols	Paper (imp.)	Steroid amines	—	Prochazka, Labler and Kotasek (1954)

Table 2.1.—continued.

Type of system	Solubility group	Stationary phase	Mobile phase	Supporting material	Main example	Non-steroid examples	References
Reversed phase (typical)	Group R Very non-polar e.g. sterols and their esters	Hydrocarbons, etc.	Aq. alcohols	Quilon-paper	Sterols, C_{19} and C_{21} steroids		Kritchevsky and Kirk (1952)
		Cellosolves	Aq. alcohols	Quilon-paper	Sterols, C_{19} and C_{21} steroids		Davis et al. (1952)
		Medicinal paraffin, Chloroform	Aq. alcohols	Siliconed Celite		Fatty acids	Howard and Martin (1950)
		Hydrocarbons Chloroform heptane	Aq. methanol	Siliconed Celite	Bile acids		Bergstrom and Sjövall (1951) Norman (1953, 1954)
		Hydrocarbon mixtures	Aq. methanol	Acetylated paper	Adrenal steroids, esters, etc.		Zijp (1955) and Micheel and Schweppe (1954) Ritter and Hartel (1958)
Reversed phase (atypical)	Group R	Medicinal paraffin–chloroform	Aq. acetic acid	Kerosened paper (imp.)	Sterols and esters		Michalec (1955) Michalec, Jirgl and Podzimek (1957)
		Paraffin–chloroform	Aq. acetic acid	Kerosened paper (imp.)			Prochazka et al. (1954)
		Paraffin	Aq. acetone	Siliconed Celite		Fatty acids	Howard and Martin (1950)
		Paraffin	30% aq. acetic acid	Paper (imp.)		Non-polar insecticides	Beroza (1956)
Straight phase (typical)	Group A Non-polar steroids, steroids and esters, etc.	Aq. methanol	n-Hexane	Celite	Progesterone, etc.		Butt, et al. (1948, 1951) Taylor (1954)
		Propylene glycol	Light petroleum	Celite	Cholesterol derivatives		Mosbach et al. (1953b)
		Aq. methanol	Cyclohexane	Celite	Cholesterol derivatives		Mosbach et al. (1953b)

Table 2.1.—continued.

Type of system	Solubility group	Stationary phase	Mobile phase	Supporting material	Main example	Non-steroid examples	References
Straight phase (typical)	Group A Non-polar sterols and steroids and esters, etc.	Formamide	Cyclohexane–benzene	Silica gel	17-Oxosteroids, etc.		Katzenellenbogen, Dobriner and Kritchevsky (1954)
		Ethanol	Methylene chloride–light petroleum (1:1)	Silica gel	$C_{19}O_2$ acetates $C_{19}O_2$, $C_{21}O_2^-$		Katzenellenbogen et al. (1954)
		Aq. ethanol	Ligroin–toluene	Paper	Oestrogen diazo dyes, some sapogenins		Heftmann (1950, 1951) Heftmann and Hayden (1952)
		Cellosolves	Heptane	Paper (imp.)	$C_{19}O_2$ acetates, $C_{19}O_2$, $C_{21}O$ compounds, etc.		Neher and Wettstein (1952) Rubin et al. (1953a) Loomeijer and Luinge (1958)
		Propylene glycol	Ligroin, etc.	Paper (imp.)	Progesterone, 17-oxosteroids, etc.		Savard, (1953, 1954) Rosenkrantz (1953)
		Aq. methanol	Light petroleum, ligroin, etc.	Paper	Progesterone, 17-oxosteroids, etc.		Bush (1952, 1954) Bush and Mahesh (1959b, c)
		Aq. methanol various concentrations	Aliphatic hydrocarbons	Paper	Progesterone, etc.		Zander, Simmer, Münsterman and Marx (1954)
		Aq. methanol	Light petroleum, etc.	Paper	Ethyl esters of bile acids		Haslewood (1954)
		Propylene glycol	Methylcyclohexane	Paper (imp.)	Less polar oestrogens		Axelrod (1953a)
		Formamide	Cyclohexane, cyclohexane–benzene	Paper (imp.)	$C_{19}O_2$ steroids, etc.		Savard (1953) Neher (1959a, p. 168)
		Benzyl alcohol	Light petroleum	Paper	Various non-polar steroids		Michalec (1955)

Table 2.1.—continued.

Type of system	Solubility group	Stationary phase	Mobile phase	Supporting material	Main example	Non-steroid examples	References
Straight phase (atypical)	Group A	70% acetic acid	Aliphatic hydrocarbons (±isopropyl ether)	Celite	Bile acids		Mosbach et al. (1954)
		Nitromethane	3% CHCl$_3$ in light petroleum	Silica gel	17-Oxosteroids	Halogenated hydrocarbons	Ramsey and Patterson (1946, 1948) Jones and Stitch (1953)
		70% acetic acid	Light petroleum	Paper (± imp.)	Bile acids		Sjövall (1952, 1954)
		70-90% acetic acid	Light petroleum, decalin, etc.	Paper (± imp.)	C$_{19}$O$_2$ and C$_{21}$O$_2$ steroids		Bush (1960b)
		Butane-1,3-diol	Methylcyclohexane, etc.	Paper (imp.)	C$_{19}$O$_2$ and C$_{21}$O$_2$ steroids		Axelrod (1953b)
		Nitromethane–methanol, (1:1)	Decalin, etc.	Paper (± imp.)	Non-polar steroids and esters		Peterson (personal communication), Bush (unpublished)
		Carbitol (diethylene glycol monoethyl ether)	Methylcyclohexane, etc.	Paper (± imp.)	Non-polar steroids and esters, etc.		Reineke (1956)
Straight phase (typical)	Group B	Ethylene glycol	Increasing CH$_2$Cl$_2$ in cyclohexane	Silica gel	Less polar adrenal steroids		Haines (1952)
		Ethanol	CH$_2$Cl$_2$–light petroleum	Silica gel	C$_{19}$O$_3$ and 17-oxosteroids, C$_{21}$O$_{4-5}$ acetates, etc.		Katzenellenbogen et al. (1954)
		Formamide	Methylcyclohexane, CH$_2$Cl$_2$, etc.	Paper (imp.)	C$_{19}$O$_3$ steroids		Axelrod (1953b)

Table 2.1.—continued.

Type of system	Solubility group	Stationary phase	Mobile phase	Supporting material	Main example	Non-steroid examples	References
Straight phase (typical)	Group B	Formamide	Benzene, cyclohexane-benzene, etc.	Paper (imp.)	$C_{19}O_3$ steroids		Neher (1959a, pp. 168, 169)
		Aq. ethanol	Ligroin	Silica gel		Dinitrophenyl amino acids	Sanger (1945)
		Aq. ethanol	Ligroin–toluene	Paper	Moderately polar sapogenins		Heftmann and Hayden (1952)
		50% aq. methanol	Carbon tetrachloride	Silica gel		Dinitrophenyl amino acids	Sanger (1945)
		70–85% aq. methanol	Light petroleum–benzene or toluene mixtures tetralin, etc.	Paper	$C_{19}O_3$, $C_{21}O_4$, $C_{21}O_5$ monoacetates, etc.		Bush (1952, 1954, 1960b) Ghillain (1956) De Courcy (1956) Bush and Willoughby (1957) Bush and Mahesh (1959b, c)
		Abs. methanol	Ligroin, etc.	Paper	Oestrogens		Mitchell (1952) Mitchell and Davies (1954) Dao (1957)
		Ethylene glycol	Benzene	Silica gel		Dinitrophenyl amino acids	Sanger (1945)
		Aq. methanol	Toluene–light petroleum	Cellulose powder	Aldosterone diacetate, etc.		Simpson et al. (1954) Ayres et al. (1957a)
Straight phase (atypical)	Group B	70% aq. acetic acid	Light petroleum–iso-propyl ether mixtures	Celite	Bile acids		Mosbach et al. (1954)
		70% acetic acid	Heptane–chloroform, etc.	Paper (±imp.)	Bile acids		Sjövall (1952, 1954)

Table 2.1.—continued.

Type of system	Solubility group	Stationary phase	Mobile phase	Supporting material	Main example	Non-steroid examples	References
Straight phase (atypical)	Group B	70-85% acetic acid	Light petroleum-toluene mixtures, decalin-xylene mixtures, etc.	Paper (± imp.)	$C_{19}O_2$, $C_{21}O_3-4$ $C_{21}O_4$-monoacetates, $C_{21}O_5$-diacetates, etc.		Bush (1960b)
		Aq. methanol (dioxan)	Cyclohexane	Paper	$C_{21}O_3-4$, $C_{19}O_3$ esters, etc.		R. Peterson (unpublished)
		(t-Butanol) water–methanol	Isooctane, isopropyl ether, etc.	Paper	Adrenal steroids and esters, $C_{19}O_3$ compounds, etc.		Eberlein and Bongiovanni (1955) Neher and Wettstein (1956b)
		(Pyridine) water	Light petroleum	Paper	Sapogenins		Callow et al. (1955)
Straight phase typical	Group C	Ethylene glycol	CH_3Cl and mixtures with cyclohexane	Silica gel	Adrenal steroids (glands)		Haines (1952)
		Propylene glycol	Toluene	Cellulose powder	Adrenal steroids (glands)		Baker, Dobson and Stroud, (1951)
		Ethylene glycol	Toluene and mixtures with light petroleum	Celite	Adrenal steroids (blood)		Morris and Williams (1953a, b)
		Aq. ethanol	Toluene	Celite	Adrenal steroids (blood)		Morris and Williams (1953a, b)
		Aq. methanol	Benzene, toluene–ethyl acetate	Celite	Adrenal steroids (cortisol, aldosterone, etc.) (gland)		Simpson and Tait (1953) Harman, Ham, De Young, Brink and Sarett (1954)

Table 2.1.—continued.

Type of system	Solubility group	Stationary phase	Mobile phase	Supporting material	Main example	Non-steroid examples	References
Straight phase (typical)	Group C	Water	Benzene and mixtures with CHCl₃ or petroleum ether	Celite	Adrenal steroids (aldosterone, etc.) (glands)		Simpson et al. (1954)
		Water	CH₂Cl₂ (gradient) in light petroleum	Silicic acid	Adrenal steroids (urine)		Heftmann and Johnson (1954) Johnson, Heftmann and Hayden (1956)
		Water	Benzene–ether	Silicic acid	Adrenal steroids		Cook et al. (1955)
		Formamide	Benzene–CHCl₃, CHCl₃, etc.	Cellulose powder	Altosterone, etc. (gland)		Mattox, Mason and Albert (1956)
		Aq. methanol	Benzene	Celite	Pregnanetriol (urine)		Cox and Marrian (1953)
		Propylene glycol	Toluene	Paper (imp.)	Adrenal steroids		Zaffaroni et al. (1950)
		Formamide	Benzene	Paper (imp.)	Adrenal steroids		Burton et al. (1951)
		Formamide	Chloroform, ± benzene, etc.	Paper (imp.)	Adrenal steroids		
		Water	Benzene, chloroform, etc.	Celite	Cardenolides, etc.		Hegedüs, Tamm and Reichstein (1953) Lichti et al. (1956)
		Aq. ethanol	Toluene, light petroleum–toluene	Paper	Sapogenins		Heftmann and Hayden (1952)
		Aq. methanol	Toluene, benzene, etc.	Paper	Adrenal steroids Cardiac aglycones		Bush (1951, 1952, 1954) Bush and Taylor (1952)

Table 2.1.—continued.

Type of system	Solubility group	Stationary phase	Mobile phase	Supporting material	Main example	Non-steroid examples	References
Straight phase (typical)	Group C	Formamide	Benzene–CHCl₃ (1:1), CHCl₃	Paper (imp.)	Cardiac aglycones some glycosides		Schindler and Reichstein (1951)
		Formamide	CHCl₃	Paper (imp.)	*Scilla* and *Bufo* sapogenins, glycosides, etc.		Schröter, Tamm, Reichstein and Deulofeu (1958) Urscheler, Tamm and Reichstein (1955)
		Propylene glycol	Benzene–CHCl₃, benzene–light petroleum, etc.	Paper (imp.)	*Scilla* and *Bufo* sapogenins, glycosides, etc.		Ruckstuhl and Meyer (1957)
		Formamide	Benzene, benzene–chloroform	Paper (imp.)	Aldosterone and its esters		Luetscher, Neher and Wettstein (1955) Mattox *et al.* (1956) Mattox and Lewbart (1958, 1959)
		Aq. methanol	Hydrocarbon mixtures, etc.	Paper	Prednisone metabolites (urine)		Gray *et al.* (1956)
		Ethylene glycol	Various hydrocarbons	Paper	Aldosterone, urinary metabolites		Nowaczynski and Koiw (1957)
		Formamide,	Benzene,	Paper (imp.)	Cardiac aglycones acetates, glycosides		Heftmann and Levant (1952)
		Propylene glycol	Toluene	Paper (imp.)			
		Aq. NaOH (0·4–3·1N)	Benzene, benzene–light petroleum, benzene–n-butanol, CHCl₃–n-butanol	Celite	Oestrogens		Swyer and Braunsberg (1951) Stern and Swyer (1952) Bitman and Sykes (1953) Haenni, Carol and Banes (1953) Braunsberg, Stern and Swyer (1954) Bauld (1955, 1956)
		Aq. 2N-NH₄OH	Benzene–CHCl₃	Paper	Oestrogens		Heusghem (1953)

Table 2.1.—continued.

Type of system	Solubility group	Stationary phase	Mobile phase	Supporting material	Main example	Non-steroid examples	References
Straight phase (typical)	Group C	80% Methanol*	Benzene, etc.	Paper	Oestrogens		Puck (1955)
		50% Methanol	Benzene	Paper	Polar oestrogens		Dao (1957)
Straight phase (atypical)	Group C	Glacial acetic acid	Light petroleum–CHCl$_3$ etc.	Paper	Sapogenins		Sannié, Heitz and Lapin (1951) Sannié and Lapin (1952a, b)
		Formamide	Xylene–ethyl methyl ketone	Paper (imp.)	Cardiac aglycones, glycosides, etc.		Kaiser (1955)
		Acetic acid (30–80%)	Toluene, benzene, etc.	Paper		Urinary phenols, etc.	Bray, Thorpe and White (1950) Boscott (1952)
		Acetic acid	Toluene, etc.	Paper (± imp.)	Oestrogens		Boscott (1952)
		Acetic acid (30–80%)	Toluene, benzene	Paper (± imp.)	Bile acids and conjugates, adrenal steroids, etc.		Beyreder and Rattenbacher-Däubner (1953) Bush (1960b)
		Acetic acid (70–90%)	Toluene, mixtures with light petroleum, etc.	Paper(± imp.)		Fatty acids Ketoacid 2,4-dinitrophenyl-hydrazones	Lieberman, Zaffaroni and Stotz (1951) Bush and Hockaday (1960)
		Acetic acid (70%)	Heptane–isopropyl ether mixtures	Paper (± imp.)	Bile acids, etc.		Sjövall (1952, 1954)
		Water–t-butanol	Isooctane	Paper	Adrenal steroids		Eberlein and Bongiovanni (1955)
		Methanol–water–(dioxan)*	Cyclohexane etc.	Paper	Adrenal steroids		R. Peterson (unpublished)
		Aq. acetic acid	Toluene	Paper	Steroid carboxylic acids		Prochazka (1954)

Table 2.1.—continued.

Type of system	Solubility group	Stationary phase	Mobile phase	Supporting material	Main example	Non-steroid examples	References
Straight phase (atypical)	Group C	Water	Light petroleum, toluene, etc.	Paper (wetted with acetone)	Adrenal steroids		Shull (1954)
		Water–(n-butanol) etc.	Toluene petroleum, etc.	Paper (wetted with water)	Adrenal steroids		Pechet (1953, 1955)
		Water–formamide	Butyl acetate	Paper (imp.)	Aldosterone and companions (urine)		Mattox and Lewbart (1958, 1959)
Reversed phase (typical)	Group C	Benzene	Aq. methanol	Rubber	Oestrogens		Nyc et al. (1951) Bosch (1953)
		Chloroform	Aq. alcohols	Siliconed paper	Oestrogens		Gunzel and Weiss (1953)
		Alcohols ($\geqslant C_5$)	Water, etc.	Paper	Corticosteroids		Schmidt, Staudinger and Bauer (1953)
		n-Butanol	Water	Paper	Cardenolides		Habermann, Müller and Schreglmann (1953)
		Octanol–pentanol, etc.	Water–formamide	Paper	Cardiac aglycones acetates, etc.		Tschesche et al. (1953) Tschesche and Seehofer (1954)
		Paraffin oil	Aq. ethanol (75%)–ammonia	Paper (imp.)	Steroid amines, etc.		Prochazka et al. (1954)
		Alcohols ($\geqslant C_5$)	Water, etc.	Paper	Corticosteroids		Schmidt and Staudinger (1954)
		Isoamyl alcohol–chloroform, n-butanol, etc.	Water	Siliconed Celite	Bile acid conjugates, etc.		Norman (1953, 1954)
		Ethanol-rich?	Aq. methanol–(ethanol)	Esterified resin	Corticoids		Seki (1959)

Table 2.1.—continued.

Type of system	Solubility group	Stationary phase	Mobile phase	Supporting material	Main example	Non-steroid examples	References
Reversed phase (typical)	Group C	Ethanol-rich?	Aq. methanol-(ethanol)	Esterified resin	Oestrogens		Seki (1958)
		Benzene	Aq. alcohols	Siliconed paper	Oestrogens		Markwardt (1954, 1955)
Straight phase (typical)	Group D	Aq. methanol	Chloroform	Paper	Cardiac glycosides and aglycones		Svendsen and Jensen (1950)
		Aq. methanol	Benzene– or toluene–ethyl acetate, benzene–chloroform, etc.	Paper	Adrenal steroids cortol, cortolones, cardiac aglycones and glycosides, etc.		Bush (1952, 1954) Simpson and Tait (1953) R. Peterson (unpublished)
		Aq. methanol	Toluene–ethyl acetate, etc.	Celite	Adrenal steroids		Simpson and Tait (1953) Ayres et al. (1957a) Simpson et al. (1954)
		Aq. methanol	Ethylene dichloride	Celite	Polar oestrogens		Bauld (1955, 1956)
		Water (± alcohols) (± salts)	Ethyl acetate, ethyl methyl ketone, chloroform	Paper	Cardiac glycosides, etc.		Jaminet (1950, 1951, 1952) Hassall and Martin (1951) Silberman and Thorp (1953, 1954)
		Water	Isobutanol	Paper	Cardiac glycosides, aglycones, etc.		Tschesche and Seehofer (1954)
		Water	n-Butanol ± toluene	Paper	Cardiac glycosides, aglycones, etc.		Schenker, Hunger and Reichstein (1954)
		Water (± methanol)	Ethyl acetate, chloroform, etc.	Silica gel, cotton linters	Scilla and Bufo glycosides, etc.		Stoll, Angliker, Barfuss, Kussmaul and Renz (1951) Stoll and Kreis (1951)

Table 2.1.—continued.

Type of system	Solubility group	Stationary phase	Mobile phase	Supporting material	Main example	Non-steroid examples	References
Straight phase (typical)	Group D	Aq. (acetic acid) (25%)	n-Butanol	Paper	Saponins and sapogenins		Tsukamoto, Kawasaki, Naraki and Yamauchi (1954) Dutta (1955)
		Aq. (acetic acid) (25%)	Isoamyl alcohol	Paper	Alkaloids (Holarrhena)		Prochazka et al. (1954)
		Aq. (acetic acid) (25%)	n-Butanol	Paper	Alkaloids		Tschesche and Peterson (1954)
		Aq. methanol (50%)	Isopropyl ether, ±benzene	Paper	Very polar cortisone analogues (e.g. triamcinolone), cortols, etc.		Bush (unpublished)
		Aq. (acetic acid) (≤ 25%)	Ethyl acetate ethanol	Paper	Solanine alkaloids, etc.		Kuhn and Löw (1954)
		Aq. acetic acid (1–2%)	Ethylene dichloride	Paper	Veratrum alkaloids		Nash & Brooker (1953)
		Aq. (pyridine) (15%)	Benzene– n-butanol mixtures	Paper	Saponins and sapogenins		Tsukamoto et al. (1954)
Straight phase (atypical)	Group D	Water	Collidine	Paper	Bile acids and their conjugates		Siperstein, Harold, Chaikoff and Dauben (1954)
		Aqueous KCl	n-Butanol	Paper	Veratrum alkaloids		Auterhoff (1954)
		(miscible)	Aqueous 20% KCl	Paper		Urinary phenols	Boscott (1951)
Straight phase (typical)	Polar glycosides, glucuronosides, sulphates, etc.	Conc. ammonia– water (3 : 2)	Isoamyl alcohol	Paper	Steroid carboxylic acids		Prochazka (1954)

Table 2.1.—*continued*.

Type of system	Solubility group	Stationary phase	Mobile phase	Supporting material	Main example	Non-steroid examples	References
Straight phase (typical)	Polar glycosides, glucuronosides, sulphates, etc.	Aq. (acetic) or formic acid (10%)	n-Butyl acetate	Paper	Glucuronosides		
		Methanol–0.1M barbiturate buffer	n-Butyl acetate (miscible)	Paper	Glucuronosides and sulphates		
		0.2% Aq. NH₄OH	Ethyl acetate–n-butanol	Paper, counter-current apparatus	Less polar sulphates		Lewbart and Schneider (1955)
		2N-Ammonium hydroxide	Ethyl acetate–toluene–n-butanol	Paper	More polar sulphates		Schneider and Lewbart (1956)
		0.1N-HCl	n-Butanol–ethyl acetate	Counter-current apparatus	More polar glucuronosides		Schneider and Lewbart (1959)
		Aq. (acetic acid) (30%)	Ethyl acetate–hexane (3:2)	Paper, silica gel, counter-current apparatus	Simple glucuronosides		
		Aq. (acetic acid) (30%)	n-Butyl ether–n-Butanol (2:1)	Paper	Polar glucuronosides		
Straight phase (typical)	Polar glycosides, sulphates, etc.	2N-Ammonium hydroxide	n-Butanol, etc.	Paper, powdered cellulose, silica gel or Celite	Group fractionation: sulphates and glucuronosides		Bush (1957)
		2N-Ammonium hydroxide	Isoamyl alcohol	Paper	Various sulphates		Cavina (1957)

Table 2.1.—continued.

Type of system	Solubility group	Stationary phase	Mobile phase	Supporting material	Main example	Non-steroid examples	References
Straight phase (typical)	Polar glycosides, sulphates, etc.	2% Aq. HCl–methanol or –ethanol	Isopropyl ether	Paper, Celite, counter-current apparatus	Less polar glucuronosides and sulphates		Bush and Mahesh (unpublished)
		Aq. methanol (70–85%)	10% Dicyclohexylamine in light petroleum, ±toluene mixtures, etc.	Paper	Sulphates		Bush (unpublished)
		Water	Ethyl acetate–dicyclohexylamine (9:1)	Paper	Less polar glucuronosides		
Straight phase (atypical)	Polar glycosides, sulphates, etc.	Acetic acid (70%)	Amyl acetate–heptane (8:2)	Paper	Bile acid conjugates		Sjövall (1955)
		(Acetic acid) (70%)	Isopropyl ether–heptane (60:40–85:15), n-butanol	Paper	Bile acid conjugates		Haslewood and Sjövall (1954)
		30% Aq. (acetic acid) ±(t-butanol)	Toluene, benzene	Paper, Celite	Less polar glucuronosides		Bush (1957) Schneider and Lewbart (1959)
		30% Aq. acetic or formic acid–(t-butanol) (5–10%)	Ethylene, dichloride, isopropyl ether, etc.	Paper, Celite	11,17-dioxysteroid 3-glucuronosides, less polar sulphates		Bush (1957, unpublished)
		Aq. (acetic acid)	n-Butanol	Paper	Saponins		Dutta (1955)
		Aq. acetic or formic acid (30%)–(t-butanol) (20–30%)	Ethylene, dichloride, isopropyl ether	Paper, Celite	Polar glucuronosides		Bush (1957, unpublished)

Table 2.1.—continued.

Type of system	Solubility group	Stationary phase	Mobile phase	Supporting material	Main example	Non-steroid examples	References
Straight phase (atypical)	Polar glycosides, sulphates, etc.	2N-Ammonium hydroxide–(t-butanol) (10–30%)	Ethylene dichloride	Paper, Celite	Less polar sulphates		Bush (1957)
		2N-Ammonium hydroxide–(t-butanol) (10–40%)	Isopropyl ether, n-butyl ether	Paper	Less polar sulphates		Schneider and Lewbart (1959) Bush and Gale (1960)
		Aq. (t-butanol) (20–30%)	Isopropyl ether–dicyclo-hexylamine (9:1)	Paper	Less polar glucuronosides		Bush (1960c)
		Aq. (t-butanol) (30–50%)	Isopropyl ether–dicyclo-hexylamine (9:1)	Paper	11,17-Dioxysteroid 3-glucuronosides		Bush (unpublished)
		Aq. (t-butanol) (10–30%)	Ethyl acetate–dicyclohexyl-amine (9:1)	Paper	Polar glucuronosides		

The first successful partition methods for steroids were those of Morris's group (Butt, Morris and Morris, 1948; Butt, Morris, Morris and Williams, 1951) using Celite columns. Soon afterwards several groups overcame the technical difficulty of running useful partition systems on paper and this led to a great increase in the use of partition chromatography for steroids. First, Zaffaroni, Burton and Keutmann (1950) described the use of paper loaded with propylene glycol and formamide as the stationary phase. Then Heftmann (1950), Heftmann and Hayden (1952), and Bush (1951, 1952) described techniques using volatile solvent systems in which equilibration by the vapour phase was responsible for the formation of the stationary phase, as in the classical method of Consden, Gordon and Martin (1944).

Ramsey and Patterson (1946, 1948a) first introduced methanol and nitromethane as stationary phases for non-polar substances (fatty acids) using silica gel as supporting medium. This type of system is the basis of many successful methods for steroids and other non-polar substances, but was unfortunately not published in a widely circulated journal.

The author used elevated temperatures originally (on general grounds and because some temperature control was obviously essential) but later showed that a reasonably well-controlled lower temperature was adequate, particularly if the original solvent mixtures were modified to contain a higher proportion of methanol (Bush, 1954a). More recently Eberlein and Bongiovanni (1955) introduced volatile systems based on t-butanol; these systems are of great interest and value as will be seen later on. Systems suitable for the more polar steroids such as glycosides, etianic acids, and bile acids are closer to those used for sugars and amino acids, and recently Lewbart and Schneider (1955) and Bush (1957) introduced systems suitable for steroid sulphates and glucuronosides. Very recently the author has tested a variety of systems in the search for different types of solvent–solute interactions and some of these have proved useful (p. 100).

The practical advantages of these different methods will be discussed later; meanwhile I shall only discuss certain features important for the theoretical part of this chapter. It was pointed out by Jermyn and Isherwood (1949) that the best separations in practice were obtained (for carbohydrates) by "overrunning" the paper for a considerable time with solvent systems in which the substances most difficult to separate had R_F values around 0·2. This is in fact the routine procedure with most of the impregnated-paper methods for steroids but is only used with the volatile systems when difficult separations are attempted, when the mixture to be separated is complicated, or when extra separation is needed for quantitative estimation or isolation of material (Bush, 1952; Bush and Willoughby, 1957). The emphasis in the earlier paper on single-length runs with volatile systems for simple mixtures has sometimes given rise to the erroneous idea that there is some essential difficulty preventing the overrunning of the volatile systems.

An important factor in partition chromatography of steroids is their low solubility. Thus very few systems based upon water as a stationary phase work well, even with columns, unless they are large (Simpson, Tait, Wettstein, Neher, von Euw, Schindler, and Reichstein, 1954). The silicic acid–water systems of Cook, Dell and Wareham (1955) are an exception but it is not entirely certain that adsorption is not playing an important part in their method. Thus the mobilities of the steroids they have published are rather lower than would be expected from their partition coefficients in the solvents used, making the assumption that the ratio of the cross-sections of the two phases is approximately 1:1. Most systems for steroids can in fact be considered as made up of three components:

(a) A main component of the mobile phase of more or less non-polar character;
(b) A main component of the stationary phase of more or less polar character such as an alcohol;
(c) A small amount of water such that two phases of sufficiently different composition separate in the presence of the supporting medium.

The last is omitted from the methods using impregnated paper and some column methods. The middle component is chosen so as to increase the solubility of the steroids in both phases and to prevent, in the case of paper (Bush, 1954a), their adsorption by the supporting material. The versatility of ternary systems has been emphasized before (Martin and Synge, 1941; Isherwood, 1946; Partridge, 1948). Simple reversed-phase systems (Winteringham, Harrison and Bridges, 1950; Mills and Werner, 1952, 1955; Spiteri, 1954; R. P. Martin, 1957) are similarly composed except that the non-polar component associates with the supporting material. Paper, glass, vulcanized rubber powder (Nyc, Maron, Garst and Friedgood, 1951), sand or carborundum can provide support for a stable stationary phase with high-boiling hydrocarbons. With other non-polar solvents a stable phase is often only secured by using rubber (Boldingh, 1948, 1950) or siliconed bases (Celite or paper; Howard and Martin, 1950), Kritchevsky and Tiselius, 1951) as supporting materials.

Certain solvent systems are more complicated in their composition and their exact mechanism is unknown. Thus the cellosolves and carbitols (Neher and Wettstein, 1952; Reineke, 1956) provide a stationary phase on paper which is probably far more like a "non-polar solvent" because the few polar groups are probably largely or entirely associated with the hydroxyl groups of the cellulose according to the amount added to the paper and the degree of swelling they cause. Nevertheless the mobile phase is itself a hydrocarbon. The same thing applies to the systems based on nitromethane (Ramsey and Patterson, 1946, 1948; Jones and Stitch, 1953) which can be used on paper as well as on silica gel (Dr. R. Peterson, personal communication). Hoffman and Staudinger (1951) used the aqueous layer of the system n-butanol—water as the mobile phase and achieved good separations of corticosteroids; the order of separation justifies (cf. Heftmann, 1955) their supposition that a reversed phase. mechanism was involved since it is known in this case that the use of the butanol-rich phase results in high R_F values and no useful separation. Similarly, higher alcohols (Schmidt and Staudinger, 1954) used in this way (i.e. with the aqueous phase for running) give slightly better separations, and it is known that the use of the alcohol-rich phase in such cases does give the opposite order of mobilities of the corticosteroids, although R_F values are too high to give useful separations (Burton, Zaffaroni and Keutmann, 1948). Tschesche, Grimmer and Seehofer (1953) however used paper impregnated with various alcohols and some more complicated mixtures and found that the order of mobility of cardiac glycosides was the same whether paper was impregnated with the aqueous phase or the alcohol phase, and the other phase used for running. This is unexplained. Various other systems of this type have been described, e.g. 25% methanol for corticosteroids (Meyerheim and Hubener, 1952) and even water with cellulose (Schmidt and Staudinger, 1954). Some of these systems have poor resolving power, since, for instance, cortisol and cortisone were not separated from each other by water–n-heptanol although aldosterone was separated from them. The system xylene–methanol (225: 75 by vol.) is an example of an unstable system of dubious mechanism: Sakal and Merrill (1953) found that R_F values were very dependent upon whether the paper was dried before use or not, being lower when undried paper was used. Judging by the scanning curves given by Tennent, Whitla and Florey (1951), who introduced this system, considerable adsorption is occurring and the separations are poor. Such "miscible" systems (p. 3) are generally poorly reproducible when suitable for the neutral steroids, since the phase composition is extremely sensitive to changes in water content.

Such "miscible" and "reversed-phase" systems have not been analysed sufficiently to allow a definite statement on their mechanism. The "miscible" systems are probably working by the sort of mechanism suggested by Martin (1950) but adsorption may be important. In the author's view there is little to commend them since their practical achievements are matched or bettered in the steroid field by systems of known mechanism, and in the absence of knowledge of their exact mechanism it is not possible to predict the behaviour of steroids or the effects of impurities in extracts.

Many types of more clear-cut reversed-phase chromatograms have been tried. On the whole they are all rather more prone to instability than "straight" systems and nearly all of them involve treatment of the supporting material with a hydrophobic substance such as silicone or rubber, or treatment with acylating agents to provide a hydrophobic layer. Siliconed Celite (Howard and Martin, 1950) was first used for separating fatty acids, and care has to be taken to use solvent systems of which the non-polar phase "sticks" firmly to the siliconed particles. A similar method with columns of siliconed Celite was used by Morris and Williams (1953b) for an elegant removal of fats and cholesterol from extracts of plasma containing adrenal steroids. Kritchevsky and Tiselius (1951) evolved a similar method with paper for steroids using the polar phase of chloroform—ethanol—water (10: 10: 6 by vol.) as mobile phase. Boldingh (1948) soaked strips of filter paper in a solution of vulcanized latex, dried them in air and then stored them under acetone. These were used for reversed-phase separation of aliphatic esters with methanol–benzene solvent systems.

Kostir and Slavik (1950) first tried acetylated paper for separating dinitrophenylhydrazones, and pointed out that paper treated in this way could be used in the "straight" or reversed-phase manner. The right conditions for the acetylation are, however, difficult to achieve and Davis, McMahon and Kalnitsky (1952) were unable to get good results with sterols. On the other hand Ritter and Hartel (1958) obtained good separations on acetylated paper using a better controlled method of preparing the paper, devised by Zijp (1955). A very simple acylation of paper was obtained by Kritchevsky and Calvin (1950) who used stearatochromyl chloride; this reagent is simply sprayed on as a 2% solution and dried at 100–110 °C, or left to dry at room temperature (Kritchevsky and Kirk, 1952b). Aqueous alcohol was used as the mobile phase and had to be more than 80% in alcohol concentration for capillary penetration of the paper.

Relatively little is known about the behaviour of steroids in such systems and they do not seem to have been applied to many practical problems.

The separations described by Ritter and Hartel (1958) do not seem to offer any advantage over the appropriate "straight" systems that are available. Indeed some of them are definitely worse; although this is almost certainly largely due to the type of solvent mixtures that these authors used and is not an essential feature of the method. Thus the ΔR_{Mg} values of the substituents of the steroids that they studied (calculated from their diagrams) are much smaller than with the equivalent "straight" systems using untreated paper (see this chapter, Section 2).

Much simpler ways of obtaining reversed-phase chromatograms on paper are available and seem, from the limited amount of work that has been done with reversed-phase systems, to be just as good as the more complicated methods described so far. These are based upon the fact that untreated paper can retain a non-polar solvent as a stable stationary phase if the latter is sufficiently viscous and relatively involatile. The first description of a method of this type was that of Winteringham et al. (1950) who used paper impregnated with Vaseline (petroleum jelly) to separate DDT derivatives. Mills and Werner (1952) found that the relatively volatile "odourless kerosene" (high-boiling light petroleum) could be used in the same way; it has the advantage that it can be removed from the paper relatively easily by drying even at room temperature. This method was very suitable for separating non-polar sterols and their esters using aqueous n-propanol or aqueous methanol as mobile phase (equilibrated with kerosene). Useful separations were obtained with this system (R. P. Martin, 1957) of various cholesterol derivatives but several important isomeric pairs were not resolved. Kodicek and Ashby (1954) applied similar methods to the separation of ergosterol and related compounds. A similar system was used by Prochazka (1954) for steroid alkaloids and acids.

At the moment it seems fair to suggest that with very non-polar steroids such as esters of steroid monohydroxymonoketones, sterols and their esters, the reversed-phase systems show little sign of competing with adsorption chromatography. On the other hand, there are reasons for hoping that much improvement of them is possible. For instance, one reason for the poor resolution of various isomeric sterols with published systems may be that in most cases the stationary non-polar phase has consisted of petroleum fractions containing a mixture of very long-chain paraffins; the Van der Waals interactions between such molecules and the hydrocarbon part of the sterol molecule will be insensitive to geometrical changes in the sterol because of the large number of conformations available to the long-chain paraffins. As discussed earlier such interactions will probably be much more specific if a solvent with less conformational flexibility is used and the successful use of tetralin as a stationary phase on untreated filter paper for separating fatty acids as their dinitrophenylhydrazides (Inouye et al., 1955) shows that there are considerable possibilities in this direction (see p. 21). Similarly, there is no reason why aqueous t-butanol and acetic acid (Prochazka, 1954; Spiteri, 1954) should not be used more often as mobile phases, now that it is known that these solvents provide differences in the relative ΔR_{Mg} values of the common substituents of steroids (see p. 99).

While the basic components of a typical system for steroids follow the pattern given on p. 61 and allow of considerable variation, it is useful to try and reach some idea of what variations are really important and why they are. The tyro may easily be confused by the profusion of systems now available and wonder whether relatively unimportant variations hold some hidden virtues known only to pundits on some distant campus. He will want to know the smallest number of systems that will be sufficient for his task, and while this is nearly always greater than one, it is often a far smaller number than might appear from the literature.

The first point is that the main types of method involve a series of solvent systems designed to provide *convenient* R_F values for important groups of steroids: *they are not unique*. These systems are usually capable of considerable modification for special purposes and some of these are suggested in Chapter 3. Neher (1958a) has also pointed out that many if not all published systems are members of an almost infinite series of possible systems based upon one or other type of stationary phase; the mobile phase is altered to provide appropriate mobilities, and ranges from aliphatic through aromatic hydrocarbons and on to chloroform, ethyl acetate, and butanol. The water content of the volatile systems is adjusted so as to provide immiscible phases in most cases, and also to cause convenient increases or reductions in ΔR_{Mg} values (see p. 125).

Two sorts of special features of the various versions can be distinguished; firstly those which improve or worsen the chances of obtaining ideal *conditions* without great elaboration of technique; secondly features which definitely improve certain *separations* under ideal conditions. The first type of variation also includes those adopted because of a particularly economical source of certain solvents, variations in local conditions, and so on. Some variations exist (Peterson, Kliman and Bollier, 1957b) which are specially designed to provide convenient R_F values for a small number of important steroids such as the diacetate and lactone-monoacetate of aldosterone.

The main advantage of the impregnated-paper systems is their capacity; particularly for extracts containing a lot of fat and other impurities (Bush, 1954a; Reineke, 1956; Neher, 1958a, 1959a). They are also better for certain separations of non-polar steroids since they give lower R_F values, thus allowing the use of a considerable degree of overrunning (p. 173). The latter difference is, however, not marked except with the cellosolve systems (Neher and Wettstein, 1952) and these and the ligroin–propylene glycol system suffer the disadvantage of serious displacement effects with certain mixtures of steroids (Savard, 1953, 1954). While these are not always a disadvantage they can sometimes be troublesome. Thus, for instance, dehydroepiandrosterone and androsterone cannot be separated in a mixture with ligroin–propylene glycol although they have different mobilities when pure; this is a nuisance with urinary extracts since a digitonin separation has to be done. The usual running times for impregnated-paper systems is also rather long but this can usually be overcome by using a more polar mobile solvent if the mixture is relatively simple (e.g. substituting benzene–chloroform or chloroform for benzene, with formamide). On the other hand, only the minimum of time is needed for equilibration.

The main disadvantages of the impregnated-paper systems are the difficulty of securing an absolutely even and reproducible deposition of the stationary phase; the protection of the prepared papers from moisture with hygroscopic stationary phases; and the removal of the rather involatile stationary phase after running the chromatogram. Even with large high-vacuum tanks propylene glycol takes about 2–3 hr to evaporate completely from filter paper (Axelrod, 1954). The aqueous methanol systems only take about three minutes for almost complete evaporation of solvent in a fume cupboard and are dry enough for all purposes after 20 min in a fume cupboard. Acetic acid and t-butanol take rather longer to remove (see p. 182).

The volatile solvent systems are rather more sensitive to temperature changes and disequilibrium as far as ordinary practical purposes of separation are concerned, and have a lower capacity than most impregnated-paper systems. The use of thick paper overcomes this difficulty to a large extent and actual comparison of zone-length in certain cases suggests that this difference is not as large as previously suggested. Their capacity for handling "dirt", however, is markedly inferior.

The higher-boiling hydrocarbons are most difficult to purify but are easier to run if temperature control is poor. Similarly, ethylene dichloride is preferable to methylene chloride or chloroform, butyl acetate to ethyl acetate, and diisopropyl ether to diethyl ether. In certain cases, however, it will be seen that separations are improved by using the lower-boiling solvents (p. 128).

Certain steroids with complex substituent groups are easier to separate in one type of system than another, and some of these are of practical importance. On the whole, however, inversions of order of mobility on changing systems are both fewer and less striking than with amino acids so that two-dimensional chromotograms are rarely useful. These differences are dealt with later (p. 102).

A few steroids have been reported as streaking in certain systems rather than giving symmetrical spots; it is not clear how regular this is and in the author's experience sporadic instances of streaking often turn out to be due to overloading or an unusually impure extract. The cases cited in Fig. 2.3 are drawn from chromatograms on which other compounds in

FIG. 2.3. Steroids prone to streak on chromatograms in particular systems

	Solvent Systems
P^5-3β,17α-ol-20-one	Methanol
αP-3β,17α-ol-20-one	Methanol
αP-3β,17α, 20β-ol	Methanol
17α-ethinyl-Δ^4-17β-ol-3-one	All methanol
βP-17α,21-ol-3,11,20-one-21-Ac	All but formamide
P^4-17α,21-ol-3,11,20-one-21-Ac	Propylene glycol
$P^{1,4}$-17α,21-ol-3,11,20-one-21-Ac	Propylene glycol
P^4-11β,17α,21-ol-3,20-one-21-Ac	Propylene glycol
$P^{1,4}$-11β,17α,21-ol-3,20-one-21-Ac	Propylene glycol: some methanol systems
2α-Me-P^4-11β,17α,21-ol-3,20-one-21-Ac	Propylene glycol
9α-F-P^4-11β,17α,21-ol-3,20-one-3,21-Ac	Propylene glycol
2α-Me-9α-F-P^4-11β,17α,21-ol-3,20-one-21-Ac	Propylene glycol: some methanol systems
P^4-6β,17α,21-ol-3,20-one	Most systems
P^4-6α-ol-3,20-one	Most systems
$P^{4,9(11)}$-17α,21-ol-3,20-one-21Ac	Most systems

similar quantity did not streak and the solvent systems were fresh; they are likely therefore to be genuinely recalcitrant compounds. Streaking is commoner with steroid acetates in impregnated-paper systems (Reineke,

1956) but probably commoner with 3β,17α-dihydroxy-5α-steroids and Δ⁵-3β,17α-dihydroxysteroids in the volatile systems (see p. 132). In some cases it is overcome by raising the temperature of operation and in most by changing to another system. Reineke's table (Reineke, 1956) is a useful guide.

Sources of Error in Determining R Values

The description of the behaviour of substances in chromatographic systems is always in terms of some function of their rate of movement through the system. With columns some volume of effluent is used such as the true retention volume (V_r) (Butt *et al.*, 1951; Van Duin, 1953) or the volume of effluent between the peak concentrations of two substances one of which is taken as a standard. On paper chromatograms, various relative mobilities or "R values" are used. Some of these measurements, indicated as subscripts to R have been rather unfortunate in that they give rise to confusion with other more conventional uses of such symbols. For single-length chromatograms the R_F value introduced by Consden *et al.* (1944) is the most important type of R value because of its relationship to the partition coefficient of the substance and to the fundamental function R_M* (p. 6). The relations between the different types of R value will be dealt with below but since all are related to R_F (or V_r) which is the only measurement which can be used for the quantitative examination of the behaviour of substances on chromatograms it is important first to describe the main sources of error in the determination of this parameter.

The first error that is common with steroids is that due to failure to obtain or maintain equilibrium in the system. The main cause of this in columns is usually a changing temperature or partial breakdown of the column, due to overloading with solute or with stationary phase or to high concentrations of material altering the surface tension between mobile and stationary phase (Howard and Martin, 1950). With paper the main causes of trouble are leaky tanks; changes of temperature; or gradients of temperature within the tanks due to a poor convecting system in the room, or to radiation causing "hot spots", or to both causes. In both types of fault the picture is the same: the spots or zones of high R_F are shortened in the direction of flow to give sharp bands of the type usually only seen with substances of very low R_F. The situation leads to what is essentially

---

* N.B. Neher and Wettstein (1956b) have most unfortunately used this term to mean "R relative to substance M" (cortisol)!

a gradient-elution overrun chromatogram of a special sort in which overrunning takes place by the evaporation of mobile phase from the lower part of the paper (in descending runs) rather than by dripping off the end. In serious cases the evaporation at some point on the paper equals the rate of entry of the new mobile phase into the front zone with the result that the front actually comes to a halt from half to two-thirds of the way down the sheet or strip. When a leaky tank is the cause there is usually no condensation of liquid on the lids or upper walls of the tank; when temperature changes or gradients are the main cause there is usually condensation of this sort which is a useful diagnostic point to look for. If the fault is not too severe the practical result in terms of separation is often quite good, and sometimes apparently better than a "good" run, because of the "sharpening" of the zones of high R_F due to the combined gradient and overrunning effects. In very mild cases it may be difficult to spot any defect in the final chromatogram; the only reliable way of detecting this sort of fault is to run a dye or standard of high R_F and a similar one of medium R_F (0·3–0·6) and check carefully the length of the zones and the R_F values against standards run under exceptionally good conditions.

The other source of error is the recognition of the front. It is not uncommon to find that the lower one-fifth of a sheet run in descending fashion is apparently "waterlogged" so that the front cannot be recognized. The front may have been easily visible with a flashlamp during the first half of the run and then gradually becomes less and less distinct as it reaches the lower part of the sheet. This phenomenon is usually associated with temperature gradients in the tank and may not affect the R_F values of slowly moving substances if the front can be found. With moderately polar systems (e.g. toluene or benzene as mobile phase) the front is accurately represented by a sharp line of blue fluorescence unless the paper is exceptionally clean. With less polar systems (e.g. light petroleum–benzene, or light petroleum itself as mobile phase) this blue line lies *behind* the front and has an R_F of about 0·93 in such systems. With unwashed paper it will be seen in ordinary light (before the mobile phase has evaporated) that the lower 1–2 cm of the paper lacks the translucency of the rest of the area over which the mobile phase has passed when the latter is of the less polar type; this zone is smaller with the more polar mobile phases. It seems very likely from the behaviour of dyes with high R_F value that this front area is occupied by hydrophobic material in the paper that is carried forward with the front but which has an R_F just less than 1·0 and is probably at a concentration giving displacement conditions in the zone occupied by it. Thus dyes and certain steroids have R_F values in light petroleum–aqueous methanol which suggest that they ought to have R_F values of 0·95–0·98 in the light petroleum–toluene–aqueous methanol systems; in fact their R_F values are not greater than 0·90–0·93 in such systems, strongly suggesting that they have been displaced backwards by the material running with the front (cf. Axelrod, 1953). This material can cause little error in the low range of R_F values but is one of the factors making high R_F values ($\geqslant 0·75$) variable and of doubtful value.

A factor which probably disturbs R_F values whatever their magnitude is the existence of unstable conditions at the solvent front (Cassidy, 1952, 1957) which are more or less independent of the presence of displacing material. This was shown very convincingly with paper sheets by Ackerman and Cassidy (1954) who observed the front very carefully and found that it "oscillated" so that small regions of the front were at times actually receding rather than moving forwards. This, however, probably varies a great deal with different solvents and different grades of paper. If the velocity of the solvent front over the first few centimetres of the sheet of paper is large this zone of turbulence may give rise to non-ideal conditions in and around the origin. Since the initial concentration of solutes at the origin is large, non-ideal conditions are very likely to occur in these early stages of the run even if the solvent flow is quite regular; even substances with low R_F values may be displaced by high concentrations of impurities of large R_F value before the latter have moved forwards out of the zone immediately around the origin.

An important factor in the running of paper chromatograms is the fact that, unlike columns, the lateral boundaries of the system are absent; they are imposed only by the surface tension between the liquid and the stationary phase, and the surface tension between the mobile phase and the vapour phase. There are thus a host of defects ascribed to this factor which are analogous, although of completely different mechanism, to the effects of "edging" in columns (Glueckauf, 1949b). One effect, a mass gradient down the paper or up the paper (Fujita, 1952; Ackerman and Cassidy, 1954) occurs even at equilibrium. The others are probably all due to disequilibrium in the system. The first effect results in a larger value for the ratio of the volumes of the phases near the origin of the chromatogram than near the front. Ackerman and Cassidy found gradients of the type shown in Fig. 2.4 for water ascending into Whatman No. 1 paper. These are probably larger than with non-polar mobile phases (see below) since, although the paper was equilibrated with the vapour additional swelling probably entered into the results of Ackerman and Cassidy; it is well known that cellulose swells more when in contact with the liquid than with water vapour (Lauer, 1944). It is important to note that these gradients are present with both ascending and descending capillary flow with water, although slightly less with the descending method. With typical

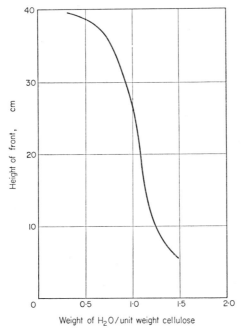

Fig. 2.4. Solvent gradients in paper sheets. Ascending run with water on Whatman No. 1 paper. Diagrammatic; after Ackerman and Cassidy (1954).

chromatographic systems, however, the gradient was large with ascending and horizontal (radial) flow and quite small with descending flow (Wood and Strain, 1954).

The other defects due to the absence of lateral boundaries of paper chromatograms are most probably due to disturbances of equilibrium. They are seen commonly with narrow strips (2·5–5 cm wide) and rarely with sheets (15–30 cm). The commonest is "edging"; that is to say zones which are concave downwards instead of straight transverse bands. This is due largely or entirely to lack of equilibration since it is diminished by increasing the time allowed for equilibration, and with short periods of equilibration is equally severe with ascending and descending runs carried out simultaneously in the same tank (unpublished experiments). It is probably due to the more rapid evaporation of mobile phase at the edges than the centre of a strip; the solvent is replaced by increased downward flow at the edge so that effectively the edge is more "overrun" than the centre (p. 69). This suggestion is confirmed by the fact that these zones are longer in the centre than at the edge (Fig. 2.5); the shortening at the edge is similar to that found with spots on sheets run in badly equilibrated tanks (p. 68).

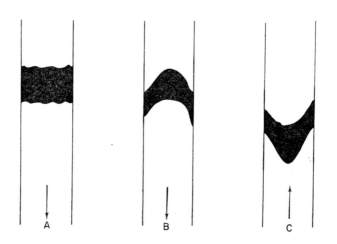

FIG. 2.5. "Edging" on paper strips.
A. Strip run by descending method with adequate equilibration.
B. Strip run by descending method with inadequate equilibration.
C. Strip run by ascending method in the same tank and at the same time as B. (Author, unpublished).

Another defect sometimes seen is diffuseness and overlap of zones after a very long run (24–48 hr) when it is known that the zones are sharp and well separated on shorter runs. The factors responsible are unknown to the author but again such defects have only occurred when there has been other evidence of disturbance of equilibrium such as the condensation of liquid on the lid of the tank (p. 175). It seems most likely that with certain solvent systems large changes of surface tension between the phases and of the intermiscibility of the components occur with small changes of temperature; these lead probably to microscopic "waterlogging" and the excess stationary phase creeps through the paper in all directions until it is taken up by the mobile phase when a rise in temperature occurs (assuming a positive temperature coefficient of intermiscibility). This is suggested since it is commonest with those systems having a very low water content and considerable intermiscibility. Unfortunately such systems are often essential and the best way out of difficulty is to run them at lower temperatures (16–20 °C); since they are all of high alcohol or acetic acid content the higher temperatures are not essential to good running (p. 60).

The phenyl cellosolve–heptane system appears to reverse its phase with rise in temperature (Rubin et al., 1953) and R_F values may be unreliable close to the critical temperature. In a careful study of the latter system, Loomeijer and Luinge (1958) found that R_F values were greatly affected by the degree of hydration of the paper before impregnation with cellosolve but not by exposure to humid air after impregnation (cf. Neher and Wettstein, 1952).

In conclusion, then, there are serious difficulties in the accurate measurement of R_F values when these are above about 0·80 which are not likely to be reduced very much by technical improvements. With volatile systems these difficulties are magnified and make it possible, though not unavoidable, to be in serious error with R_F values of all magnitudes. These difficulties are, however, amenable to improvements in technique and R_F values should only be reported when measured by methods known to be adequate in this respect. The variations of R_F values in day-to-day use reported by Reineke (1956) are acceptable in situations where standards are run in parallel with unknowns and where identification is helped by circumstantial evidence; such large variations ($\pm 10\%$) however, should not be accepted for the reporting of R_F values which should always be based on results obtained with fresh systems and with special precautions to avoid disequilibrium in the system.

Variation of R_F values with the size of the tank, with the velocity of the mobile phase and with the length of run has been reported for amino acids (Clayton, 1950; Kowkabanky and Cassidy, 1950, 1952; Saifer and Oreskes, 1953) and other substances. These effects must also be ascribed to lack of equilibrium in the system; either due to insufficient equilibration, temperature changes or to slow swelling of the cellulose during the run. A useful study of the need for equilibration was made by Shute (see Cassidy, 1957).

Description of Results: Relations between the Different R Functions

The R_F is the ratio of the velocity of the point of maximum concentration of a solute to the velocity of the mobile phase. On paper chromatograms, this is the same as the ratio of the distance moved by the centre of a zone to that moved by the front. For columns the retention volume (Butt et al., 1951) is the most usual function of R_F that is used because it is easily measured. The function usually employed is V_r, the "true retention volume" (Van Duin, 1952, 1953) in which the volume of mobile phase in the interstices of the column is not included in the measurement (Butt et al., 1951). These two parameters are both directly related to the partition coefficient of the solute in question (p. 7) by the equations

$$\alpha_S = k \cdot \left(\frac{1}{R_F} - 1\right) \qquad \text{(p. 7)}$$

$$\alpha_S = k' V_r \qquad \text{(Van Duin, 1952)}$$

As shown in Chapter 1, the logarithm of the partition coefficient is directly related to the sum of the free energies of transporting the constituent groups of a substance from one phase to the other so that the comparison of the chromatographic behaviour of different solutes is based upon consideration of the logarithms of the functions given above: for columns $\log V_r$ is used (Van Duin, 1952) and for paper, $R_M [= \log(1/R_F - 1)]$ (Bate-Smith and Westall, 1950). Since the information available for paper chromatograms is much more extensive than that for columns the discussion will be mainly centred upon R_M values; the arguments based on these values however are directly applicable to columns by substituting $\log V_r$ values (Van Duin, 1952).

Since the quantitative treatment of chromatographic behaviour depends entirely upon the measurement of R_F values it is important to see how the other types of R value are related to it. The most commonly used measurement in the steroid field is probably the R_S value where the movement of a steroid is expressed as a ratio of the movement of a standard steroid S rather than by comparison with the front. This is commonly done because an overrun system is being used and no front is available for measuring R_F values; it is most common with the impregnated-paper methods where overrun chromatograms are almost invariably used. In order to obtain R_F values from R_S values the R_F of the standard must be known; because R_F values vary considerably with variations in the technique of impregnation, a great disadvantage of this type of system from the point of view of a theoretical treatment is that there is no check on the R_F of the standard compound on the chromatograms used to measure the R_S values of slow-moving compounds. Another difficulty with R_S values is that it is commonly supposed that they afford an internal control on variations of technique, and that R_S values remain constant even if the R_F value of the standard is different on two different chromatograms. For qualitative purposes this is more or less true, but for the quantitative comparison of chromatographic behaviour it is far from the truth and small differences in R_S values may be in considerable error if the two compounds showing such a difference have not actually been compared on the same sheet of paper. The only way of obtaining valid comparisons by correcting small errors in R_F value is by making corrections of the R_M values (Bate-Smith and Westall, 1950) which are linear and additive. Errors in R_F and R_S follow the sigmoid relation shown in Fig. 2.6 and cannot be corrected accurately by any simple linear relationship.

In most other fields a relatively limited number of compounds have to be separated, and hence a solvent system is sought which gives the series of compounds conveniently placed on one sheet of paper. The steroids, however, cover such a wide range of polarity and involve so many closely related structures that it is usual to devise solvent systems that deal separately with a limited group of compounds on each run. Furthermore, many separations require the use of overrun chromatograms whatever type of solvent system is used. It is therefore useful to obtain equations by which R_S values can be related to R_F and R_M values since it is often necessary to calculate what solvent system will provide the most convenient and efficient separation of a set of compounds differing by known R_M or R_F values on chromatograms overrun by a certain factor. Similarly, R_F values smaller than 0·1 are best measured accurately by overrunning a chromatogram under very carefully controlled conditions and obtaining R_S values with respect to a compound of a known R_F value lying in the range 0·2–0·4.

It will be shown later that steroids do in fact behave approximately as predicted by Martin's theory and that steroids differing by a particular structure do show approximately constant differences in R_M values. These ΔR_M values depend mainly upon the nature of the stationary phase and the intermiscibility of the two phases. By changing the nature of the mobile phase, and adjusting the water content of the stationary phase to maintain approximately the same degree of intermiscibility, it is possible therefore to devise systems giving very approximately constant ΔR_M values for

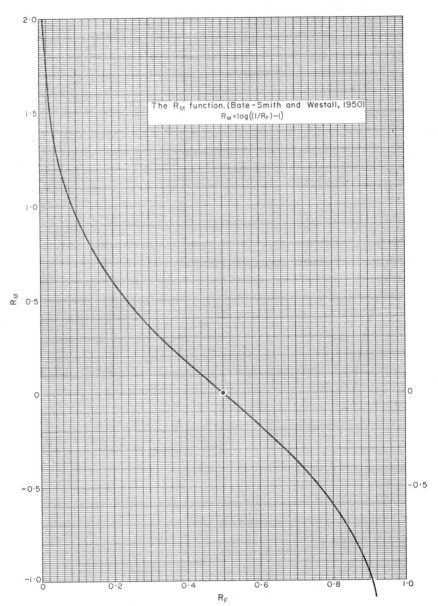

Fig. 2.6. The R_M function (Bate-Smith and Westall, 1950).
Note the almost linear part of the curve in the R_F range 0·3–0·7 and the rapidly increasing change of R_M with R_F outside this range. Reichl (1955) has proposed an alternative form which avoids negative values of R_M.

polar groups while altering the R_M value of any given compound. Suppose, then, that we wish to calculate the R_M values needed to provide a given R_S value for two substances with a known ΔR_M, so as to arrive at a solvent system giving a convenient separation of the pair on an overrun chromatogram. If the first substance is taken as the standard and has the smaller R_F value ($R_{F2} > R_{F1}$) we have

$$R_S = R_{F2}/R_{F1} \tag{2.1}$$

and

$$\Delta R_M = \log\left(\frac{1}{R_{F1}} - 1\right) - \log\left(\frac{1}{R_{F2}} - 1\right) \tag{2.2}$$

Hence

$$\Delta R_M = \log\left\{\frac{\left(\frac{1}{R_{F1}} - 1\right)}{\left(\frac{1}{R_{F2}} - 1\right)}\right\} \tag{2.3}$$

Substituting equation 2.1 and simplifying to eliminate R_{F2}

$$\Delta R_M = \log\left\{\frac{(R_S - R_S R_{F1})}{(1 - R_S R_{F1})}\right\} \tag{2.4}$$

Putting $k =$ antilog ΔR_M we obtain

$$k = \frac{(R_S - R_S R_{F1})}{(1 - R_S R_{F1})} \tag{2.5}$$

Whence

$$R_S = \frac{k}{(k-1)R_{F1} + 1} \tag{2.6}$$

If the first substance is to be used as standard and $R_{F2} < R_{F1}$, the analogous expression is

$$R_{S2} = \frac{k}{k + R_{F1}(1-k)} \tag{2.6a}$$

These equations can be used to predict the relative mobilities of two steroids in terms of the ratio of their R_F values when the R_M value of one of them is known and their ΔR_M is known for the type of system employed. A family of curves for different ΔR_M values is given in Fig. 2.7.

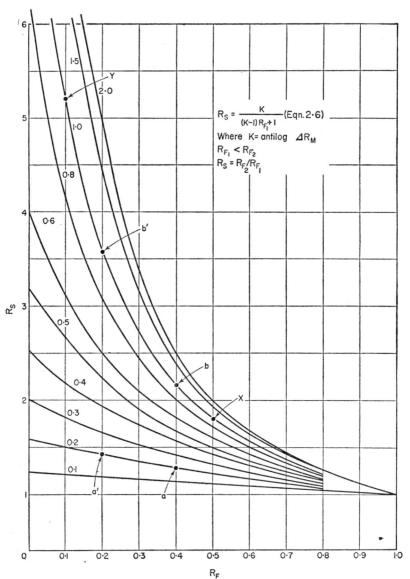

Fig. 2.7. R_S values for varying values of the R_F of the standard and different values of ΔR_M: (i.e. standard − test or unknown steroid).

ΔR_M values are alongside each curve. Suppose the unknown is calculated to have an R_M which is 1·0 less than that of the standard. Then by following the curve "1·0" we find that with a mobile phase giving the standard an R_{F1} of 0·5 the R_S is 1·8 (point X) and hence R_{F2} will be 0·9. If the mobile phase is adjusted to give $R_{F1} = 0·10$ $R_S = 5·2$(Y) and R_{F2} will be 0·52—a much more useful value. The curves illustrate the increased resolution afforded by using over-run systems giving low R_F values, and the difficulty of increasing resolutions if the ΔR_M value is small.

The reverse operation is used for obtaining R_M (or ΔR_M) values on overrun chromatograms. In this case, the R_F value of one of the compounds must be known. Since the point is to obtain R_M values for slow-moving substances, for which the geometrical error of measuring the R_F values directly is large, the standard will usually be the substance of higher R_F value. The easiest procedure is to measure R_S on the chromatogram and to calculate the R_F of the slower compound (R_{F2}) by equation 2.1; R_{M1} and R_{M2} are then read from a plot of R_M and R_F (Fig. 2.6). An alternative is to use the nomogram given by Lederer (1959).

Brooks, Hunt, Long and Mooney (1957) use the approximation

$$\Delta R_M = -\log_{10} R_T$$

where R_T is proportional to R_S in Equations 2.1 to 2.6. This is valid where

$$1/R_S \gg R_{FS} \ll 1\cdot 0$$

In the case where substances 1 and 2 differ by a single substituent this ΔR_M is, of course, an ordinary ΔR_{Mg} value. This approximation is thus only useful for moderately large ΔR_M values and considerably overrun chromatograms. In addition, it can be rendered invalid by any large differences in velocity of the mobile phase, so that for instance results on Whatman No 3MM paper (fast) cannot always safely be compared with results on Whatman No 2 paper (slow): this is because $R_T = K.R_S$ where K is the *velocity* of the standard in cm/hr (Savard, 1954). On the whole, it seems better to use the full calculation which is simple enough and is applicable to all values of R_F and R_S.

Equations 2.6 and 2.6a are merely tautologous rearrangements of equations 1.2, 2.1, and 2.2, and contain nothing new. They are, however, convenient in calculating ΔR_M values and predicting R_S values in the steroid field where overrun chromatograms and low R_F values are used extremely frequently. Suppose, for instance, it was necessary to separate three compounds with $R_{M1} < R_{M2} < R_{M3}$, and $\Delta R_{M12} = 1\cdot 0$ and $\Delta R_{M23} = 0\cdot 2$, preferably on one sheet. Then it is quickly seen from Fig. 2.7 that if a system is chosen giving the slowest substance an R_{F3} of $0\cdot 4$ then R_{S23} is $1\cdot 29(a$, Fig. 2.7) and R_{F2} will be $0\cdot 51$. R_{S12}, similarly, is $2\cdot 17$ (b, Fig. 2.7) and R_{F1} will be $0\cdot 87$, placing the fastest substance in a position close to the front and with a considerable probability of errors in its R_F value and of contamination with fast-moving impurities. If a system giving an R_{F3} of $0\cdot 2$ is used ($a,'$ Fig. 2.7), R_{F2} will be $0\cdot 286$ and (b', Fig. 2.7) R_{F1} will be

0·715: this gives a reasonably useful R_{F1} and a separation of 3·4 cm between the peaks of the two slowest compounds on a 40-cm run. With R_{F3} set at 0·1, R_{F2} will be 0·15 and R_{F1}, 0·525; for a separation between the two slowest peaks of 3·4 cm, a 40-cm sheet will have to be overrun by a factor of 1·7 and the fastest compound will be placed $40 - (1·7 \times 0·525 \times 40)$ = 4·3 cm from the front. For routine work this will be difficult to control unless a convenient marker dye running close to the fastest compound is available, and it is clear that the best compromise for getting a reasonable separation of all three compounds on one sheet will be to select a system giving an R_{F3} in the range 0·15–0·20 and using a slightly overrun or single-length chromatogram.

Separation and Resolving Power

The above expressions are useful for calculating the relative positions of the peak concentrations of different compounds but give no indication of "resolving power" which depends also upon the width of the zones; or more accurately the distribution of concentration of each zone, since every substance is in fact spread throughout the chromatogram, albeit at very low concentrations. This problem has been studied by Mayer and Tompkins (1947) for ion-exchange columns and along similar lines by Dixon and Stack-Dunne (1955). Suitable expressions for comparing the performance of paper strips and partition columns were derived by Tait and Tait (1960) using the treatment of the last authors. These treatments, however, are all based upon the discontinuous method of approach; that is to say they depend upon assuming the existence of small finite theoretical plates in the system at each one of which equilibrium is achieved. This approach is entirely accurate for counter-current separations of the type achieved by Craig's machine, in which the transfers are indeed discontinuous, and work well enough for deriving R values in chromatographic systems. Glueckauf (1955) has shown however that this approach is not good enough for calculating the overlap, or mutual contamination, of two zones separated by chromatographic systems. For such calculations the assumption of discontinuous small transfers in the system leads to considerable errors and Glueckauf has derived alternative expressions using a mathematical approach which avoids the discontinuous treatment. According to Glueckauf the original equations of Mayer and Tompkins underestimate the degree of overlap of two zones running relatively close to one another ($\alpha_S^1/\alpha_S^2 =$ ca. 1·2) by a factor of three, which is unacceptable for many purposes. The error is still large with values of 1·5 for α_S^1/α_S^2 and very much larger when $\alpha_S < 1$ (i.e. $R_F > 0·5$).

The mutual contamination of two zones is given by Glueckauf (1955) as

$$\eta = \frac{2m_1m_2}{m_1{}^2+m_2{}^2}\left(0\cdot5 - A_\epsilon\left\{\frac{\sqrt{N}(\sqrt{\bar{V}_{r2}}-\sqrt{\bar{V}_{r1}})}{4\sqrt{V_{r1}V_{r2}}}\right\}\right) \qquad (2.7)$$

where η is the percentage of m_1 and m_2 contaminating each other if the fractions are cut so as to obtain equal contamination of the two peaks; m_x = m-equiv. of solute x put on column; N = number of effective theoretical plates of column; V_{rx} = true retention volume of solute x; $A_\epsilon(t)$ = area under curve of normal error given for t in *Handbook of Chemistry and Physics*, 13th ed., p. 204 (1947).

Glueckauf's theory should be studied in the original papers since it is too long to reproduce here. The distinctive feature of his approach is to set up a mass balance equation which uses the "theoretical plate" concept but which is based on differentials rather than an expansion of a set of finite terms. The differentials correspond mathematically with a model based on the continuous flow of the mobile phase through a column; the binomial expansion due originally to Martin and Synge (1941) corresponds to a model based on discontinuous contact between the elements of the mobile and stationary phases, as in the Craig counter-current machine. Many of Glueckauf's points are of more relevance to ion-exchange separations where particle-diffusion and displacement conditions are commonly important factors in practical work. These are usually absent in partition chromatography as applied to steroids although they may become important in heavily loaded columns used for preparative work. Other conclusions however are of great use in partition chromatography.

Two convenient equations are derived from such treatments for calculating the effective number (N') of theoretical plates. The first is useful (Glueckauf, 1955, Equation 28) for columns:

$$N' = 8\left(\frac{V_r}{\beta}\right)^2 \qquad (2.8)$$

where V_r is the true retention volume and β is the width of the zone in ml of effluent at the concentration $c_{\max}/e$. A similar equation which can be used for elution curves from columns or for scanning curves on paper is given by Keulemans (1956; p. 112) as

$$N' = 16(V_r/w)^2 \qquad (2.9)$$

where w is the base (in ml) of a triangle formed by the tangents to the points of inflection of the peak in the effluent concentration curve. The analogous expression for scanning curves on paper strips is obtained by equating V_r in ml to D/R_F in cm (Tait and Tait, 1960), since the material is as if "still on the column":

$$N' = 16\left(\frac{D}{R_F W}\right)^2 \qquad (2.10)$$

where D is the distance from the origin to the peak of the zone in cm, and W is measured in centimetres. Glueckauf's equation can also be transformed to an approximation suitable for scanning records on paper.

The analogous equations based on Mayer and Tompkins's (1947) treatment due to Dixon and Stack-Dunne (1955), and to Tait and Tait (1960) for columns and paper are:

$$N' = \frac{V_r(V_r + V_m)}{(s.d.)^2} \qquad \text{(column effluents)} \qquad (2.11)$$

and

$$N' = \frac{D^2(1 - R_F)}{(s.d.)^2} \qquad \text{(band on paper)} \qquad (2.12)$$

Where V_m is the volume of the mobile phase of the column (i.e. its effective part) and $s.d.$ is the standard deviation of the concentration profile in ml of effluent from columns and in cm on paper. These equations are simpler to use than the original ones of Martin and Synge (1941) although derived by essentially the same treatment.

With very slow runs and the elution of substances with very large retention volumes, Glueckauf's equations for mutual contamination of two zones give very similar results to those of Martin and Synge (1941) and of Mayer and Tompkins (1947). This similarity, however, is only achieved with conditions that are rarely practicable; thus Glueckauf's calculation of a discrepancy by a factor of three between the two types of treatment was for a column of 1000 theoretical plates, and a mean α_S of about 5·0 for the two substances. Since this is in a common range of α_S for convenient separations on columns it is probable that the expressions of Tait and Tait (1960) underestimate the mutual contamination of zones in partition systems.

SECTION II: QUANTITATIVE TREATMENT OF STEROID BEHAVIOUR ON CHROMATOGRAMS

In the first chapter it was seen that quantitative comparisons of R values can only be made conveniently by using the function of the R_F value that is directly proportional to the partition coefficient of the solute in question, namely $(1/R_F) - 1$, and that since this in turn is related to the chemical potential of a substance by its logarithm (Martin, 1949), it follows that the logarithm of this function is the parameter we have to work with (Bate-Smith and Westall, 1950). Furthermore this parameter should be determined approximately by the sum of a set of terms which correspond to the constituent parts of the solute (Martin, 1949) and which are constants for the different types of group of which the molecule of the solute is composed. The purpose of this section is to show that this treatment gives a reasonably adequate account of the behaviour of steroids on chromatograms if the restrictions mentioned in Chapter 1 are adopted. This account is far from complete because much of the available data have probably not been obtained under ideal conditions, and for reasons of convenience many important steroids have not been run in enough solvent systems to provide all the data that would be required for a full treatment. However, the results of applying this method to the steroids are sufficiently promising to be worth describing, if only to encourage others to undertake the careful experiments which will be needed to establish the full picture.

It was found in the early stages of this investigation that most of the common solvent systems used for steroids fell obviously into one single group, and that most of the important differences in ΔR_M values were determined largely by the more polar components of the systems. The systems of this group are all based on stationary phases which are hydroxylic solvents or which can donate protons to form hydrogen bonds; for convenience these systems will be called "typical" systems and in general they can be thought of as "water-like" solvents. The other group of systems contained appreciable proportions of solvents less capable of forming hydrogen bonds by donating a proton, or known to have particularly strong interactions with certain groups in steroids, although otherwise apparently typical in their ability to form hydrogen bonds. These solvents will be called "atypical" for convenience although they are much more complicated and various in their effects than the "typical" group and will undoubtedly have to be subdivided and reclassified as more information becomes available.

The "typical" systems will be discussed first because they are better known and there are more data available for them. They comprise all the systems based largely or completely upon water, methanol, ethanol, formamide, propylene glycol, and ethylene glycol. Isopropanol (Bush, 1952) probably also falls into this class. It will be seen that except for certain complex functions and strongly hindered groups, these "typical" systems all show very much the same differences in the effects of different groups upon R_M values. It is to be regretted that much of the data available for the impregnated-paper systems could not be used for this investigation because it was in the form of R_S values and no reliable R_F values were available for making the necessary conversion to R_M values; enough data were collected however to calculate the R_M values of certain series of compounds and these were sufficient to justify the placing of most of these systems in the class of "typical" solvent systems. A valuable treatment by Kabasakalian and Basch (1960) is now available for those systems. The results are very similar to those reported in this book and represent a useful and independent confirmation of these general conclusions.

Brooks et al. (1957) were the first to publish ΔR_M values of ketone and hydroxyl groups in steroids using their own data and those of Savard (1953, 1954). The author was unfortunately unaware of this paper until after the main part of this book was written and an earlier review (Bush, 1960b) regrettably lacked a reference to it. Their study showed that ΔR_{Mg} values in two propylene glycol systems were very similar to those found in two systems based on aqueous methanol and aqueous ethanol. The values were very similar to those given here and are particularly valuable in that they include several types of sapogenin. Their results thus confirm independently the classification of propylene glycol systems among the "typical" systems which is made above, and show that the behaviour of a wide range of steroids is indeed satisfactorily accounted for on the basis of Martin's theory.

Before reviewing these calculations the definitions of the different types of ΔR_M values that are given in the Glossary (p. xiv) must be explained. The symbol ΔR_M means "difference in R_M" and was used originally by Bate-Smith and Westall (1950) both to describe differences between compounds differing in numerous features of their structure, and also to describe the contribution of single substituents to the R_M value of a given substance. It is useful however to apply this symbol to many different kinds of "difference of R_M", and in order to specify these different uses of the term it has been found convenient to add a second subscript letter, although sometimes the kind of use is obvious from the context. When the difference in the R_M values of a pair of compounds is implied without special reference to particular groups or particular solvent systems, or when the meaning is quite clear from the context, the original term without an extra subscript will be used. When a difference in R_M value due to the substitution of hydrogen atoms by a definite group is implied the term ΔR_{Mg} will be used (Fig. 2.8a). This term will also be used for complex substituents or parts of a molecule when necessary. Some substitutions and other alterations are most conveniently thought of as due to a particular reaction, e.g. oxidation of hydroxyl groups to ketone groups or esterification of the former, and the change in R_M value caused by such operations will be called ΔR_{Mr} values, although there is no fundamental difference between them and ΔR_{Mg} values which can also be thought of as due to particular reactions (Fig. 2.8b). Finally, the difference in R_M values of a substance in two different solvent systems will be called the ΔR_{Ms} value (Fig. 2.8c).

These usages will sometimes be slightly modified, or combined to describe certain observations, but the sense should be obvious from the context. The only important thing to note about these different terms is that if Martin's theory is obeyed they can all be added to one another irrespective of *type*, as long as the *order* of operations is taken into account. Thus if we add a group to a given steroid and then compare the R_M value of the *original* steroid in one solvent system with the R_M value of its *derivative* in another solvent system, the overall difference can be calculated either by adding the ΔR_{Ms} of the original steroid to the ΔR_{Mg} of the group in the *second* solvent system; or by adding the ΔR_{Ms} of the added group to the ΔR_{Mg} of the group in the *first* solvent system. In the second method of calculation the group is considered as if it were subject to its own characteristic ΔR_{Ms} value on changing the type of solvent system: this is a direct consequence of Martin's postulate.

1. ΔR_{Mg} Values in Typical Solvent Systems

In Table 2.2 the ΔR_{Mg} values for hydroxyl and ketone groups are given for four volatile systems and one impregnated-paper system. The values for the chloroform–formamide system ($CHCl_3/F$) and the benzene–methanol–water (2:1:1 by vol.) "B5" system are calculated from the figures of Neher and Wettstein (1956b) using average R_F values for cortisol

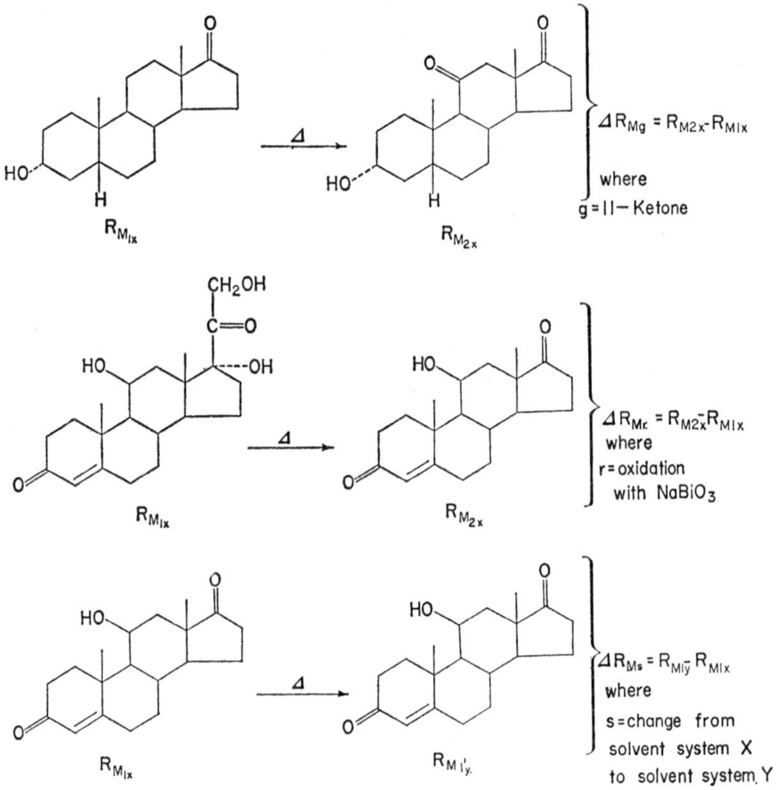

FIG. 2.8. Different types of ΔR_M value

ΔR_{Mg} may be considered as the partial contribution of the group g to the "polarity" of substance 2, with respect to solvent system X.

ΔR_{Mr} represents the change in R_M due to conversions other than the substitution of hydrogen, particularly geometrical isomerizations and complicated reactions: also with respect to a given solvent system.

ΔR_{Mr} is the change in R_M of *one* substance on changing solvent systems. It may also be broken down to the ΔR_{Ms}, value for parts of the molecule i.e. to the changes in ΔR_{Mg} values that can be measured.

(Neher, 1958a). Since the R_S values were not corrected using R_M values (p. 75) there is some uncertainty about these figures although they were measured under conditions which have been well standardized by many years of experience with the method. In many cases these groups are part

Table 2.2. ΔR_{Mg} Values of Ketone and Hydroxyl Groups

Group, position, orientation	Substance in which group is substituted	ΔR_{Mg} values in solvent systems				
		L/85 25·5° C	LT21/85 25·5° C	LB21/80 25·5° C	B/50 22° C	CHCl$_3$/F 22° C
6-Ketone	P^4-3,20-one (progesterone)		0·31			
1-Ketone	P^4-3,20-one (progesterone)		0·44			
	A^4-3,17-one		0·56			
	P^4-21-ol-3,20-one-21-Ac		0·68			
	P^4-21-ol-3,20-one		0·67	0·67		
	P^4-17α,21-ol-3,20-one		0·67	0·62	0·79	
	P^4-17α,21-ol-3,20-one-21-Ac		0·68			
	βA-3α-ol-17-one		0·64			
	βP-3α,17α,21-ol-20-one				0·82	1·0
1-OH	P^4-3,20-one		0·73			
	P^4-11β-ol-3,20-one		0·82	1·0		
	P^4-3,11,20-one		1·06			
	P^4-6β,17α-ol-3,20-one			1·0		
7α-OH	P^4-3,20-one		0·64			
	P^4-21-ol-3,20-one			0·86		
	P^4-6β-ol-3,20-one		1·11			
	P^4-11β,21-ol-3,20-one			0·94		
	P^4-11β,21-ol-3,20-one-21-Ac			0·72	1·20	1·11
	P^4-21-ol-3,11,20-one			0·81	1·04	0·70
	βP-3α-ol-20-one			0·77		
1β-OH	P^4-3,20-one	1·31	0·90			
	A^4-3,17-one		0·93			
	P^4-21-ol-3,20-one		1·07	0·86		
	P^4-17α,21-ol-3,20-one		0·99		1·20	1·16
	P^4-21-ol-3,20-one-21-Ac		0·81			
	βA-3α-ol-17-one	1·34	1·06			
	αA-3α-ol-17-one	1·27	1·08			
β-OH	P^4-3,20-one		1·03			
	A^4-17β-ol-3-one		1·35	1·17		
	P^4-17α-ol-3,20-one		1·48			
	P^4-17α,21-ol-3,20-one			1·40	1·61	1·47
	P^4-11β,21-ol-3,20-one				1·77	1·50
	P^4-21-ol-3,11,20-one				1·50	1·12
	P^4-21-ol-3,20-one		1·50*			

Table 2.2. ΔR_{Mg} Values of Ketone and Hydroxyl Groups—*continued*

Group, position, orientation	Substance in which group is substituted	ΔR_{Mg} values in solvent systems				
		L/85 25·5° C	LT21/85 25·5° C	LB21/80 25·5° C	B/50 22° C	CHCl$_3$/F 22° C
11α-OH	P^4-3,20-one A^4-17β-ol-3-one P^4-17α,21-ol-3,20-one		1·39 1·88	1·50		2·6
7α-OH	P^4-21-ol-3,20-one		1·70	1·29		
15α-OH	P^4-21-ol-3,20-one		1·77	1·57		
15β-OH	P^4-21-ol-3,20-one		1·57	1·38		
2α-OH	A^4-17β-ol-3-one		0·75			
19-OH	P^4-17α,21-ol-3,20-one P^4-11β,21-ol-3,20-one				1·98 1·32	1·71 0·90

Systems L/85, LT21/85, LB21/80 calculated from the author's figures: Systems B/50 and chloroform–formamide calculated from the data of Neher and Wettstein (1956).

* Value from Neher (1958) System BL$_1$.

of complex functions and in some cases the ΔR_{MS} values are obviously affected by neighbouring substituents. It is clear, however, that groups well separated from others have relatively constant and characteristic ΔR_{Mg} values, and that groups substituted close to others (e.g. the 17α-hydroxyl group which is adjacent to the 20-ketone group in all the quoted C$_{21}$ steroids) have approximately similar ΔR_{Mg} values as long as the neighbouring substituents are the same. In each case an equatorial hydroxyl group has a larger ΔR_{Mg} than the axial group at the same position (Savard, 1953), and the least hindered hydroxyl groups such as those at positions 15 and 16 (see also Table 2.3) have the largest values. It will be noticed also that the deviations of particular compounds from the mean ΔR_{Mg} of the group in question are in the same direction and usually similar in magnitude in the formamide–chloroform and benzene–aqueous methanol systems.

In Table 2.3, the data have been grouped according to the different structures in which the various substitutions have been made. The figures were calculated from the data of Reineke (1956). It is probable that some

Table 2.3. ΔR_{Mg} VALUES FOR DIFFERENT SERIES OF ROOT SUBSTANCES

Unsubstituted root substance	Substituent	ΔR_{Mg} values in solvent system			
		L/80	LT11/70	LB21/80	B/50
P⁴-3,20-one (progesterone)	11-ketone	1·0	0·64	1·03	
	21-OH(p)	1·02	0·67	1·26	
	17α-OH(t)	1·22	0·75	1·26	
	11β-OH(a)	1·4	0·91	1·42	
	6β-OH(a)	1·46	0·95	1·58	
	11α-OH(e)	1·93	1·35	1·93	
	14α-OH(t)	1·57	0·95	1·58	
	15α-OH(e')	2·30	1·47	2·12	
	15β-OH(a')	1·90	1·32	1·9	
	16α-OH	2·10	1·42	2·1	
	21-OAc	0·41	—	0·38	
	11α-OAc	0·98	—	0·91	
A⁴-3,17-one (androsteredione)	6-ketone	—	0·29	0·51	
	11-ketone	0·90	0·68	0·67	
	11β-OH(a)		0·86	1·0	
	11α-OH(e)		1·39	1·75	
	14α-OH(t)		1·21	1·41	
	6β-OH(a)		1·03	1·17	
	15α-OH(e')		1·46	—	
A⁴-17β-ol-3-one (testosterone)	11-ketone		0·80	0·82	
	11β-OH(a)		1·03		
	11α-OH(e)		1·58		
	14α-OH(t)		1·32		
	6β-OH(a)		1·21		
P⁴-21-ol-3,20-one (deoxycorticosterone)	11-ketone		0·75	0·86	0·65
	11β-OH(a)		0·93		0·87
	11α-OH(e)		1·58		1·50
	14α-OH(t)		1·20	1·20	
	6β-OH(a)		1·20	1·22	
	15α-OH(e)		1·83	1·65	
	17α-OH(t)		0·93	1·03	

Table 2.3. ΔR_{Mg} Values for Different Series of Root Substances—*continued*

Unsubstituted root substances	Substituent	ΔR_{Mg} values in solvent system			
		L/80	LT11/70	LB21/80	B/50
P^4-17α-ol-3,20-one (17α-hydroxyprogesterone)	11-ketone 11β-OH(*a*) 11α-OH(*a*) 21-OH(p) 6β-OH(*a*)		0·60 1·07 1·58 0·85 1·17	0·67	
P^4-17α,21-ol-3,20-one	11-ketone 11-OH(*a*) 11α-OH(*e*) 6β-OH(*a*)		0·70 1·07 1·65 1·30		
P^4-6β-ol-3,20-one	11-ketone 11α-OH(*e*) 21-OH(p)		0·75 1·71 0·92		
P^4-3,11,20-one	6β-OH(*a*) 21-OH(p)		1·06 0·78		
βP-3,20-one	11-ketone 11β-OH(*a*) 11α-OH(*e*)		0·83 1·29 1·55		
19-*nor*A^4-3,17-one	11α-OH(*e*)			1·74	

p = primary; t = tertiary; *a* = axial; *e* = equatorial; *e'* = quasi-equatorial, etc. Calculated largely from data of Reineke (1956).

of the higher R_F values are subject to the errors described earlier and these have been corrected where reasonable by using R_F values from my own experiments. While these figures are less accurate, due probably to operating at 34° C, than those of Table 2.2, they are of considerable value because of the wide range of compounds studied by Reineke. In Table 2.4, the ΔR_{Mg} values for the three types of 11-oxygen function are collected (aqueous methanol system). This position is well isolated from all the other substituents and so it is interesting that the ΔR_{Mg} values show less variation over a wide range of steroids than for groups at other positions.

Much less information is available for other substituents and fewer of these are really isolated groups. The effect of a 9α-fluorine atom in the presence of an 11β-hydroxyl group is shown in Table 2.5 to be relatively constant. Methyl groups have also been studied only in positions where they undoubtedly influence neighbouring groups both by their positive inductive effect and their steric hindrance. However, it is seen in Table 2.6 that their ΔR_{Mg} values are of the same order for similar positions and neighbouring groups, in any one solvent system.

Table 2.4. ΔR_{Mg} Values for Substituents at C-11

Root substance	ΔR_{Mg} of substituents		
	11-ketone	11β-OH	11α-OH
P⁴-3,20-one	0·64	0·91	1·35
A⁴-3,17-one	0·68	0·86	1·39
A⁴-17β-ol-3-one	0·80	1·03	1·58
P⁴-21-ol-3,20-one	0·75	0·93	1·58
P⁴-17α-ol-3,20-one	0·60	1·07	1·58
P⁴-17α,21-ol-3,20-one	0·70	1·07	1·65
P⁴-6β-ol-3,20-one	0·75	—	1·71
19-norA⁴-3,17-one	—	—	1·74
βP-3,20-one	0·83	1·29	1·55
Mean ΔR_{Mg}	0·72	1·02	1·57
Range	0·60–0·83	0·86–1·29	1·35–1·74

These have been listed for system LT11/70 at 34° C (calculated from the data of Reineke (1956)).

With the exception of the 19-hydroxyl group, for which few examples are available, it is clear that characteristic differences in ΔR_{Mg} values are found for different positions and orientations of hydroxyl groups in steroids. On the other hand, the equatorial (less hindered) secondary hydroxyl groups have values which differ much less among themselves than the axial groups, and the less hindered they are the more their $\Delta_M R_g$ values approach a limiting figure of about 1·6–1·8 in different "typical" systems. Only one exception to the general rule that the ΔR_{Mg} value for an equatorial hydroxyl group is larger than that of the axial group at the same position is known at present, namely the 7-hydroxyl group in

Table 2.5. ΔR_{Mg} of 9α-Fluoro Group in 11-Substituted Steroids

Root substance	ΔR_{Mg} for substitution with 9α-fluorine	
	LT21/85	B/50
P^4-11β,17α,21-ol-3,20-one		+0·13
2α-Me-P^4-11β,17α,21-ol-3,20-one	+0·25	+0·13
P^4-11β,17α,21-ol-3,20-one-21-Ac		+0·10
2α-Me-P^4-11β,17α,21-ol-3,20-one-21-Ac		+0·10

Solvent systems at 25·5° C (author's data, unpublished).
Both are "typical" methanol systems.

Δ^5-3β-hydroxysteroids. On both columns and paper (Mosbach, Nierenberg and Kendall, 1953; Haslewood, 1954; Prochazka, 1945; Neher, 1958a, p. 69, but see p. 88) the 7α-hydroxysteroid (axial) is more polar than the 7β-epimer. This, however, is accounted for by the fact that in this position the equatorial group is *more* hindered than the axial, not by the atoms of the B-ring, but by the D-ring (Barton, 1950; Barton and Cookson, 1956). This position appears to be highly sensitive to steric factors and the relative mobilities of epimeric 7-hydroxysteroids are often inverted on changing solvent systems (Reineke, personal communication).

The ΔR_{Mg} for the 14α-hydroxyl group is the only one available for an isolated tertiary hydroxyl group. It appears to be relatively constant in four families of compounds (Table 2.3) with a mean value of 1·17 (range 0·95–1·32) which is close to that for hindered axial secondary hydroxyl groups such as the 11β-hydroxyl. Under Reineke's conditions (34 °C) there is no significant difference in the values for this group and the 6β-hydroxyl group although in any one family of compounds the 6β- and 14α-hydroxy-derivatives can usually be separated by overrun chromatograms: the difference between the values for these two groups and for the 11β-hydroxyl group is also rather small when compared with the figures of Neher and Wettstein (Table 2.2) and of the author. The latter two groups of figures, however, are taken from chromatograms run at 22° and 25·5 °C, respectively. This effect of temperature, which has been mentioned briefly before, is to be expected and is due to two principal factors which are known to occur with widely different classes of organic compounds. First the higher

Table 2.6. ΔR_{Mg} Values of Methyl Groups

Root substance	Methyl group (conformation)	ΔR_{Mg} in solvent system				Source
		34° C LT11/70	34° C LB21/80	25.5° C LT21/85	25.5° C B/50	
P⁴-17α,21-ol-3,11,20-one	2α(e)	−0.41				Author
P⁴-11β,17α,21-ol-3,20-one	,,				−0.41	Reineke (1956)
P⁴-11β,17α,21-ol-3,20-one-21-Ac	,,	−0.30			−0.51	Author
9α-F-P⁴-11β,17α,21-ol-3,20-one-21-Ac	,,			−0.25		Author
P⁴-11β,17α,21-ol-3,20-one	16α(−)				−0.52	Author
P¹,⁴-11β,17α,21-ol-3,20-one	,,				−0.68	Author
P¹,⁴-17α,21-ol-3,11,20-one	,,				−0.70	Author
19-norA⁴-3,17-one	10β(a)		−0.09			Reineke (1956)
19-norA⁴-17β-ol-3-one	,,		−0.16			Reineke (1956)
A⁴-17β-ol-3-one	17α(a)	−0.20	−0.21			Reineke (1956)
A⁴-17β-ol-3,11-one	,,	−0.44	−0.25			Reineke (1956)

Solvent systems all "typical" (methanol–water–hydrocarbon).

temperatures produce in most cases a greater intermiscibility of the two phases so that they become more alike and hence all differences in chemical potential in the two phases become smaller. Second the excess free energy or activity coefficients of nearly all solutes decrease in non-electrolyte solvents with increases in temperature and the differences between different solutes in this respect becomes less (see p. 126).

ΔR_{Mg} values for isolated primary hydroxyl groups are only available for two 19-hydroxysteroids (Table 2.2) and in one of these the 11β-hydroxyl group is also present and very close to it, although on the γ-carbon atom. Judging by the figures given by Loke, Marrian and Watson (1959) for 18-hydroxyoestrone, however, the 18-hydroxyl group has a high ΔR_{Mg} value of around 1·8–2·0. In the latter compound, the 18-hydroxyl and 17-ketone group are relatively far apart and the former group will probably be polarized only to a small extent by the negative inductive effect of the β-ketone group. The value of 1·8–2·0 is therefore probably typical for primary hydroxyl groups and the ΔR_{Mg} value of the 19-hydroxyl group in 19-hydroxycorticosterone should be considered exceptionally low (see p. 284). This effect of close hydroxyl groups is seen in simpler compounds. Thus the R_M values of propane-1,3-diol and propane-1,2-diol are -0.09 and -0.31 in chloroform–ethanol (3:2 by vol.), and 0·78 and 0·50 in ether–water (Bergner and Sperlich, 1953), the ΔR_M values thus being 0·22 and 0·28 in favour of the 1,3-diol. The rather larger difference between the ΔR_{Mg} values of the 19-hydroxyl group in 17α,19-dihydroxy-11-deoxycorticosterone and 19-hydroxycorticosterone (approximately 0·7) is probably due to the conformational rigidity of the steroid nucleus which stabilizes the mutual hindrance of the two hydroxyl groups (11β and 19) and any hydrogen bonding between them.

The relatively low ΔR_{Mg} values of the 17α- and 21-hydroxyl groups in 20-oxosteroids must also be considered atypical and due to interactions with adjacent groups in the side-chain and D-ring of the steroids (cf. the 14α-hydroxyl group with the tertiary 17α-hydroxyl; and the 18- and 19-hydroxyl groups with 21-hydroxyl groups: Tables 2·2, 2·3). Internal hydrogen bonding between these hydroxyl groups and the 20-ketone group is a possible explanation of these findings (Brooks et al., 1957) but simple hindrance by neighbouring polar groups is probably enough to account for the effects.

Certain 1,2-diol groups have been studied and it is clear that the *trans*-diol has a larger ΔR_{Mg} value than the *cis*-diol (e.g. oestriol R_M 0·95, and epi-oestriol R_M 0·72 in benzene–methanol–water (2:1:1 by vol. B/50)). In both cases, however, the ΔR_{Mg} of the diol group is less than the sum of the ΔR_{Mg} values for the separate hydroxyl groups of the appropriate configuration (see p. 121). In the case of 11,12-ketols and diols of the sapogenin series, Brooks et al. (1957) observed particularly strong interactions, shown by ΔR_{Mg} values for the diol or ketol group which were less than the sum of the values for the individual groups by 0·4–1·0 R_M units. This area of the steroid molecule is, however, subject to far more subtle and extensive steric influences than others and the deficits in ΔR_{Mg} values are probably exceptionally large (see. p. 121).

This account of ΔR_{Mg} values in steroids is far less complete and accurate than desirable for the reasons mentioned earlier, and many of the deviations from agreement with the theory are believed to be due to experimental error and the extra difficulty of obtaining accurate values with these solvent systems. For instance, the ΔR_{Mg} values all tend to be lower for the family based on progesterone (P^4-3,20-one) than for other more polar steroids. This is probably largely due to the high R_F value of progesterone in the solvent systems favourable for studying its substituted derivatives; this is not only subject to the large errors of such high R_F values but also gives rise to much higher proportionate errors in the R_M value due to the shape of the R_F/R_M function (p. 76). It is not unreasonable, however, to suggest that the main suppositions of this theoretical approach are confirmed by the available data. They thus support the basic theory of Martin (1950) and justify the restricted definition of "group" proposed in Chapter 1. More important still, the approximate constancy of ΔR_{Mg} values for substituents in close proximity to other substituents, despite their difference from the values given by the same substituents when isolated, provides a sound basis for applying the R_M treatment to complex functional groups and to the use of chromatographic results in the elucidation and interpretation of structural formulae (Chapter 5). These steric considerations are very similar to the vicinal corrections required in the additive treatment of molecular rotations (Barton and Cox, 1948).

It has been suggested that the ΔR_{Mg} values described for steroid substituents can reasonably be considered as the result of general rules modified by the special steric features of the steroid hydrocarbon skeleton. This is seen by considering the angular 10-methyl group which, in the absence of neighbouring substituents (e.g. at C-11), has a ΔR_{Mg} of $-0·1$ to $-0·15$ in typical systems (Table 2.13). The value for methylene groups in short- and long-chain alkyl derivatives of various types in similar solvent systems is usually about $-0\cdot 2$ in "straight" systems (p. 20). Similarly, the numerous studies of two- and three-ring compounds by Prelog's group confirm this supposition. Thus the ΔR_{Mg} values for epimeric hydroxyl groups in several dihydroxy-*trans*-decalin derivatives differ by $0\cdot 18$–$0\cdot 2$ in the sense ΔR_{Mg} (equatorial-OH) $> R_{Mg}$ (axial-OH) when run in aqueous methanol systems (e.g. Baumann and Prelog, 1958a, b). This epimeric difference is close to that found for epimeric hydroxyl groups in steroids at positions 3, 17 and 20; that is to say in those positions at the extreme ends of the molecule which are already known to show the smallest differences in steric hindrance (Barton and Cookson, 1956). Again the differences in the ΔR_{Mg} values of isolated hydroxyl groups of the same conformation but at different positions in decalin and indane derivatives are small to vanishing, as would be expected in the absence of additional hindering groups. (Calculated from data of Acklin and Prelog, 1959; Baumann and Prelog, 1959; Acklin, Prelog and Prieto, 1958; Baumann and Prelog, 1958a, b; Prelog and Zäch, 1959.) The values obtained in aqueous methanol systems are indeed close to those obtained for isolated hydroxyl groups of a given conformation at positions 3, 16 and 20 in the steroids.

2. ΔR_{Mg} Values in Atypical Solvent Systems

The most interesting of these systems are the ones based on t-butanol that were introduced by Eberlein and Bongiovanni (1955). These authors noted that the relative "polarities" of ketone groups and hydroxyl groups at positions 3 and 11 were usually the reverse of those found with the typical solvent systems. These systems were mainly useful because of the important reversal of the relative mobilities of cortisol and aldosterone. With the work of Neher and Wettstein (1956b) and some unpublished experiments by the author there are now sufficient data available to suggest that these systems show a difference with respect to hydroxyl groups in steroids that is quite general. Despite lack of data, it is possible to suggest from ΔR_{Mr} values that the ΔR_{Mg} values of ketone groups are, in contrast, relatively similar to those found with typical systems.

As originally published the systems were used under non-equilibrium conditions, the tanks being heated during equilibration to 32 °C and allowed to cool during the run. The mobilities obtained by Eberlein and Bongiovanni (1955) cannot therefore be analyzed in terms of R_M values. In Table 2.7 the figures of Neher and Wettstein (1956a), obtained at 40 °C with one of these systems, and the figures of the author, obtained with a modification of the original E_4 system at 20 °C, are compared with figures from two comparable "typical" systems taken from Table 2.2. It is clear that the ΔR_{Mg} values for all the hydroxyl groups that have been studied are from one-quarter to one-half of those found in typical systems. The few values available for ketone groups vary in system E_4 from one-half to two-thirds the values for typical systems, and the one value available in system E_2B is slightly greater than the value in typical systems. It will be seen later, however, that a comparison of ΔR_{Mr} and ΔR_{Ms} values for the two types of system suggests that ketone groups at positions 3, 17 and 20 have similar ΔR_{Mg} values in the t-butanol systems and in typical systems.

The interpretation of these figures is made rather difficult by the fact that system E_2B contains a large proportion of water, and that system E_4

Table 2.7. A Comparison of ΔR_{Mg} Values of Typical and Atypical Solvent Systems

Group, position, orientation	Substance in which group is substituted	LT21/85 25·5° C 1	B/50 22° C 2	E$_4$ 20° C 3	E$_2$B 40° C 4
-Ketone	A^4-3,17-one	0·56	—	0·39	—
	P^{4}21-ol-3,20-one-21-Ac	0·68	—	0·30	—
	P^4-17α,21-ol-3,20-one	0·67	0·79	—	0·37
	βA-3α-ol-17-one	0·64	—	0·31	—
	βP-3α,17α,21-ol-20-one	—	0·82	—	0·97
-OH	P^4-3,20-one	0·73	—	0·27	—
	P^4-11β-ol-3,20-one	0·82	—	0·42	—
α-OH	P^4-11β,21-ol-3,20-one-21-Ac	—	1·20	0·07	0·36
	P^4-21-ol-3,11,20-one	—	1·04	0·30	0·07
	P^4-11α,21-ol-3,20-one	—	—	—	0·35
β-OH	P^4-3,20-one	0·90	—	0·16	—
	A^4-3,17-one	0·93	—	0·22	—
	P^4-21-ol-3,20-one	—	—	0·31	—
	P$_4$-17α,21-ol-3,20-one	—	1·20	—	0·36
-OH	P^4-3,20-one	1·03	—	0·19	—
	A^4-17β-ol-3-one	1·35	—	0·34	—
	P^4-17α,21-ol-3,20-one	—	1·61	—	0·63
	P^4-11β,21-ol-3,20-one	—	1·77	—	0·88
	P^4-21-ol-3,11,20-one	—	1·50	—	0·68

Table 2.7. A Comparison of ΔR_{Mg} Values of Typical and Atypical Solvent Systems—*continued*

Group, position, orientation	Substance in which group is substituted	LT21/85 25·5° C 1	B/50 22° C 2	E₄ 20° C 3	E₂B 40° C 4
11α-OH	P⁴-3,20-one A⁴-17β-ol-3-one P⁴-17α,21-ol-3,20-one	1·39 1·88 —	— — 1·70	0·35 0·52 —	— — 0·49
7α-OH	P⁴-21-ol-3,20-one	1·70	—	0·57	—
15α-OH 15β-OH	P⁴-21-ol-3,20-one P⁴-21-ol-3,20-one	1·77 1·57	— —	0·65 0·65	— —
19-OH	P⁴-17α,21-ol-3,20-one P⁴-11β,21-ol-3,20-one	— —	1·98 1·32	— —	0·84 0·48
2α-OH	A⁴-17β-ol-3-one	—	0·75	0·39	—

1 and 2 are "typical" systems. (methanol–water–hydrocarbon) 3 is a modification of Eberlein and Bongiovanni's (1955) original E₄ system and had the composition methylcyclohexane–t-butanol–methanol–water (100: 45: 45: 10 by vol.). E₂B is the original system of these authors as run by Neher and Wettstein (1956). Both the t-butanol systems were run under equilibrium conditions at the temperatures stated: i.e. not with the falling temperature of the original authors. Systems 1 and 3: author's figures (unpublished). Systems 2 and 4 calculated from figures of Neher and Wettstein (1956).

contains very little water but equal proportions of t-butanol and methanol (by vol.). Also no figures are available for the phase composition; it is probable however that the components of E₄ are much more intermiscible than those of E₂B (Table 2.7). The fact that the relative magnitudes of the ΔR_{Mg} values for different hydroxyl groups are in the same order in t-butanol systems as in typical systems is probably largely due to solute interactions with the methanol in E₄ and with the water in E₂B. Further discussion of the nature of the solvent–solute interactions in these systems will be deferred until a later section.

Another atypical solvent system introduced by Peterson *et al.* (1958) is based on dioxan–methanol–water. Preliminary tests of this system suggest that the ΔR_{Mg} values of hydroxyl groups are slightly reduced and those of ketone groups largely unchanged relative to typical systems. These systems also give very nearly the same ΔR_{Mg} value for the complex hemiacetal group of aldosterone diacetate, as is obtained with typical systems. Some figures from the author's experiments are given in Table 2.8; they were obtained at approximately 22 °C.

Table 2.8. ΔR_{Mg} Values in Another Atypical System

Group, position, orientation	Substance into which group is substituted	ΔR_{Mg}
6-Ketone	P⁴–3,20–one	0·51
11-Ketone	P⁴–3,20–one A⁴–3,17–one βA-3 α-ol-17-one	0·70 0·71 0·62
11β-OH	P⁴–3,20–one A⁴–3,17–one βP-3 α-ol-17-one αA-3 α-ol-17-one	0·75 0·79 0·81 0·79
11α-OH	P⁴–3,20–one	1·23
17α-OH	P⁴–3,20–one	0·61

Composition: methylcyclohexane – dioxan – methanol – water (4: 4: 2: 1 by vol.). After R. Peterson: author's results (unpublished).

Systems based on concentrated ($\geqslant 70\%$, v/v) aqueous acetic acid as stationary phase were used for separating bile acids in the counter-current apparatus by Ahrens and Craig (1952), on paper by Sjövall (1952, 1954) and on Celite columns by Mosbach et al. (1953a, b; Mosbach, Zomzely and Kendall 1954). The ΔR_{Mg} (or ΔV_r) values for the groups present in the compounds they studied seemed to be approximately similar to those found in typical solvent systems. On examination, however, it was found that such systems give exceptionally high R_M values with a variety of saturated and unsaturated 3-oxosteroids, while the R_M values for 3-hydroxysteroids were rather larger than those obtained in systems having the same main components of the mobile phase but with methanol substituted for acetic acid (Bush, 1960). At present, some of these findings can only be discussed in terms of ΔR_{Mr} values but it is convenient to note here that ketone and hydroxyl groups at position 11 have ΔR_{Mg} values in these systems which are very close to those obtained with typical systems, although slightly, and probably significantly, larger in the case of the 11-ketone group. These latter differences are small, however, compared with those seen with the 3-, 17- and 20-ketones, and it seems probable that these systems will give similar ΔR_{Mg} values for hydroxyl groups and ketone groups in other positions (Table 2.12). That a highly specific interaction occurs with terminal ketone groups[1] is also suggested by the fact that this type of system is the only partition system giving large enough differences in the R_M values of saturated 3-ketones epimeric at C-5 to provide really good separations of such pairs of epimers, apart from the system methylcyclohexane—butane-1,3-diol (Axelrod, 1953). The acetic acid systems are useful in that low R_F values can be obtained for steroid diketones which are otherwise not well separated in volatile systems (e.g. progesterone, R_F 0·2 in methylcyclohexane–acetic acid–water (10: 9: 1 by vol.); and 0·50 in methylcyclohexane–methanol–water (10: 9: 1 by vol.); both run at approximately 22 °C). They are also useful in giving a much *lower* ΔR_{Mg} value for the hemiacetal group of aldosterone (see p. 296).

Other atypical systems have been tried out by the author in a search for general changes of the relative ΔR_{Mg} values of common substituents, the main aim being to find systems showing atypical ΔR_{Mg} values of ketone groups in general. Systems based on pyridine–water–hydrocarbon (Callow, Dickson, Elks, Evans, James, Long, Oughton and Page, 1955) show a general reduction in the ΔR_{Mg} values of both ketone and hydroxyl groups (author, unpublished). They do not seem to have any special merits except that the ΔR_{Mg} of the 11β-hydroxyl group is greater than that of the 6β-hydroxyl group, an inversion of the usual order. Systems based on aqueous acetone (Howard and Martin, 1950) show even greater reductions in the ΔR_{Mg} values of these two groups and no useful distinction between them. These solvents are so soluble in aromatic hydrocarbons that only a limited range of mixtures is feasible as solvent systems. As soon as the mobile phase is composed of a hydrocarbon mixture more polar than light petroleum or decalin a very large proportion of acetone or pyridine goes over into the mobile phase and leaves a stationary phase largely made up of water.

The figures for systems based on butane-1,3-diol (Axelrod, 1953) are inadequate for calculations of R_M values but deserve special attention because of their excellent resolution of saturated 3-oxosteroids epimeric at C-5. These separations are better than obtained with acetic acid systems

---

[1] Note added in proof: Dr. J. Sjövall points out that these results agree with the partition coefficients given by Ahvens & Craig (1952).

and appear to be at least as good as are obtained with adsorption columns.[1] These two types of partition system are at present unique in giving differences in R_M value which are sufficient for good separation of this type of epimeric pair. With most "typical" systems, the differences in R_F (or R_M) value between epimeric 3-ketones are not even measurable. In all systems, however, the R_F value of the 5α-epimer is *lower* than that of the 5β-epimer if there is a significant difference.

It was mentioned earlier that more reproducible results would be expected with more viscous and less volatile mobile phases than have been usual in steroid chromatography. Decalin, tetralin and xylene have been found excellent for this purpose and are very useful for obtaining more accurate measurements of R_F values in the higher ranges (Bush, 1960b). It has been possible recently to obtain values of reasonable reproducibility for monosubstituted steroids and thus to obtain ΔR_{Mg} values for substituents at positions 3, 17 and 20. Some typical values obtained with these systems are given in Table 2.12. It is seen that the majority of polar groups have the same ΔR_{Mg} values as are obtained with the more usual systems based on aqueous methanol. On the other hand, the androstanediones epimeric at C-5 are partially separated in these systems, an effect which must be attributed to the change in the mobile phase. As suggested in the first chapter, the use of solvents having rather rigid and assymmetrical molecules leads to a greater sensitivity of non-polar interactions to steric differences in the solutes.

Some of the ΔR_{Mg} values given in Table 2.12 are in doubt by about ± 0.20 R_M units because of the nature of the available monosubstituted steroids. However they are precisely what would be expected: thus, for instance, for three sets of compounds the ΔR_{Mg} values for the equatorial 5α(H)-3β-OH group are 1·63, 1·64 and 1·55 allowing +0·05 units for each double bond in the monosubstituted compounds, a value found for the isolated 9(11)-double bond in several disubstituted C_{19} steroids. The value for less hindered positions is probably slightly larger, and in $\Delta^{3,5}$-17-one is probably increased by conjugation so that the actual values for the 5α(H)-3β-OH group may be as much as 1·7–1·8. This value is in fact close to the limiting value to which ΔR_{Mg} values for equatorial hydroxyl groups in steroids tend as progressively less hindered positions are considered, and in particular is close to those for the 15α- and 17β-hydroxyl groups.

[1] Note added in proof: Dr. W. Taylor has been unable to confirm this (personal communication).

The value for the 17-ketone group is in the range 1·0–1·2 for five different compounds, and for the 5α(H)-3-ketone group of one compound is 1·1. The results at these extreme ends of the steroid molecule where steric differences are at a minimum lend considerable support to the theories of Martin (1949) and Bate-Smith and Westall (1950).

3. ΔR_{Ms} AND ΔR_{Mr} VALUES

If the R_M theory is obeyed by two solvent systems with respect to ΔR_{Mg} values then it follows that the difference in the R_M values of a given solute in the two systems should also be an additive property of the solute. If we call this difference the ΔR_{Ms} of the substance with respect to the particular pair of solvent systems then it follows from the above arguments that it will be equal to the sum of the differences of the ΔR_{Mg} values of its constituent parts in the two systems. Similarly, the difference in the ΔR_{Mg} value of any substituent or part of the molecule in the two solvent systems can be considered as the ΔR_{Ms} value for that particular substituent or part (p. 85).

For certain positions in the steroid nucleus it is difficult to find many examples in which these positions are unoccupied by a substituent. Until further values are obtained for groups at these positions it is useful therefore to compare the differences in the R_M values of pairs of compounds differing only by the nature of their substituents at positions 3, 17 and 20 (i.e. ΔR_{Mr} values). Again it is often useful to consider the ΔR_M caused by extensive modifications to molecular structure such as the opening of rings to form seco-dicarboxylic acids, or oxidations followed by ring-closure. As emphasized previously, these structural differences are merely more general examples of the substitution of hydrogen by simple functional groups such as hydroxyl or a ketonic oxygen (p. 85).

These two types of ΔR_M value are necessary consequences of the R_M theory and an example of the former type (ΔR_{Ms}) was given for methylene groups in Chapter 1 (p. 22). Tables 2.9 and 2.11, which give some typical values of ΔR_{Ms} and ΔR_{Mr}, are therefore given merely as an illustration of the treatment rather than as additional proof of its validity. The ΔR_{Mr} and ΔR_{Ms} values for substituents at positions 3, 17 and 20 are collected separately in Tables 2.10 and 2.12, since few ΔR_{Mg} values are available and the figures do therefore lend extra support to the theory as applied to steroids.

Two types of effect enter into the ΔR_{Mg} values of compounds. The

Table 2.9 Some Typical ΔR_{Mr} Values for Conversions at Positions Other than C-3, C-17 and C-20

Conversion	Root substance	Solvent system		ΔR_{Mr}	Source
11-Ketone → 11β-OH(a)	P⁴-3,20-one	LT11/70	34° C	+0.27	Reineke (1956)
	A⁴-3,17-one	,,	,,	+0.18	,,
	A⁴-17β-ol-3-one	,,	,,	+0.23	,,
	P⁴-21-ol-3,20-one	,,	,,	+0.18	,,
	P⁴-17α-ol-3,20-one	,,	,,	+0.47	, ,,
	P⁴-17α,21-ol-3,20-one	,,	,,	+0.37	,,
	βP-3,20-one	,,	,,	+0.46	,, ,,
	P⁴-17α,21-ol-3,20-one	LDT111/85	25.5° C	+0.35	Author (unpublished)
	P⁴-21-ol-3,20-one	,,	,,	+0.22	,,
11-Ketone → 11α-OH(e)	P⁴-3,20-one	LT11/70	34° C	+0.41	Reineke (1956)
	A⁴-3,17-one	,,	,,	+0.71	,,
	A⁴-17β-ol-3-one	,,	,,	+0.78	,,
	P⁴-21-ol-3,20-one	,,	,,	+0.83	,,
	P⁴-17α-ol-3,20-one	,,	,,	+0.98	,,
	P⁴-17α,21-ol-3,20-one	,,	,,	+0.95	,,
	P⁴-6β-ol-3,20-one	,,	,,	+0.96	,,
	βP-3,20-one	,,	,,	+0.72	,,
21-OH → 21-OAc	P⁴-21-ol-3,20-one	LB21/80	,,	−0.88	Reineke (1956)
	P⁴-17α,21-ol-3,20-one	LT11/70	,,	−0.73	,,
	P¹,⁴-17α,21-ol-3,11,20-one	B/50	,,	−0.84	,,
	βP-17α,21-ol-3,11,20-one	LT11/70	,,	−0.84	,,
	P⁴-11β,17α,21-ol-3,20-one	,,	,,	−1.08	,,
	2α-Me-P⁴-11β,17α,21-ol-3,20-one	,,	,,	−0.97	,,
	P⁴-11β,17α,20α,21-ol-3-one	B/50	,,	−1.20	,,

Table 2.9—continued.

Conversion	Root substance	Solvent system		ΔR_{Mr}	Source
17β-OH → 17β-OAc	A^4-17β-ol-3-one	L/80	34° C	−1·24	Reineke (1956)
19-OH → 19-OAc	P^4-19,21-ol-3,20-one	LB21/80	22° C	−1·27	Neher (1958a)
11α-OH → 11α-OAc	P^4-11α-ol-3,20-one	LB21/80	34° C	−1·02	Reineke (1956)

Table 2.10. ΔR_{Mr} Values for Conversions at C-3, C-5, C-17 and C-20 in "Typical" Systems

Conversion	Root substance or skeleton	Solvent system		ΔR_{Mr}	Source
5β(H)-3α-OH → 5α(H)-3β-OH	P-3,11β,17α,21-ol-20-one	B/50	22° C	+0·07	Neher and Wettstein (1956b)
	P-3,17α,21-ol-11,20-one	,,	,,	+0·25	,, ,, ,,
	P-3,21-ol-11,20-one	,,	,,	+0·29	,, ,, ,,
	P-3,11β,21-ol-20-one	,,	,,	+0·28	,, ,, ,,
	A-3-ol-17-one	LD11/85	20° C	+0·04	Author (unpublished)
	P-3-ol-20-one	,,	,,	+0·04	,, ,,
5β(H)-3α-OH → 5α(H)-3α-OH	A-3-ol-17-one	LD11/85	20° C	−0·16	Author (unpublished)
	P-3-ol-20-one	,,	,,	−0·08	,,
	A-3,11β-ol-17-one	,,	,,	−0·11	,,

Table 2.10—continued.

Conversion	Root substance or skeleton	Solvent system		ΔR_{Mr}	Source
$5\beta(H)$-3α-OH $\rightarrow$ Δ^5-3β-OH	A-3-ol-17-one	LD11/85	20° C	+0.11	Author (unpublished)
	P-3-ol-20-one	,,	,,	+0.13	,,
5β(H)-3α-OH $\rightarrow$ Δ^4-3-Ketone	P-3,11β,17α,21-ol-20-one	B/50	22° C	−0.48	Neher and Wettstein (1956b)
	P-3,17α,21-ol-11,20-one	,,	,,	−0.47	,,
	P-3,11β,21-ol-20-one	,,	,,	−0.20	,,
	P-3,17α,21-ol-20-one	,,	,,	−0.44	,,
	P-3,21-ol-11,20-one	,,	,,	−0.24	,,
	A-3-ol-17-one	LD11/85	20° C	−0.29	Author (unpublished)
	A-3-ol-11,17-one	,,	,,	−0.33	,,
	A-3,11β-ol-17-one	,,	,,	−0.43	,,
5β(H)-3α-OH $\rightarrow$ Phenolic-3-OH	βA-3α-ol-17-one	LD11/85	20° C	+0.40	Author (unpublished)
	βA-3α,17β-ol	,,	,,	+0.38	,,
5β(H)-3α-OH $\rightarrow$ 5β(H)-3-ketone	P-3,11β,17α,21-ol-20-one	B/50	22° C	−0.93	Neher and Wettstein (1956b)
	P-3,17α,21-ol-11,20-one	,,	,,	−0.83	,,
	A-3-ol-17-one	LD11/82	20° C	−0.65	Author (unpublished)
17-Ketone $\rightarrow$ 17β-OH	Δ^4-3,17-one	LD11/85	20° C	+0.71	Author (unpublished)
	Δ^5-3β-ol-17-one	,,	,,	+0.78	,,
	0-3-ol-17-one	,,	,,	+0.78	,,
	βA-3α,11β-ol-17-one	BL$_1$	22° C	+0.60	Neher (1958a)
	Δ^5-3β,11β-ol-17-one	,,	,,	+0.67	,,
	Δ^5-3β-ol-11,17-one	,,	,,	+0.57	,,

Table 2.10—continued.

Conversion	Root substance or skeleton	Solvent system		ΔR_{Mr}	Source
17-Ketone $\rightarrow$ 17β-OH	αA-3α,11β-ol-17-one A⁴-11β-ol-3,17-one	BL₁ LT11/70	22° C 34° C	+0·60 +0·62	Neher (1958a) Reineke (1956)
20α-OH $\rightarrow$ 20β-OH	αP-3β,20α-ol αP-3β,17α,20α-ol P⁴-17α,20α-ol-3,11-one P⁴-11β,17α,20α,21-ol-3-one	L/85 ,, ,, B/50	34° C ,, ,, ,,	−0·22 −0·17 −0·25 −0·13	Author (unpublished) ,, ,, Reineke (1956)
20-Ketone $\rightarrow$ 20β-OH	Cortisol Cortisone P⁴-21-ol-l3,11,20-one	B/50 ,, ,,	22° C ,, ,,	+0·78 +0·90 +1·05	Neher and Wettstein (1956b) ,, ,, ,, ,,

Table 2.11. ΔR_{Ms} Values for changes from the nearest Equivalent "Typical" Solvent System to Three Types of "Atypical" System

Group	Root substance	ΔR_{Ms} for change to:			Source as in Fig. 2.7
		t-Butanol	Dioxan–Methanol	Acetic acid	
11β-OH	P⁴-11β-ol-3,20-one A⁴-11β-ol-3,20-one αA-3α,11β-ol-17-one βA-3α,11β-ol-17-one	−0·74 −0·71 −0·51 —	−0·15 −0·12 −0·38 −0·23	−0·28 −0·16 −0·28 −0·25	
11-Ketone	P⁴-3,11,20-one A⁴-3,11,17-one βA-3,11,17-one	— −0·14 −0·31	+0·16 +0·18 0·0	−0·03 −0·05 −0·02	
11α-OH	P⁴-11α-ol-3,20-one A⁴-11α,17β-ol-3,one P⁴-11α,17α,21:ol-3,20-one	−1·04 −1·36 −1·21	−0·16 −1·08 —	−0·34 −0·20 —	
6β-OH	P⁴-6β-ol-3,20-one A⁴-6β,17β-ol-3-one P⁴-6β,17α,21-ol-3,20-one P⁴-6β,11β,21-ol-3,20-one P⁴-6β,21-ol-3,11,20-one	−0·84 −1·01 −0·98 −0·89 −0–82		−0·31 +0·04 	
7α–OH	P⁴-7α,21-ol-3,20-one	−1·13	—	—	
15α-OH	P⁴-15α,21-ol-3,20-one	−1·12	—	—	
15β-OH	P⁴-15β,21-ol-3,20-one	−0·92	—	—	
19-OH	P⁴-17α,19,21-ol-3,20-one P⁴-11β,21-ol-3,20-one	−1·14 −0·84	— —	— —	

It should be emphasized that these are *not* the ΔR_{Ms} values for the *compounds* in column 2. The values are obtained by subtracting the ΔR_{Mg} values of the group in the first column in the "typical" system from that of the group in the "atypical" system. It will be noted that there is a large and generally negative ΔR_{Ms} for all hydroxyl groups examined in the change to t-butanol systems (column 3). Other solvent changes give small or variable changes probably related to changes in intermiscibility of the solvent phases.

Table 2.12. ΔR_{Mg} Values in More Reproducible Systems

Structure or group	Root compound	ΔR_{Mg} values in				ΔR_{Ms} value	
		1 D/85	2 D/A90	3 DX21/85	4 DX21/A90	$1 \to 2$	$3 \to 4$
C_{19} skeleton	(mean of 4)	−2·3	−2·4	−2·5	−3·0	−0·1	−0·5
11-Ketone	Progesterone	0·8	0·88	0·74	0·71	+0·08	−0·03
	Androst-4-enedione			0·61	0·56		−0·05
	Aetiocholanolone			0·65	0·63		−0·02
11β-OH	Progesterone	1·18	0·98	1·17	0·89	−0·20	−0·28
	Androst-4-enedione			0·93	0·77		−0·16
	Aetiocholanolone			1·07	0·82		−0·25
	Androsterone			1·12	0·84		−0·28
6β-OH	Progesterone			1·33	1·02		−0·21
	Testosterone			1·19	1·23		+0·04
11α-OH	Progesterone			1·68	1·34		−0·34
	Testosterone			1·50			
	Androst-4-enedione			1·60	1·43		−0·07
17α-OH	Progesterone	1·04	0·84	1·07	0·71		−0·36
6β-OH-17α-OH	Progesterone	—	—	2·41 (theory 2·40)	1·80 (theory 1·73)		−0·61 (theory −0·67)
17-Ketone	Androstan-3β-ol	1·2	1·73			0·53	

CHROMATOGRAPHIC SEPARATION OF STEROIDS

Table 2.12—continued.

Structure or group	Root compound	ΔR_{Mg} values in 1 D/85	ΔR_{Mg} values in 2 DA/90	ΔR_{Ms} value 1→2
17-Ketone	Aetiocholan-3α-ol	1·17	—	—
	Androstan-3α-ol	1·16	1·56	0·40
	Aetiocholan-3β-ol	1·22	1·55	0·33
Δ⁴-3-Ketone	Androstan-17-one	1·4	2·0	0·60
	Androstan-17β-ol	1·6	2·1	0·50
5β(H)-3α-OH	Androstan-17-one	1·84	1·56	−0·28
5α(H)-3α-OH	Androstan-17-one	1·71	1·46	−0·25
Δ⁵-3β-OH	Androstan-17-one	2·0	1·70	−0·24
	Androstan-17β-ol	2·6	1·79	−0·81
5α-3-Ketone	Androstan-17-one	1·24	1·51	0·27
5β-3-Ketone	Androstan-17-one	1·23	1·43	0·20

Solvent systems at 22 to 23 °C.
1. Decalin–methanol–water (100: 85: 15 by vol.).
2. Decalin–acetic acid–water (100: 90: 10 by vol.).

The small to moderate falls in ΔR_{Mg} for hydroxyl groups on changing to the equivalent acetic acid system are probably due to differences in intermiscibility of the phases. Note, however, the moderate to large increases in ΔR_{Mg} of 3- and 17-ketone groups.

first is due to specific changes in the interactions between the solvent phases and the compound's constituent parts. The second is due to more general changes in the phase composition of the system. As mentioned previously, increased intermiscibility of the solvent phases leads to a reduction of all ΔR_{Mg} values and a tendency of all R_M values to become closer to zero. Within one type of system (say aqueous methanol or t-butanol) it is usually possible to adjust either or both the water or alcohol content of the system to provide almost constant ΔR_{Mg} values for the polar parts of the molecule. Thus the principal methanol systems described in this book give ΔR_{Mg} values for the 11β-hydroxyl group in the range 0·9–1·2 at 22–25 °C.

Apart from the ΔR_{Mg} values of the polar groups it is also useful to calculate the ΔR_{Mg} values of the non-polar parts of the molecule for each system. At present it is possible only to give approximate values for the steroids in this respect because of the paucity of results on monosubstituted steroids. With the accumulation of more data a more extensive type of treatment such as that of Reichl (1955, 1956) should be possible. At present, therefore, approximate figures for the ΔR_{Mg} value of the C_{19} and C_{21} steroid skeleton must be used and these are given in Table 2.13. These are useful for calculating the very large ΔR_{Mr} values caused by extensive modifications (e.g. the formation of a triacetate); values which are too large to be actually measured in one solvent system. Suppose for instance we had a steroid with an R_M value of 1·0 in system B_5 and which gave a triacetate. The ΔR_{Mr} should be of the order of $-3·3$ and the triacetate would have an unmeasurable R_F of more than 0·99 in the B_5 system. On changing to an A system, however, an R_M value of 0·0 (R_F = 0·50) would be expected which would be measured with reasonable accuracy in, say, a decalin system. This is calculated as follows:

ΔR_{Mr} for triacetylation in methanol systems	approx. $-3·3$
ΔR_{Ms} for C_{21} steroid skeleton	approx. $+2·3$
R_M of original steroid triol in B5-type system	$+1·0$
R_M of triacetate in A-type system	approx. $0·0$

The basis for this type of calculation is the predominance of the polar interactions and the relative constancy of the ΔR_{Mg} values of the polar groups in solvent systems based on the same polar phase. Similar calculations can be carried out for changes of a more complicated nature such as the combination of a change from typical to atypical solvent systems with a modification of structure. The more steps involved, however, the less precise the end result will be (see Table 2.13).

The main point of defining these types of ΔR_M values and drawing attention to them is to show the ease with which features of molecular structure and changes due to chemical reactions may be identified purely on the basis of chromatographic behaviour, even in the absence of colour reactions or other specific photometric methods. This use of ΔR_{Ms} and ΔR_{Mg} values will be dealt with more fully in Chapter 5 but an example will be given here to illustrate the argument.

Suppose we have a substance extracted from urine and detected by a

Table 2.13. Approximate ΔR_{Mg} Values for the $5\alpha(H)$-C_{19} Steroid Hydrocarbon Skeleton in Various Systems (22–28 °C)

	Group A					Group B			Group C	
	1 L/96	2 L/85	3 D/85	4 D/A90	1 LT21/85	2 DT21/85	3 DT21/A70	1 T/75	2 B/50
Skeleton	−2.2	−2.2	−2.3	−2.5	−2.7	−2.5	−3.0	−3.9	−4.5
Methylene	−0.08	−0.12	—	—	—	—	—	—	—
Methyl	−0.10	−0.15	—	—	—	—	—	—	—
Double bond	+0.03	+0.05	—	—	+0.03	—	—	—	—

These values are in doubt because of insufficient data but are good enough for deciding which solvent systems will be best for unknown compounds.

Values for double bonds, methyl and methylene groups are for isolated groups. They will be considerably larger if substituted close to polar groups due to steric hindrance by methyl groups, and to conjugation with double bonds.

non-specific colour reaction, or by non-specific labelling with a radioactive isotope, which has an R_M of 0·55 in light petroleum–toluene–methanol–water (67:33:85:15 by vol.) at 25·5 °C and −0·45 in methylcyclohexane–t-butanol–methanol–water (100:45:45:10 by vol.) at 22 °C. The ΔR_{Ms} for this solvent change is −1·0 while the ΔR_{Ms} for the hydrocarbon skeleton of a C_{19} or C_{21} steroid is about +1·0. The over-all ΔR_{Ms} value of the compound suggests therefore that the sum of the ΔR_{Ms} values of its polar groups is about −2·0 which could be made up of a variety of combinations, as seen from Tables 2.7 and 2.11. Acetylation gives a ΔR_{Mr} in the first solvent system of −1·2; and reduction with borohydride gives two products with ΔR_{Mr} values of +0·8 and +1·05. Making reasonable assumptions, the first ΔR_{Mr} value shows clearly that the compound has one esterifiable hydroxyl group, and the second that it has either one reducible ketone group giving epimeric hydroxyl groups on reduction, or two reducible ketone groups one of which is less completely reduced than the other. The latter possibility, while reasonable on qualitative, is, however, unreasonable on quantitative grounds since the ΔR_{Mr} for the second ketone → hydroxyl conversion would be only +0·25, while the smallest known value (for the 11-ketone → 11β-hydroxyl conversion) is in the range +0·37 to +0·46 for this solvent system. On the other hand, the difference in the two ΔR_{Mr} values for the reduction with borohydride is too small for the difference in ΔR_{Mg} values of all known epimeric equatorial and axial hydroxyl groups on the steroid nucleus other than for epimeric hydroxyl groups at C-3, C-16 or C-17 for which it is too large (0 −0·13). But it is typical of the difference between epimeric 20-hydroxysteroids (0·17–0·25 for three pairs). The original compound is therefore most probably a 20-ketone with an esterifiable hydroxyl group which, on circumstantial grounds, is most probably at position 3.

These two groups now account for about −1·3 of the ΔR_{Ms} value of −2·0 due to polar groups. The residual −0·7 could be due to one hydroxyl or two ketone groups. Since the only position at which ketone groups are reduced with considerable difficulty in steroids is at C-7, a hydroxyl group is most probably the source of this residual ΔR_{Ms}. Since this is not esterifiable it could be tertiary, or at position 11, or in some cases, 12 (p. 291). Since the commonest tertiary hydroxyl group in urinary steroids is at C-17 and has a ΔR_{Ms} of about −0·8 in other 20-oxosteroids we should explore this possibility first. Oxidation of the borohydride reduction products with sodium bismuthate (Appleby and Norymberski, 1955) would be found to give a substance from both of them with an R_M

of -0.05; the over-all ΔR_{Mr} for the reduction plus oxidation with bismuthate is therefore -0.60 which is close to the typical ΔR_{Mr} for the conversion 17α-hydroxy-20-ketone (C_{21}) → 17-ketone (C_{19}). Oxidation of the acetate of the original substance with chromic acid, but not with bismuthate, would give the same ΔR_{Mr}. Finally the acetate would be saponified and the ΔR_{Mr} measured very accurately to distinguish between the value of 1·3 typical for the hydrolysis of 5β-3α-acetoxysteroids and that of 1·1 of 5α-3α-acetoxysteroids and the former value would be found. On oxidation of the hydrolysed product a ΔR_{Mr} of -0.4 would be obtained (ΔR_{Mr} -0.35 to -0.45; 5β(H)-3α-OH → 5ξ(H)-3-ketone) and this final derivative would show a ΔR_{Ms} of about $+0.3$ on changing from a methanol to an acetic acid system (ΔR_{Ms} for 5β(H)-3-ketones $+0.2$ to $+0.35$; for 5α(H)-3-ketones $+0.4$ to $+0.5$; for Δ^4-3-ketones $+0.55$ to $+0.60$; for 17-ketone $+0.33$ to $+0.53$; for C_{19} skeleton -0.2 to -0.3).

The probable formula of our unknown in this example is therefore 17α-hydroxypregnanolone (βP-3α,17α-ol-20-one) and the whole series of reactions and solvent changes would now be gone through with the suspected reference steroid. Two points must be emphasized about this example. First that in the present state of our knowledge and technique, the ΔR_{Ms} and ΔR_{Mr} values are subject to errors and variations which are too large to give *exact* identifications of the structural changes involved, except in the final derivatives of such a scheme in which the structural features are relatively simple. Second that the quantitative estimation of the changes in chromatographic mobilities in terms of ΔR_M values is similar in principle to the qualitative examination of such changes that have been used previously (Burton et al., 1951a, b; Bush and Sandberg, 1953; Bush and Willoughby, 1957) but much more powerful.

In practice, the degradation of a complicated steroid involves such large over-all changes of mobility that several solvent systems of each type are needed for the complete sequence. It is useful in practice to compare such large ΔR_{Mr} values with standard compounds giving similar ΔR_{Mr} values. Thus the difference in the ΔR_{Mr} values for the acetylation of a 5α(H)-3α,21-dihydroxysteroid and its 5β(H) epimer will be in the range 0·1–0·17, well within the range of variation for the ΔR_{Mr} of acetylation of the 21-hydroxyl group in different steroids (-0.73 to -1.20). If, however, there is good evidence that the steroid concerned has, say, a dihydroxyacetone side chain, then a pair of 3α-hydroxysteroids containing this group and epimeric at C-5 can be acetylated and run in parallel with the diacetate of the unknown. The difference, ΔR_M {(unknown)−(5α(H)-

reference compound)} $-\Delta R_M$ {(unknown diacetate)$-$(5α(H)-reference compound diacetate)} will be approximately zero if the unknown is a 5α-steroid and positive if it is a 5β-steroid (see Fig. 2.9). Since this difference is small, however, it is essential to confirm this supposition by partial

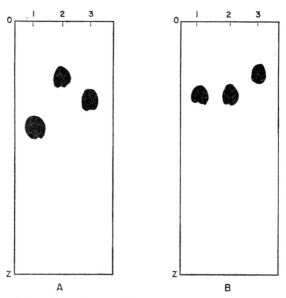

FIG. 2.9. ΔR_{Mr} of acetylating epimeric 3α-hydroxysteroids.

A. Polar system (e.g. B/50) with 1. Unknown 3α-hydroxy-20,21-ketol; 2. Standard 5β(H)-3α-hydroxy-20,21-ketol; 3. Standard 5α(H)-3α-hydrosy-20,21-ketol.

B. Less polar system of same type (e.g. LT21/85) with diacetates of the above three steroids. The ΔR_M (3$-$1) is the same in the two cases while $\Delta R_M(2-1)_A > \Delta R_M(2-1)_B$. For a 5β(H)-3β-hydroxysteroid (x), $\Delta R_M(x-1)_A > \Delta R_M(x-1)_B$ and for a 5α(H)-3β-hydroxysteroid (y), $\Delta R_M(y-1)_A < \Delta R_M(x-1)_B$. The four types of epimer are distinguishable but the differences in ΔR_{Mr} are small compared with variations in the very large negative ΔR_{Mr} of forming a diacetate. In this case the unknown is probably a 5α(H)-3α-hydroxy-20,21-ketol. These results would be confirmed by acetylation of the 17-oxosteroid or methyl etianate derived from the unknown ketol, and by the behaviour of the 3-ketone in an acetic acid system.

hydrolysis of the diacetates to the 3-monoacetates of the unknown and reference substances. If the ΔR_M {(unknown monoacetate)$-$(reference monoacetate)} is the same as the ΔR_M of the two *diacetates*, then the difference between the latter can be ascribed entirely to the configuration at C-5 with reasonable confidence (cf. Bush and Willoughby, 1957).

In principle, it should be possible to avoid this last concession to qualitative methods of argument by obtaining more accurate values for the ΔR_{Ms} values of the non-polar parts of the steroid molecule (Table 2.13). Large ΔR_{Mr} values could then be measured really accurately by changes of solvent system within one type (i.e. change of mobile phase only).

Much of the variation in the ΔR_{Mg} values that have been calculated from the figures in the literature is probably due to errors in the measurement of R_F values. This is not a serious criticism of these measurements since the error acceptable in routine screening work is far larger than that acceptable for the measurement of R_M and ΔR_M values (Bate-Smith and Westall, 1950) and the R_M approach to steroid behaviour on chromatograms has not been possible until recently when figures for large numbers of steroids became available. Some of the variation is, however, almost certainly due to complicated steric and electronic interactions within the molecule (see e.g. Barton, 1955) and it will be well worth seeing whether these obey general rules when really accurate measurements of R_M values are made. The range of application of this approach will then be greatly extended. Despite the errors of the R_M values used in this discussion small ΔR_{Mr} and ΔR_{Ms} values are sufficiently regularly observed to be regarded as significant and worth commenting upon. Since some of these depend upon small changes in the water content of the stationary phase or upon the hydrocarbon components of the mobile phase they are important when considering the finer details of ΔR_{Mr} values obtained with a change of solvent system within the same type. It will be seen in the next section that they can be related quite reasonably to known solvent–solute interactions derived from other types of study.

One of these is the conversion 20-ketone → 20ξ-hydroxyl group. The ΔR_{Mr} value of this conversion is much smaller for the formamide, and probably also for the propylene glycol system (Neher, 1958a, p. 89) than with aqueous methanol systems. This difference is so large that the ΔR_{MS} for 20-hydroxysteroids is quite large on changing from methanol to the usual impregnated-paper systems. Since the ΔR_{MS} for other hydroxyl groups between these systems is small and variable in sign this is a useful way of characterizing this class of steroids.

A very useful set of ΔR_{Ms} values is obtained with the various types of 3-oxygenated steroids. In some cases, these must be dealt with in terms of ΔR_{Mr} values in the different solvent systems (Table 2.12). It was noticed early on (Savard, 1953) that androstenedione (Δ^4-3,17-one) had a lower R_F value in light-petroleum or ligroin than androsterone (αA-3α-ol-17-one). All other pairs of this type showed the inverse order of their R_F or R_S values. This is a definite exception to the R_M rule but is marginal and with certain solvent systems can be shown to be a general phenomenon for Δ^4-3-ketones.

In light petroleum–methanol–water (100:85:15 by vol.) the ΔR_{Mr} values for the conversion 5β(H)-3α-hydroxyl group → Δ^4-3-ketone group are all negative but much greater for polar compounds like 11β-hydroxyandrostenedione (Δ^4-11β-ol-3,17-one) than adrenosterone (Δ^4-3,11,17-one) and androstenedione. With the same methanol–water composition but light petroleum–toluene as mobile phase the values are all negative and with this solvent system sufficiently so to give androstenedione a higher R_F value than androsterone. When acetic acid systems are substituted for aqueous methanol systems the ΔR_{Mr} values all become positive and all three Δ^4-3-ketones have lower R_F values than the related 5β(H)-3α-hydroxysteroids. Since the ΔR_{Mg} of the 11-ketone and 11β-hydroxyl groups are similar in the two systems these changes must be attributed to the structures in ring A, and the changes in the sign of the ΔR_{Mr} values for the three compounds in the two types of system must also be due to complicated differences in the interactions of these structures with the stationary phase.

The ΔR_{Mr} for the conversion 5α(H)-3-ketone → 5β(H)-3-ketone is sensitive not only to the main organic components of both phases but also to the water content of the system. Thus the epimeric 3-ketones are separated well with systems based on 90–95% acetic acid but poorly with the same hydrocarbon and 70% acetic acid. Little or no separation occurs in aqueous methanol systems using light petroleum or mixtures with toluene as mobile phase. Better, though incomplete, separations are obtained with decalin or xylene as mobile phase with the same aqueous methanol stationary phase. Other ΔR_{Mr} differences are sensitive to changes in the main components of the mobile phase, although it is not always easy to say how much of these differences are due to changes in phase composition due to differences in the intermiscibility of the components. Thus the inversion of the relative mobilities of androsterone and androstenedione on changing from light petroleum–85% methanol to light petroleum–toluene (67:33)–85% methanol is most reasonably ascribed to the addition of a cyclic component to the system.

In light petroleum (100–120 °C fraction)—95% methanol there is no separation of 11β-hydroxyandrosterone (5α(H)-3α-hydroxyl) and 11β-hydroxyaetiocholanolone (5β(H)-3α-hydroxyl); slight separation in the same boiling fraction with 85% methanol; and good separation with light petroleum (60–80 °C fraction)—85% methanol.* Even with the high-boiling petroleum, however, this pair becomes separable with a mixture such as light petroleum–toluene (67:33); and they are also well separated with methylcyclohexane or decalin with 90–95% methanol. Although slight improvements in the separation of this type of epimeric pair is obtained by increasing the water content of the system, much larger improvements are obtained by using mobile phases based on smaller aliphatic molecules or on more rigid assymetrical alicyclic molecules such as methylcylohexane, xylene, decalin, or tetralin.

Much more work needs to be done to find solvent systems giving selective changes in ΔR_{Mg} values for different classes of substituents, and also to find solvent systems more sensitive to the steric properties of hydrocarbon skeletons. Unfortunately many of these will involve involatile components which will have to be used by the impregnated-paper method. Nevertheless their successful use in gas–liquid chromatography (e.g. the use of phenanthrene or 7,8-benzoquinoline to separate m- and p-xylenes (Desty, Goldup and Swanton, 1959)) is an encouragement to use similar components for liquid–liquid separations of steroids.

Schauer and Burlisch (1955) have pushed the treatment of ΔR_{Mg} values to the theoretical limit by what is implicitly the same argument as the above. They suggest that for a set of substances containing a maximum of $(n-1)$ different constituent groups, n solvent systems should be set up. This will enable one to obtain $n^2 \Delta R_M$ values, each one of which will be the sum of the ΔR_{Mg} values of the constituents. The sums are then set down as the linear equations of a matrix and solved for the constituents' ΔR_{Mg} values, for each solvent system. They give as an example a set of values for the system n-butanol–acetic acid–water (40:10:50 by vol.) at room temperature. They imply that an unambiguous identification will be achieved by this *number* of solvent systems. However, they make no mention of the important fact that this will only hold if the solvent systems are chosen so that a significant ΔR_{Ms} effect is obtained for *each* constituent with at least one pair of solvent systems in the set. At present, a choice of solvent systems must be made largely on the basis of empirical findings

---

* Dr. R. V. Brooks, personal communication.

that certain systems do give such ΔR_{Ms} effects. It seems better therefore to try and devise systems which give special ΔR_{Ms} effects with one type of constituent group at a time while minimising any ΔR_{Ms} effects with other types of group. This is not difficult to do and will in any event be far more economical of time, space and material.

4. Solvent–Solute Interactions and ΔR_{Mg} Values

While it is usually easy to suggest reasonable mechanisms to explain the regularities and anomalies of solubilities and partition coefficients, it is usually difficult to prove that a particular explanation is the correct one. The phase-composition of most of the systems used in steroid work is not only unknown but on paper is subject to unknown modifications by the cellulose supporting material. Nevertheless, it is useful to try and find a mechanistic explanation for the regularities observed in the previous sections, if only to find thereby reasonable explanations of the deviations from exact correspondence with Martin's theory that occur and to help predict the behaviour of other steroids likely to show deviations. Some of these are clearly steric in origin and were anticipated by him. Others are equally clearly due to polar interactions. In this section I shall attempt to give an account of some of the outstanding interactions which probably determine the partition coefficients and hence the R_M values of the steroids in the solvent systems discussed above.

One useful way of looking at solubilities is to consider the interactions between the parts of the solute and parts of the solvent molecules as competing with the interactions between the solvent molecules (p. 12). In dilute solutions we can usually neglect the interactions between the solute molecules themselves, which also compete with solvent–solvent interactions in concentrated solutions. The tendency of a polar solvent to dissolve polar substances is thus seen as an ability of the polar groups of the solute to attract parts of the solvent molecules with a force comparable to that between the solvent molecules themselves. A non-polar solute however cannot exert this attraction and if it finds itself in a polar solvent it is, as it were, squeezed out of the solvent as the molecules of the latter re-establish their positions in the relatively ordered lattice formed by polar solvents. Non-polar solvents exhibit far less short-range order than polar solvents and their molecules, unless constrained, tend to adopt linear or spheroidal shapes at the usual temperatures (Glasstone, 1947). The molecules of such solvents are only attracted to one another by relatively weak dispersion forces unless they contain heavy atoms. With highly unsaturated molecules stronger interactions probably occur due to delocalization of electrons and consequent increase in the dispersion forces (London, 1946). Although the theoretical basis of such interactions is still incomplete it is known empirically that unsaturated molecules tend to attract one another more strongly than they attract saturated molecules. Similarly, unsaturated groups attract the molecules of polar substances much more strongly than saturated groups. Unsaturated groups may in fact be thought of as weakly polar in nature.

One important feature of a dissolved solute has already been mentioned and is probably the basis of the influence of steric factors. This is that liquids possess what is called short-range order: over small distances a lattice similar to that found in crystals is formed, the best example being that of water (Bernal and Fowler, 1933). It is not enough, therefore, for a polar group in an organic compound to be able to associate with one molecule of the solvent; there must be room around the group to fit in several molecules of solvent if it is to interact strongly with the lattice of the solvent. This is probably the reason for the low ΔR_{Mg} values of hydroxyl groups added to phenols in positions *ortho* to an existing hydroxyl group (Bate-Smith and Westall, 1950) and for the differences between certain *cis*- and *trans*-diols noted previously. In the *ortho*-dihydroxyflavonoids the phenolic hydroxyl groups are strictly equatorial in conformation and there is not room between them for part of the water lattice. Each hydroxyl group, therefore, presents effectively only about two-thirds of its "surfaces" for interaction with the water lattice; isolated equatorial secondary hydroxyl groups can be considered crudely as presenting three co-ordination positions to the water lattice.

In the steroid 16,17-diols (p. 95) the 16α,17β-diol has the 17-hydroxyl group in the equatorial conformation (less hindered) while the 16α-hydroxyl group is in a conformation intermediate between axial and equatorial. The 16β,17β-*cis*-diol however also has an equatorial and intermediate conformation of the 17β- and 16β-hydroxyl groups which should give the same ΔR_{Mg} value as its epimer if the groups were behaving as they do when isolated; in fact the *trans*-diol group has a higher ΔR_{Mg} than the *cis*-diol (R_M of oestriol (*trans*) 0·95, and of 16-epi-oesteriol (*cis*) 0·72 in benzene–methanol–water (2: 1: 1 by vol.)).

Another diol group showing an interesting deviation from the expected ΔR_{Mg} value is the 6β,11β-diol group in 6β-hydroxycorticosterone. The ΔR_{Mg} value for the 6β-hydroxyl group in this compound is +1·77 in benzene–methanol–water (2:1:1 by vol.) (Table 2.2); the figures for 6β,17α-dihydroxy-11-deoxycorticosterone and for 6β-hydroxy-11-dehydrocorticosterone are +1·61 and +1·50 respectively. Similar differences, though smaller, are seen with chloroform–formamide and with t-butanol systems, so that these small differences are probably significant. It seems likely here that the 6β,11β-diol group is favourably placed to fit into the water lattice (Fig. 2.10).

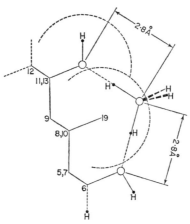

FIG. 2.10. Possible mode of interaction of 6β,11β-diols with water.

The figure is a projection of the steroid skeleton (Rings B and C only, for clarity) on to a plane perpendicular to the line through carbon atoms 8 and 10. The dashed arcs represent the two angular methyl groups (large van der Waals radius). Dotted lines represent hydrogen bonds. The hydrogens of the 6β- and 11β-hydroxyl groups are placed very nearly in the same positions that would be occupied by those of a tetrahedral (ice) lattice extending from the centrally placed water-molecule. Although 2·8 Å is about the limit for the length of known hydrogen bonds (Barrer, 1957) the stability of such a double co-ordination would be reasonably large as in chelates.

While these examples are useful in showing relatively unequivocally the effect of steric factors influencing the ΔR_{Mg} values of close hydroxyl groups in suitable conformations it is not sufficient to suggest how *cis-* and *trans-α*-diols will behave in general. In a cyclohexane ring the substitution of two adjacent hydroxyl groups can occur in three sterically distinguishable ways or conformations (Klyne, 1954). These are shown in Fig. 2.11 and can be referred to as *trans-(e,e)*, *trans-(a,a)*, and *cis-(a,e)*. It is clear from a variety of studies that the interaction between the hydroxyls of the *trans*-(e,e)-conformation are larger than the interactions of the other two conformations. The ΔR_{Mg} value of the 16α-hydroxyl group in epi-oestriol is 0·25 units less than its probable value when present as an isolated group. If we assume that this deficit is of steric origin it will be slightly larger in a typical *trans*-(e,e)-diol, rather less for a typical *cis*-(e,a)-diol, and negligible in a *trans*-(a,a)-diol (Reeves, 1951). It is then possible

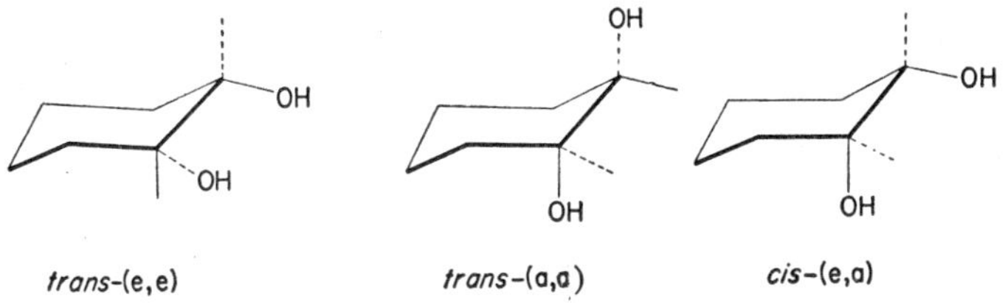

trans-(e,e)   *trans*-(a,a)   *cis*-(e,a)

FIG. 2.11. Conformations of alicyclic 1,2-diols.
Repulsive interactions are strong in the *e,e* form, moderately strong in the *e,a* and negligible in the *a,a* form (see Klyne, 1954).

to make a reasonable guess at the ΔR_{Mg} values for the three types of α-diol group assuming that the group is placed at an otherwise unhindered position, for example a 2,3-diol or a 3,4-diol. It is seen from Table 2.14 that the differences are small and comparable with the variation in the ΔR_{Mg} values of hydroxyl groups of either conformation in different steroids; in view of the uncertainty of the deficits due to steric hindrance it is clear that the separation of epimeric α-diols in general may be difficult in "typical" solvent systems unless the position is hindered by other parts of the molecule.

The ΔR_{Mg} deficits for the four epimeric 11,12-dihydroxytigogenins which were measured by Brooks *et al.* (1957) are much larger than expected from the above calculation. Although these authors attribute most of this to hydrogen bonding between the two polar groups, it seems that other factors are at least equally important. For instance the ΔR_{Mg} deficit for the 11β,12α-diol group was 0·53. This diol group has the *trans*-diaxial conformation which in simpler molecules has been shown to behave as if little or no interaction occurred between the hydroxyl groups (Reeves, 1951). It seems much more likely that the strong interactions between the 11β-hydroxyl and the angular methyl groups at C-10 and C-13 deform the C-ring and modify the conformation of the 12α-hydroxyl group and

Table 2.14. Attempt at Calculating ΔR_{Mg} Values for Different Types of α-Diol Group

Typical equatorial OH, $\Delta R_{Mg} = 1\cdot 8$
Typical axial OH, $\Delta R_{Mg} = 1\cdot 5$

Conformation of α-diol group	Sum of ΔR_{Mg} values	Probable hindrance deficit	Probable ΔR_{Mg} of diol group
trans (e, e')	3·6	−0·4	3·2
trans (a, a)	3·0	0·0	3·0
cis (e, a)	3·3	−0·2	3·1

The probable hindrance deficits are assessed on the basis of the results with oestriol and 16-epi-oestriol in system B/50 and studies on the interactions of typical 1,2-diol groups in sugars (Reeves, 1951).

It is clear that whatever the exact figures the R_M values of epimeric α-diols are going to be uncomfortably similar in "typical" solvent systems unless other hindering groups complicate the picture (cf. e.g. steroid 1,2-diols with 6,7-diols).

E-ring (Barton, 1955). Brooks et al. (1957) themselves point out that the anomalously low ΔR_{Mg} value of the 12β-hydroxyl group (equatorial) in rockogenin (αaS-3β,12β-ol) is correlated with an abnormal rotational contribution of this group ($\Delta M_D - 9$), and suggest that hindrance and interaction with groups in rings E and F may be involved. In the sapogenins, the methyl group at C-20 introduces a considerable steric complication not found with the C_{19} and C_{21} series of steroids. This will probably cause considerable deformation of the C- and D-rings and hinder the 12-position. A fair degree of hindrance of this position is already well known in the bile acids (C_{24}) in which the 20-methyl group is nevertheless not directly linked to a fused ring. It is probably safer at present to consider the figures of Brooks et al. (1957) as atypical, although extremely interesting, and take the less hindered steroid 16,17-diol group as a guide to the probable behaviour of α-diols in general, although a study of the epimeric steroid 2,3-diols would be more valuable.

Another case in which it seems that "lattice-fitting" of the solute or the lack of it is an important factor is the exceptionally low ΔR_{Mg} value of steroid hydroxyl groups in the t-butanol systems. In some of these systems this is probably partly due to an increase in the intermiscibility of the two phases as indicated by an atypically small ΔR_{Mg} of ketone groups. However, in all of them the ΔR_{Ms} (relative to "typical" systems) for hydroxyl groups is far larger than for ketone groups and in the system E$_2$B (isooctane—t-butanol—water (50:25:45 by vol.) in which the situation is not complicated by the presence of methanol the ΔR_{Mg} for ketone groups seems to be little changed (cf. for instance the ΔR_{Ms} values for hydroxyl groups in Table 2.11 with the ΔR_{Mr} values for the conversion 5β(H)-3α-hydroxyl → 5β(H)-3-ketone in Table 2.10). Water and primary alcohols probably form hydrogen bonds with steroid hydroxyl groups by acting as proton donors as well as acceptors of protons; with steroid ketone groups they must act as donors. In the case of t-butanol, however, it is probable that hydrogen bonding with ketones will be very weak because of the large charge on the oxygen atom of this alcohol molecule due to the positive inductive effect of the three methyl groups. Hydrogen bonding between t-butanol and steroid hydroxyl groups will occur mainly if not entirely by the steroid hydroxyl group donating the proton. Furthermore once a hydrogen bond has been formed between t-butanol and a steroid hydroxyl group the methyl groups of this alcohol will hinder the approach of other polar molecules such as water or methanol and to a large extent prevent the formation of other hydrogen bonds. The steroid hydroxyl group will thus be prevented from forming the secondary and tertiary hydrogen bonds (i.e. to the oxygen atoms involved) needed to incorporate it in the solvent lattice. It is probably justifiable to consider the situation in terms of a short-lived steroid-t-butanol complex; instead of a hydroxyl group linked to the polar solvent lattice we have a t-butoxyl group presenting its non-polar methyl groups to the solvent (cf. Brooks *et al.*, 1957).

While this type of interaction would be sufficient to explain qualitatively the atypical ΔR_{Mg} values of hydroxyl groups in t-butanol systems it is probably not the whole story. The other factor is the distribution of t-butanol itself between the two phases of such solvent systems. The activity coefficients for t-butanol in hydrocarbons and in water are very different from those of methanol (Pierotti *et al.*, 1959) and it is likely that the larger part of the t-butanol in such systems is in the mobile phase; in contrast very much less methanol is present in the mobile phase of the "typical" solvent systems. This is probably the basis of the temperature sensitivity of the E$_2$B system; the solubility of t-butanol in water has a large negative temperature coefficient while its solubility in paraffins has the more usual positive coefficient. Thus the higher the temperature the more the mobile phase of such systems will be relatively enriched in t-butanol and the more the stationary phase will be enriched in water.

In conclusion, two factors probably combine to decrease the ΔR_{Mg} values of hydroxyl groups in this type of system. First of all, hydrogen bonding with t-butanol in the stationary phase competes with methanol or water or both to prevent the hydroxyl group from being incorporated into the polar solvent "lattice". Second, unlike methanol, water, or other typical polar solvents, an appreciable amount of t-butanol is contained in the mobile phase where it is also available for forming hydrogen bonds with steroid hydroxyl groups. The much smaller reduction in ΔR_{Mg} values for ketone groups is due to the fact that t-butanol is less able to form strong hydrogen bonds by using its tightly bound and hindered proton; the normal interactions of ketone groups with water and typical hydroxylic solvents, therefore, dominate the picture. It should be noticed that the findings are not explained entirely by the effect of a large proportion of t-butanol in the mobile phase. If this were the only effect occurring then it would be much larger with less hindered and equatorial, than with hindered and axial, hydroxyl groups; the effect observed is of such magnitude that one would expect a reversal of the sign of the ΔR_{Mr} value of the conversion, axial → equatorial hydroxyl group. In fact the order of magnitude of the ΔR_{Mg} values of the various epimeric hydroxyl groups that have been studied is the same as in typical systems although the differences are reduced.

The effect of changes of temperature has been mentioned at various times and needs further clarification. The most important effect of temperature is to increase the intermiscibility of most pairs of solvent phases. Except in a very few instances involving particularly the polar solvents isopropanol and t-butanol, temperature coefficients of mutual solubility of polar and non-polar liquids are positive. The two phases of most systems used for chromatography thus become more alike and partition coefficients become closer to unity with a rise in temperature. This should result in all R_F values becoming closer to 0·5; in R_M values becoming closer to zero; and in ΔR_{Mg} values closer to zero. In general, therefore, substances with R_F values greater than 0·5 will run more slowly, those with values less than 0·5 more rapidly. The importance of phase composition was first shown clearly for amino acids by Consden et al. (1944), and for sugars by Isherwood and Jermyn (1951); the latter authors showed that the R_M values of sugars showed a linear relation to the logarithm of the water content of the mobile phase. Consden et al. (1944) had shown earlier that the main effect of raising the temperature of operation was through an increase in the water content of the mobile phase. With the more polar compounds such as sugars and amino acids the effect is usually to raise all the R_F values in a given system. Most of these systems, however, contain an alcohol or third component with a negative temperature coefficient of solubility in water. The aqueous methanol systems seem to change their phase composition more symmetrically thus producing more often the tendency of all R_F values to approach 0·5 ($\pm$0·2). The effect is shown well in the R_S values given by Neher and Wettstein (1956) for toluene–ethyl acetate–methanol–water (9: 1: 5: 5 by vol.) run at 22 °C and 40 °C.

The practical value of using raised temperatures has been mentioned before (p. 31) but it is useful to re-emphasize here that all separation will be diminished unless tailing, caused by adsorption or poor solubility, is the main cause of poor separation. Apart from the unavoidable effect of increased similarity of the two phases a second effect also diminishes ΔR_{Mg} values by the same means that intermiscibility is increased. Just as non-electrolyte solvents become more intermiscible with increase in temperature so solutes become more "miscible" with solvents and the differences between them become less. This is well shown by some of the figures given by Pierotti *et al.* (1959) which are assembled in Table 2.15. Since the most selective effects of solute–solvent interactions are due to relatively weak and short-range forces it is understandable that these effects, which are especially sensitive to steric factors, should be diminished rapidly as thermal agitation increases. Wherever possible, therefore, low temperatures should be used and most of the volatile systems described in this book are all based on the modified systems described earlier with this in view (Bush, 1954a); they can be used at 18–25 °C for almost all purposes. If it is necessary to raise the temperature it is preferable to use the earlier type of system with a higher water content so as to reduce somewhat the intermiscibility of the phases (Bush, 1952).

Similar rules will govern the results obtained with any other means of increasing the similarity of the two phases. Thus raising the methanol content of a "typical" system at low temperatures tends to reduce the R_F values of steroids with high R_F values and increase those of low R_F values. This effect, however, is not easily predictable or symmetrically distributed about an R_F value of 0·5 because of the low and variable solubility of methanol in the different non-polar solvents used for mobile phases. With paraffins the ratio of activity coefficients of methanol in the hydrocarbon to that in water is of the order of 60: 1 at 25 °C (Pierotti *et al.*,

Table 2.15. ACTIVITY COEFFICIENTS AT INFINITE DILUTION OF VARIOUS SOLUTES AT DIFFERENT TEMPERATURES

Solute	Solvent	Temp. °C	γ°
Heptane	Ethanol	25	16
		50	13
		70	11·8
Benzene	Ethanol	25	6·6
		50	6·0
		70	5·4
Butylcyclohexane	Ethanol	25	22·5
		50	18·0
		70	16·0
Hexadecylcyclohexane	Ethanol	25	165
		50	97
		70	67
Heptane	Ethyl methyl ketone	25	3·65
		50	3·1
		70	2·75
Butylcyclohexane	Ethyl methyl ketone	25	6·6
		50	5·1
		70	4·2
Dodecylbenzene	Ethyl methyl ketone	25	4·7
		50	3·5
		70	2·7
Benzene	n-Heptane	25	1·55
		50	1·43
		70	1·35
Toluene	n-Heptane	25	1·45
		50	1·32
		70	1·33
Ethylbenzene	n-Heptane	25	1·65
		50	1·52
		70	1·43
Isopropanol	n-Heptane	25	32·4
		50	12·1
		70	5·1
n-Butanol	n-Heptane	25	37·1
		50	13·6
		70	6·2
t-Butanol	n-Heptane	25	18·0
		60	7·6
		100	3·8

Data from Pierotti *et al.*, (1959.)

1959) so that raising the methanol content of methanol–water–paraffin systems increases the intermiscibility of the phases by very little; the R_F values in most of the measurable range are all decreased. At the extreme of absolute methanol–light petroleum (Mitchell, 1952), however, a steep increase in intermiscibility occurs with the result that even cortisol, which in light petroleum–96% methanol has an R_F of zero, has an R_F of the order of 0·05 in Mitchell's system (Short, 1959). With s-butanol, however, the ratio of activity coefficients in short chain paraffins to that in water at 25 °C is close to unity. The addition of such an alcohol to a paraffin–aqueous methanol system would be expected to increase the intermiscibility of the phases considerably and to distribute between them more or less equally. An effect similar to a rise in temperature should be obtained. (Precisely this was found in counter-current systems by Carstensen (1956)). However, with mobile phases based on more polar solvents such as benzene or chloroform, increasing the methanol content of the system causes a more appreciable rise in the methanol and water content of the mobile phase. With benzene, R_F values greater than about 0·4 tend to decrease and those less than 0·4 to increase; with toluene the dividing point comes at an R_F value of about 0·2.

These general rules are useful in making up special modifications of the published solvent systems. More accurate predictions can only be made by careful study of the phase compositions of the mixtures under consideration. In general, systems very close to complete miscibility are likely to be very sensitive to temperature changes, e.g. collidine–water for amino acids (Dr L. Woolf, personal communication) and to show small ΔR_{Mg} values. They are unlikely to be useful in the steroid field except for sub-fractionation for instance, when it may sometimes be useful to obtain a wide range of steroids all placed in the R_F range 0·05–0·80.

Little attention has been given so far to the main non-polar component of the solvent systems used for steroids; that is to say the main component of the mobile phase. In most cases, this must be considered as a mere "vehicle" for the solute molecules with rather weak and non-specific interactions with them. The activity coefficients of hydrocarbons in n-heptane for instance only vary between 0·50 and 3·0 at 25 °C for a variety of monocyclic and bicyclic aromatic hydrocarbons with alkyl side chains varying from one to twelve carbons long (Pierotti *et al.*, 1959). With the more polar and asymmetrical aromatic compounds such as phenanthrene, chrysene, and anthracene, however, the activity coefficients rise to the range 30–50 (Table 2.15). Alicyclic hydrocarbons have coefficients in the lowest ranges. The aliphatic solvents therefore show weak interactions with alicyclic structures, such as the hydrocarbon skeletons of the majority of steroids, which are probably relatively insensitive to steric changes in the solute molecule. Such solvents can be thought of as merely "receivers" of the solute molecules that are squeezed out of the polar phase.

Thus the difference in chemical potential of the hydrocarbon parts of the steroid in the two phases of a typical solvent system depends largely upon the nature of the stationary phase with aliphatic mobile phases. Additional methyl groups in the solute, for instance, increase the volume of the solute and hence the number of bonds in the lattice of the more polar phase which have to be broken to accommodate the solute molecule. Even with aliphatic solvents, however, small but definite effects are seen which can only be attributed to the steric properties of the mobile phase. Thus the separation of $5\beta(H)$-3α-hydroxysteroids from their $5\alpha(H)$-epimers is relatively little affected by changing the mobile phase from light petroleum to methylcyclohexane or decalin in the $C_{19}O_2$ series. However the separation of the 11β-hydroxy-analogues of these epimers (e.g., 11β-hydroxyaetiocholanolone from 11β-hydroxyandrosterone) is negligible with light petroleum of b.p. 100–120 °C, quite good with light petroleum of b.p. 60–80 °C, and even better with methylcyclohexane, xylene, decalin, or tetralin (i.e. in terms of ΔR_M values). With light petroleum (b.p. 100–120 °C)–toluene–methanol–water separation is good with a petroleum: toluene ratio of 2: 1 but poor with a ratio of 3: 1. These systems all contain very similar proportions of water and methanol and the separation is not correlated with the changes in the amount of water and methanol in the mobile phase. It is clear that the ΔR_{Mr} values increase as soon as a cyclic molecule is used as a main component of the mobile phase. There is also a considerable increase on changing from high-boiling petroleum fractions to lower-boiling fractions.

When the mobile phase is largely or entirely based on aromatic hydrocarbons or chloroform (whose hydrogen atom has a significant power to form hydrogen bonds (Moelwyn-Hughes and Sherman, 1936)) polar interactions become significant in both phases of the system, and the methanol and water content of the mobile phase becomes greater. In addition the mutual attraction of unsaturated hydrocarbon structures comes into play. Such effects, however, are still small with the usual range of steroids compared to the role of polar interactions with the stationary phase. In hydrocarbons themselves these effects are more easily seen, as in the selective retardation of aromatic hydrocarbons on gas–liquid columns with aromatic or unsaturated oils as stationary phase (James and Martin, 1954). These effects should, however, be worth exploring with very non-polar steroids such as the sterols and their esters; and with hydrocarbon degradation products of steroids such as androstane, pregnane, and chrysene (see Chapter 5).

In conclusion, then, the changes of ΔR_{Mg} values produced by simply changing the nature of the non-polar components of the mobile phase will be small in most cases when the intermiscibility of the two phases is not grossly altered. A few separations may be improved by changing from high-boiling petroleum fractions to lower-boiling fractions, or to alicyclic or aromatic solvents.

Finally we come to the important question of what role, if any, is played in paper chromatography by interactions between the steroids and the polysaccharide molecules of the supporting material. This is important because the solvent systems used for steroids (with the exception of formamide) probably swell the cellulose fibres very much less than the solvents generally used for amino acids and sugars, which usually have a much higher water content. Suggestions that "true partition" is not the basis of separations in the latter type of solvent system on starch, or cellulose (reviewed by Martin, 1950) where the polysaccharide gel is fully expanded, are likely to be far more pertinent to the less swollen gel that is probably the basis of the stationary phase in most of the systems for steroids.

In most cases it seems that the influence of steroid–polysaccharide interactions is in fact negligible. Thus the ΔR_M values of several adrenal steroids were identical on Whatman No. 2 paper and Celite columns using identical aqueous methanol systems (Tait and Tait, 1960). In these systems the water content was relatively high (approximately 25% of the total volume of the phases). A more stringent test was made by the author who compared a limited number of ΔR_M values in the system tetralin–90% aqueous methanol on Whatman No. 2 paper at 22 °C using it both in the "straight" and reversed-phase manner. For the latter use the paper was impregnated first with a 20% (by volume) solution of tetralin in methylene chloride and the latter solvent allowed to evaporate. The results are given in Table 2.16. It is seen that the ΔR_{Mg} and ΔR_{Mr} values are of the same relative magnitude but of the expected opposite sign in the two ways of running this system. They are all slightly greater in the reversed-phase method. It seems likely that this is due to a difference in phase composition

Table 2.16. Comparison of ΔR_M Values in the same Solvent System Run in "Straight" and "Reversed-phase" Manner

Solvent system: tetralin–methanol–water (100:90:10 by vol.) 22 °C Whatman No. 2 paper.
Straight: **equilibrated** via vapour phase.
Reversed-phase: impregnated with 20% tetralin in CH_2Cl_2 (v/v) and hung 5 min in fume cupboard before putting in tank.

Group or change	Root substance	"Straight"		"Reversed phase"	
		ΔR_{Mg}	ΔR_{Mr}	ΔR_{Mg}	ΔR_{Mr}
11-Ketone	Progesterone	0·38			
	Androst-4-enedione	0·38			
	Aetiocholanolone	0·40		−0·54	
11β-OH	Progesterone	0·77			
	Androst-4-enedione	0·69			
	Aetiocholanolone	0·80		−1·15	
Δ^4-3-Ketone ↓ 5α-3-Ketone	Androstan-17-one		−0·17		+0·31
5β(H)-3α-OH ↓ Δ^4-3-Ketone	Aetiocholan-17-one		−0·34		
	Aetiocholan-11,17-dione		−0·36		
	11β-Hydroxyaetiocholan-17-one		−0·45		
17β-OAc ↓ 17-Ketone	Progesterone		+0·23		−0·39

and volume due to the cellulose; in the reversed phase method the latter is probably completely unswollen and unavailable to the polar solvents. In the "straight" method it is reasonable to expect that during equilibration with the vapour phase the swelling cellulose can associate appreciably with tetralin thus enriching the polar phase with tetralin: in this method the two phases will be more alike than in the reversed phase method thus reducing ΔR_{Mg} and ΔR_{Mr} values in the usual way (see above). If steroid–polysaccharide interactions were of any serious magnitude steric factors would be particularly strong with the 11-position (cf. the relative mobilities of 11β-hydroxysteroids and their 11-ketone analogues on alumina, p. 37) and, if anything, ΔR_{Mg} values for polar groups should be less in the reversed-phase method; no such effects are seen and it is more reasonable to attribute the small general reduction of ΔR_{Mg} values in the "straight" run to an alteration of the volume ratio of the phases or to an influence of the cellulose on phase composition. Even better agreement between "straight" and "reversed-phase" runs of the same solvent system was obtained with sapogenins in entirely similar experiments by Brooks et al. (1957) using an aqueous methanol system. The values of log V_r for a variety of steroids run on Celite columns agree closely with those calculated from R_F values on paper run with analogous systems (Tait and Tait, 1960). In particular the 17-oxosteroid glucuronosides behave almost identically in identical solvent systems used on Celite columns, Whatman No. 2 and 3MM paper, and in the Craig counter-current machine (Bush, 1957; Schneider and Lewbart, 1959; Bush and V. B. Mahesh, unpublished).

While the majority of steroids behave in this way there are a few cases where adsorption to the cellulose appears to be important. All steroids are adsorbed strongly to cellulose from water or non-polar organic solvents if methanol or some similar polar organic solvent is absent (Bush, 1952). Some steroids are still adsorbed strongly (as shown by streaking) under conditions where other steroids on the same sheet have not streaked (p. 67) in aqueous methanol systems and impregnated-paper systems. The 21-acetates of typical adrenal steroids streak slightly if equilibration is limited to 3 hours; this is abolished by prolonging equilibration to 16 hours. Certain types of steroid however streak even after prolonged equilibration. It is interesting that these compounds are nearly all C_{21} steroids containing equatorial 3β-hydroxyl groups, a 17α-hydroxyl group, and a *trans* A-ring/B-ring junction (5α-H), or the approximately flat Δ^5-3β-hydroxyl structure. The α-surface of these steroids is approximately flat (cf. Fig. 1.8) and the hydroxyl groups at positions 3 and 17 are about 10·5 Å apart in models. It is an old empirical observation that the "fastest" (i.e. most firmly adsorbed) dyes for cellulose are those with molecules that are flat and that possess two groups capable of forming hydrogen bonds about 10·3 Å apart (Hodgson and Marsden, 1944). It is probably more than a coincidence that the cellobiose unit repeats every 10·3 Å (Ott, Spurlin and Grafflin, 1954) with a glycosidic oxygen available for hydrogen bonding; a possible association of 3β,17α-dihydroxysteroids is given in Fig. 2.12. Numerous other possibilities exist but this one gives the best fit of the hydrocarbon skeleton of the steroid and hence the maximum energy of dispersion

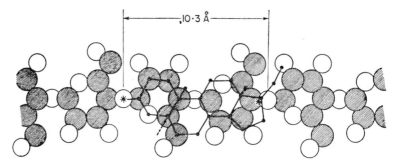

FIG. 2.12. Possible mode of adsorption of 5α(H)-3β,17α-dihydroxysteroids to cellulose. Other positions are possible but this one gives the best fit of the chair-rings to the cellulose angulations (note e.g. carbons, 1, 4, 10 and 12 in the steroid which is drawn with its α-side apposed to the cellulose chain).
● carbons of cellulose
○ oxygens of cellulose
* oxygens of steroid 3β- and 17α-hydroxyl groups

interactions. Furthermore, the group in cellulose that takes part in hydrogen-bonding with the oxygen at C-3 probably acts as a proton acceptor since no such streaking is seen with analogous 3-ketones.

In conclusion, it seems that the only influence, if any, of cellulose on the chromatographic behaviour of the great majority of steroids is exerted by a general effect on the actual composition of the effective stationary phase of paper chromatograms. In a few cases, streaking occurs under ideal conditions: in many of these the molecular structure of the steroid is analogous to that of strongly adsorbed dyes in respect of the two features believed to be essential for their adsorption by cellulose.

CHAPTER 3

TECHNIQUES AND APPARATUS

In this chapter I shall attempt to describe the main technical features upon which steroid chromatography depends for useful results. As far as possible these technical details will be correlated with the specific theoretical points which are relevant and they should always be considered in this way. Some of the details described in this chapter may seem unduly sophisticated, but the steroids are in many ways more difficult to deal with in chromatographic systems than other classes of compounds. While the methods suggested are not unique solutions of the problems considered here, it should be recognized that each detail deals with some point that must be taken into account if good results are to be obtained. In many cases I have described the techniques with which I am most familiar but in others I am very grateful to the many colleagues who have given personal accounts or demonstrations of techniques with which I am less familiar. Detailed prescriptions are best gleaned from original papers but some useful routine procedures are collected in the appendices.

1. Transfer Operations and Volumetric Error

Steroids are usually handled in solutions in organic solvents, and for convenience these are usually volatile and mobile liquids. This gives rise to special problems not found in the chromatography of water-soluble substances, especially when quantitative results are desired. The author has found the following techniques necessary for good quantitative methods but less rigorous standards suffice for qualitative work.

Standard solutions of steroid are conveniently made up in ethanol and are usually quite stable for at least six months when kept in the dark in a refrigerator. Owing to the appreciable temperature coefficient of expansion of ethanol they must be brought to room temperature before use, and care must be taken that vessels and micropipettes are not warmed by radiation or by handling. This is best achieved by using marked pipettes held in a short length of plastic or rubber tubing connected to a well-greased tuberculin syringe; a light grasp with two fingers is then sufficient to guide the pipette while delivery is smoothly controlled by the syringe plunger. Alternatively the self-filling type of pipette can be used which is mounted within its own glass extension. A micrometer-syringe pipette (as in the Conway microburette (Conway, 1947)) is sometimes useful but needs great care if accuracy is to be obtained with ethanolic solutions.

With rare compounds a standard solution may have to be used relatively infrequently over a period of many months. This is difficult because of evaporation of the solvent in glass-stoppered vessels. For routine work this is partly overcome by using ribbed plastic stoppers such as the E-mil type. Even with such stoppers, however, ethanolic solutions with a surface area of 0·5 cm^2 lose about 0·2 ml per month in the refrigerator, probably as a result of gradual loosening of the stopper owing to its higher temperature coefficient of expansion than glass. To avoid this it is useful to make up standard solutions in finely graduated measuring cylinders (10–25 ml) and to record the volume remaining after each use. Any loss by evaporation can then be made good by adding ethanol, and the solution is usable even after many months. If the solution is rejected for quantitative work after 6 ml have been used of an original 10 ml stock, the error with a good cylinder is about 0·5–1·0%. With the orthodox volumetric flask the original standard solution will be more accurately known than with a cylinder but this state of affairs will not last long as the surface area for evaporation increases rapidly after use.

The first operation in chromatography itself is the transfer of a solution to the system in such a way that the solutes are distributed evenly across it in as small a length of the system as possible. In many cases it is essential to carry out this transfer quantitatively. Whatever the particular method used, the main problem with solutions in volatile and mobile solvents is to obtain quantitative transfer while not lengthening the original zone of distribution in the system. It is best to consider this problem in connexion with the different methods.

(a) Transfer to Adsorption Columns

This is usually not difficult but with a mixture of steroids varying greatly in their solubilities it may be difficult to get them all into solution in a suitable solvent. The general procedure is to dissolve the mixture in the least polar solvent possible and pour this on to the column using a fairly generous volume (up to one interstitial volume). The solutes are then adsorbed in a narrow zone at the top of the column when the solution is allowed to run through the column. A few washes of pure solvent are then allowed to run into the column so that the amount of solute remaining in the thin layer of solvent above the adsorbent, which must not be allowed to become dry, is negligible. The chromatogram is then "developed" with more polar solvent mixtures capable of eluting the solutes.

Three procedures can be used if all the material is not dissolved by the least polar solvent which can be used. The first is to dissolve the mixture in a small volume of a more polar solvent (e.g. chloroform or ethyl acetate) and then dilute this with about ten times the volume of a less polar solvent such as light petroleum or benzene. The filtered solution is run into the column and any precipitate on the filter dissolved again with the same sequence of solvents. The process is repeated until all the material has been transferred in solution to the column. This procedure can sometimes be very tedious and often the less polar steroids of the mixture are eluted in mixed bands because the final solvent mixture applied to the column is still too polar to allow their adsorption at the top of the column. The second possibility is to proceed in the same way but to distil off the polar solvent from an approximately one-to-one mixture with a suitable less polar solvent so that the precipitate forms gradually and does not carry down the non-polar steroids with it. The filtered solution is then put on the column and "developed" with the usual series of solvents. The residue on the filter is dissolved in the least polar solvent possible and added to the column when the "development" has reached the stage of using this particular solvent. This method is more convenient but the precipitation is not always clean and non-polar steroids may be carried down with the polar steroids and thus emerge late from the column. This, however, is usually only a minor difficulty since they will usually be eluted rapidly from the column, and confusion with the more polar steroids can be avoided by prolonging slightly the stage of development with the solvent mixture used for the final transfer. These non-polar steroids will then constitute a slight loss from the non-polar fractions and can be separated and recovered on another smaller column if necessary. By far the most satisfactory method in the author's experience is to proceed in the opposite direction; namely to dissolve the mixture completely in the least polar solvent possible and put this straight on the column. The eluate is collected immediately and "development" with this solvent continued until all the non-polar steroids have been eluted. The polar steroids on the column are then separated by continuing with the usual sequence of solvents. The first eluates are then evaporated and run on another column using the less polar solvents of the usual series. This method needs two columns but avoids the time and uncertainties of precipitation; there are no losses; and the overall time taken is shorter since the second column can be run as soon as the first eluates have been evaporated.

Whatever the technique adopted it is always unwise to apply a suspension to the column in the hope that the precipitate will be dissolved at an appropriate state in the sequence of elution. This hope is almost always vain.

(b) Transfer to Partition Columns

This is more difficult than the case of adsorption columns because the material must usually be applied to the column either in solution in the stationary phase or in solution in the mobile phase itself. There is usually no way of putting the material on the column in a solvent which gives a sharp concentrated band at the top of the column, but sometimes it is possible to use a variation of the technique used with adsorption columns.

In favourable cases the mixture can be dissolved in a minimal volume ($\leqslant 0.05$ interstitial volume) of the mobile phase and put directly on the column. The mixture is washed in very carefully by running two or three very small volumes (about 0.005 interstitial volume) of mobile phase into the column using a fine pipette to rinse the walls of the column near the top of the stationary phase. More usually the mixture is only easily dissolved in small volumes of the stationary phase. At least four possible techniques can then be used.

In the first (Hegedüs, 1953) the mixture is dissolved in a good volatile solvent and mixed well with a small weight ($\leqslant 0.05$ the weight of the total column-supporting material) of dry supporting material (e.g. Celite) using a round-ended glass rod or pestle. This solvent is then evaporated in a vacuum desiccator and the dry powder ground up well with 0.5 to 1.0 parts by weight of the stationary phase. It is then transferred to the top of the column with a spatula and pressed down well. The residual powder is mixed with small amounts of supporting material and stationary phase and again transferred to the column. A very small volume of mobile phase is left above the column before starting this transfer operation and in order to avoid spreading of the starting zone by having too large a volume or the trapping of bubbles due to too small a volume, it needs to be carefully measured to match the amount of supporting material that is going to be used for the transfer.

A second method is very similar to the first. Here the mixture is dissolved in the stationary phase itself, ground up with the calculated amount of supporting material and transferred to the column as above.

The third method is to pack about 0·05 column-volume of *dry* supporting material on top of the column proper and then run in a solution of the mixture dissolved in the stationary phase (Butt *et al.*, 1951). This top zone is then mixed gently with a glass rod and a small volume of mobile phase added in the final stages to remove bubbles. Again the relative amounts of supporting material and stationary phase need careful calculation before starting the manoeuvre and it is best to err on the side of underloading this starting zone. This method when carefully performed is probably the best in securing quantitative transfer and a small sharp starting zone.

The fourth possible method can only be used when the more polar steroids are to be separated and the solvent system is stable to fairly rapid changes of mobile phase. This holds for most columns using silica gel and for Celite used in the "straight" manner, but not for siliconed Celite used in the reversed-phase manner. In this method the steroid mixture is dissolved in a mixture of $\leqslant$ 0·05 interstitial volume of stationary phase and about four times this amount of light petroleum. The supporting material (0·05 column weight) is now added and ground in with a glass rod. The slurry is mixed by sucking in and out with a pipette on a syringe and is then transferred with the pipette to the column. The residual slurry is transferred with further washes of petroleum and then packed down with a Martin packer and secured with two disks of filter paper. The petroleum is then run into the column and followed by the mobile phase (e.g. toluene or benzene). This method has the advantage of allowing the best packing of the starting zone to be achieved using a packer and avoids the very serious danger of trapped bubbles in the starting zone. It also enables the transfer to be made in the form of a semi-fluid slurry which is pipettable rather than having to use a spatula to scrape up a damp powder. It is however limited to steroids which are sufficiently polar to have partition coefficients giving negligible concentrations in the petroleum phase: this is run through the starting zone before applying the mobile phase itself, which will usually be a more polar solvent such as benzene or toluene.

Another method which has not been used a great deal is to apply the mixture dissolved in the stationary phase to a few disks of filter paper and to place these on top of the rest of the column. This method is more appropriately used with the "chromatopile" technique (Mitchell and Haskins, 1949, p. 186) in which the whole column is made up of a pack of filter paper disks under pressure. Owing to the strong adsorption of steroids by cellulose from aqueous solutions this method will probably only work with systems based on an alcohol or glycol stationary phase. Such adsorption is less serious with cardiac glycosides which may be run on cellulose columns with aqueous systems (e.g. Lichti *et al.*, 1956).

These transfer operations are difficult with partition columns unless the mobile phase is such that the least polar steroid in the mixture to be separated has a low R_F value in the system. Wherever possible it is a great help to transfer the extract on the supporting material using petroleum to give a pipettable slurry, and then change to the more polar solvent to be used as the mobile phase of the actual system. This technique is thus often conveniently combined with gradient elution methods (p. 153).

(c) Transfer of Material to Paper Chromatograms

The technique is generally easier than those needed for partition columns but presents several special difficulties to be overcome. Two general cases need consideration. In the first a known *fraction* of an extract or solution has to be transferred: in the second the *whole* of an extract must be quantitatively deposited on the necessarily limited area of the starting zone. It is difficult to transfer such an extract quantitatively with volatile solvents in less than 0·1–0·3 ml, followed by at least two washings of the original vessel with 0·05–0·2 ml of the solvent, if serious volumetric errors are to be avoided. With less than 1·0 μg of any single component in the mixture and a total weight of solids less than about 40 μg it is possible to use the conventional method of starting with a small spot at the origin of 2–4 mm diameter. With amounts in the range 4–100 μg it is best to use a starting zone in the form of a small "bar" perpendicular to the direction of the eventual solvent flow measuring about 15 × 1–2 mm. For quantitative estimation by the scanning method (p. 222) this bar must be extremely even in concentration of solutes, and deposited across the full width of a strip 2·5 or 5 cm wide.

Even with volatile solvents it is tedious to obtain such starting zones by trying to fill them directly with a pipette and a strong draught of air or nitrogen. Unless very small volumes can be used (5–20 μl as in the deposition of a fraction of a solution) it takes almost superhuman patience and skill to avoid the formation of an annular deposit (Fig. 3.1) since evaporation is maximal at the edge of the zone and the process is effectively that of a minute "radial chromatogram" in which the solutes have an R_F close to unity and are "overrun". Furthermore, it is uncommon in the steroid field to be able to use less than 10 μl of solvent even for fractional transfers and it should be remembered that the orthodox technique of direct

pipetting was evolved with amino acids in which a less volatile solvent (water or pyridine) is used and volumes usually lie in the range 1–10 μl.
For these reasons the author evolved the technique of applying the

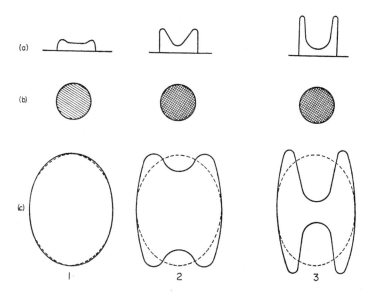

FIG. 3.1. Annular "spots" with large volumes of solution by orthodox techniques
(a) Concentration profile.
(b) Appearance.
(c) Final "spot" after running the chromatogram. Magnification approx. ×2.
1. One deposition of small volume (1 –10μl).
2. and 3. Successive stages towards an annulus. If the original "spot" is also allowed to enlarge it is possible to get "rings" instead of "spots" on the final chromatogram (C. S. Hanes, personal communication. Woolf, 1959).

material over a wide zone with generous volumes of solvent and then concentrating the material to a fine band at the origin by a preliminary run as far as the origin line with a solvent in which the solutes have an R_F of unity (Bush, 1952). This has been extended (Bush and Mahesh, 1959a) to concentration in the lateral direction when very small starting spots are desired. The two procedures are referred to as "running up" and "running in". The latter technique has also been used successfully by Brooks (1960).

Quantitative transfer is far more easily achieved by this method and it is relatively easy, once the knack has been learned, to achieve absolutely even deposition across the width of the starting zone. Furthermore, the time taken is scarcely longer than by the more orthodox method because the speed of pipetting is far greater. When the orthodox method is used, care must be taken to hold the pipette square to the paper so that solvent does not creep on to the outside of the tip of the pipette and evaporate. A useful pipette for this sort of work is described by von Arx and Neher (1956).

If a draught of air is used to hasten evaporation of the solvent it must be directed at the lower surface of the paper so as to avoid evaporation of solvent on the outside of the extreme tip of the pipette. Furthermore, the draught must be of cold and not warm air. With warm air, adsorption of steroids occurs very rapidly and is usually more or less irreversible so that material is left on the origin after the chromatogram has been run.

The "running up" technique is illustrated in Fig. 3.2 and the only precaution needed is to ensure that the rate of traverse of the pipette and the rate of delivery of solvent on to the paper are such that the tip of the pipette moves faster than the solvent creeps radially through the paper, thus avoiding the formation of miniature "radial chromatograms" with annular deposition. This is especially important at the edges of the "lane" over which the original deposition is made. It is best to carry the pipette to about 2 mm within the edge of this lane during the first deposition of the extract. The first washing of the vessel is then deposited (after drying off the first lot of solvent) carrying the tip of the pipette at each traverse about 1 mm beyond the area convered by the first deposition. The final wash of the vessel is then deposited over the same area as the second deposition, or, even better, over a smaller fresh area at the bottom of the lane which is left uncovered in the first two depositions. In this way, piling up of material at the edge of the lane is avoided.

With the usual technique of using impregnated papers developed by Zaffaroni and his colleagues it is not possible to use the "running up" technique, and even the ordinary method of direct pipetting into the starting zone involves special problems. If the extract or steroid is deposited in ethanol or methanol the stationary phase (e.g. propylene glycol) is washed away from the centre of the zone of deposition leaving a partly or totally denuded area of paper. If the amount of stationary phase on the paper is sufficient, and if a short time is allowed for equilibration in the tank before the running strip, the displaced stationary phase creeps back to fill the denuded area and also concentrates the extract by what is essentially a miniature "radial chromatogram" running in reverse. The problem can be avoided by depositing the extract dissolved in approximately 5% propylene glycol in methanol or ethanol, and is not serious if the deposition is carried out in a less polar solvent such as benzene or chloroform (R. Peterson, personal communication). If the strip is run before such a denuded area has been "refilled" with stationary phase, much of the

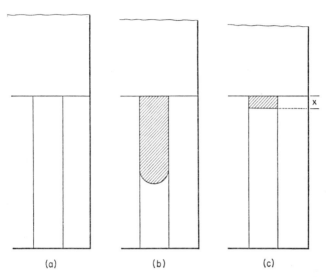

Fig. 3.2. Concentration of 0·02–0·2 ml of extract for chromatography by a preliminary chromatogram. "Running up" technique. (a) Marked sheet. (b) After depositing extract. (c) After running up to the origin. (By kind permission of the Editors of the *Biochemical Journal*.)

steroid will be adsorbed at the origin and either remain there during the run or be eluted gradually to give streaks or artefact spots.

These difficulties are also avoided by using the modified technique of impregnating paper described below (p. 166). This technique is also useful in allowing the "running up" method to be used, thus removing one of its previous limitations.

The concentration of the material to starting zones in the form of small spots is achieved by "running up" an extract deposited in the usual manner, and then driving the solutes into the centre of the strip by delivering the "running up" solvent to the edges of the strip with strokes of a pipette

(Fig. 3.3). A concave front is used to avoid radial dispersion and loss of the solutes, and in the final stages the spot is concentrated by "ringing" it with the pipette in the presence of a draught of air (Bush and Mahesh, 1959a). These concentrating techniques are sometimes upset by extracts containing a large amount of difficultly soluble material. This is usually more soluble in dry ethyl acetate than in the ethyl acetate–methanol mixture that we use (2: 1–3: 1 by vol.) and can be run in successfully by using

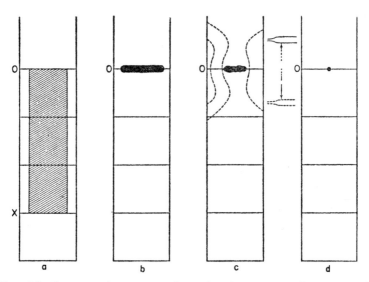

FIG. 3.3. Concentrating extracts for estimating very small amounts of Δ^4-3-oxosteroids with the NaOH fluorescence reaction. (a) The extract is deposited as usual over the shaded area between O and X, and (b) is run up to O with ethyl acetate: methanol. (c) The band is compressed to a spot by "running in" with the same solvent delivered by hand pipette moving over the course shown by the arrow; the two sides of the strip are irrigated with solvent in rapid succession. (d) The process completed.
The drawing is to scale for a 5 cm wide strip of filter-paper.
(By kind permission of the Editors of the *Journal of Endocrinology*.)

ethyl acetate or methylene chloride, followed by the usual mixture of one of these solvents with methanol. Great difficulty will be experienced if the solvents are not quite dry due to the waterlogging of the zone as successive lots of solvent are evaporated during the process of washing the material into the central spot.

The "running up" operation can also be used to achieve some purification of the material. With extracts of urinary conjugates which commonly contain salts, urea, and other polar contaminants, we usually deposit the extract over an area extending to within 2–3 cm of the origin rather than up to the origin itself. Much polar material remains in the original area of deposition, leaving the origin "clean" and avoiding salt-effects or water-logging during the critical first minutes when the mobile phase reaches and passes through it (p. 70).

Similarly, fat and cholesterol esters can be "run up" beyond the origin with dry hexane or light petroleum leaving moderately polar steroids behind. This may fail with large amounts of fat and one must then run a properly equilibrated chromatogram in a group-A solvent system (e.g. Bondy and Upton, 1956) to carry the fatty material off the paper. The paper is then dried and run again directly in the appropriate solvent system for the steroids it is desired to separate. Alternatively the fat can be treated as if it were an intentional non-polar stationary phase, and the polar steroids can be "run up" beyond the fat-laden area with 85% aqueous methanol (Bush, 1957b). The original area of deposition is now cut off and rejected and the polar steroids "run up" in the usual way to the origin (Fig. 3.4). Lewis (1957) was unable to succeed with this technique without siliconing the area for the original deposition but the author has had no difficulty as long as the aqueous methanol run is repeated if the extract is excessively oily, and the area for the original deposition is generous.

Whether orthodox or more complicated methods of depositing material on paper chromatograms are used it is our experience that for amounts of steroid over about 5 μg, results are better if the starting zone is in the form of a thin bar across the strip rather than in the form of a round spot. This is probably a simple consequence of the limited solubility of most steroids in most solvents: ideal conditions are only achieved with lower concentrations than those feasible with other classes of compounds and this is most easily achieved by spreading the material to be separated over a larger width of the paper.

I have dealt with these transfer operations in some detail since my own experience is that chromatograms are made or marred by this first step in the method to a greater extent than by any other single feature of technique. With columns the quality of packing is almost equally important; with paper no other factor approaches this one in its importance for good results.

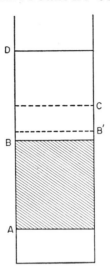

Fig. 3.4. Concentration of fatty extracts. The extract is deposited between A and B and the strip then run up to C with 85 per cent aqueous methanol. It is then cut at B′ and the lower part rejected. The part above B′ is then run up to C with ethyl acetate–methanol. B′C is a length sufficient for the folds needed for hanging the strip from the trough, etc.

2. Column Chromatography

(a) Preparation of Adsorption Columns

This is a simple technique which has been fully described in many books and papers (e.g. Reichstein and Shoppee, 1949). There are no features specific to work with steroids apart from the nature of the adsorbent which will be dealt with below. It is usual to fill the column with the first solvent of the elution series and add the adsorbent as a fine stream while tapping and rotating the column. The solvent is run through the column until its surface is just above that of the top of the adsorbent (which must always be kept covered with solvent) and then fresh solvent is added and run through the column. This is repeated until the top of the adsorbent no longer settles down as the solvent runs through. The column of adsorbent should be an exact and properly squared-off cylinder. This is achieved by having a sintered glass plate at the bottom of the column covered with a few accurately fitting disks of filter paper; or by packing with glass wool, a filter plate and filter paper, followed by a layer of about 1cm of fine, washed sand. The top of the column is levelled by vibrating and rotating the column, and then covered with sand or a tight-fitting disk of filter paper in order to prevent disturbance of the top layers by entering solvent. Rapid mechanical vibration of the column during its filling is a great help in removing bubbles and avoiding uneven packing. This is easily secured by holding the column against a rapidly rotating eccentric of small diameter: this can be run in the chuck of a stirring motor and even a bent nail is quite efficient.

Bubbles of air must be removed from the solvent and the region of the tap and stem of the glass tube before adding the adsorbent. This is much easier with a tube fitted with a sintered glass plate at the bottom than with the load of glass wool, filters and sand that are used with a tapered tube outlet. It is often best to fill the column from the bottom via the stem and tap if the latter arrangement is used. If there is any danger of a rise in temperature during the run, either due to heat of mixing of solvents or to a poorly controlled room temperature, it is a good idea to "de-gas" the solvents by a brief exposure to reduced pressure before adding the adsorbent to the tube. If aqueous alcohols are to be used, this is particularly important since the solubility of air in alcohols is much greater than in water: on raising the water content of the eluant bubbles of gas will be released unless the alcohol has been partially de-gassed before use. It is not, however, common to use aqueous solvents with adsorption columns.

(b) *Preparation of Partition Columns*

This is more difficult than the preparation of adsorption columns and it is not usual to rely upon gravity for good packing. This, however, is not a fundamental difference but due to the popularity of Celite as a supporting medium. Silica gel columns are readily packed under gravity (Martin and Synge, 1941) because of the high density of this material. With kieselguhr, or "Celite" (as the type used for partition columns is usually called), however, the low density of the material makes it essential to pack the column with mechanical pressure (Martin, 1949).

With Celite and silica gel the stationary phase is mixed with the supporting material by grinding up well with a pestle in a tube until the mixture has the appearance of an almost dry powder, absolutely free of lumps. The only sign that it has been wetted at all is that the powder is no longer free running. With both materials the amount of stationary phase must be within the limits of 0·5 to 1·0 parts of solvent per part of supporting material by weight. The best figure must be found by practice and depends upon the nature of both phases of the solvent system and the temperature of operation. A larger amount of solvent is usually held satisfactorily by silica gel than by Celite (see p. 152) but this is not a great difference with the solvents commonly used for neutral steroids because their swelling capacity with silica gel is relatively small.

The powder is then shaken up with a good excess of the mobile phase to form a slurry and poured into the glass tube which has been prepared in the same way as for adsorption columns. With Celite it is then essential to pack the column by the method described by Martin (1949) and Howard and Martin (1950). (Silica gel can be packed in the same way or simply by the method used for adsorption columns.) A perforated packing disk (Fig. 3.5) on the end of a stainless steel rod is plunged up and down the column through the slurry with a rapid motion until all bubbles have emerged and the slurry is quite even. The packer is then brought smoothly down until it lies about 0·02 column length (i.e. that length occupied by the stationary phase in the final column) above the filter disk at the bottom of the tube. It is then moved down slowly and firmly so that the particles are packed down beneath it instead of rushing through the perforations as occurs with the rapid strokes. The slurry is again stirred up by a few rapid strokes and again a small section pressed down with a slow firm stroke. This is repeated until all the powder is packed down tightly leaving clear mobile phase above the column paper. The smaller the section packed at each firm stroke the more even and efficient the column will be. For a 1cm diameter column, good results are usually obtained if the packer obtains a section 0·2–0·3 cm deep with each firm stroke. A typical column would, therefore, have about 50 to 100 zones packed in this way. For crude separations into groups a less stringent standard can be adopted and columns packed in ten to twenty zones are quite adequate.

Cellulose powder columns have not been used much, but are sometimes useful in providing more stable columns with certain solvent systems. They are much more difficult to pack and indeed it is almost impossible to obtain a really even and channel-free column by ordinary methods. Dr. P. N. Campbell (personal communication) found that the best method was to dehydrate the powder with acetone and pack the column under gravity in acetone as for adsorption columns. The column is then run with large volumes of mobile phase (i.e. previously equilibrated with stationary phase) so that the cellulose takes up the stationary phase and swells to pack itself tightly in the column. The effluent mobile phase is re-equilibrated with the stationary phase and the process continued until the effluent mobile phase emerges fully saturated with the stationary phase. This is estimated by titration when acid or alkaline systems are used, or by conductivity measurements with neutral or buffered systems. Alternatively, density or

Fig. 3.5. The Martin packer. Made of stainless steel. Perforations must be of a diameter determined by the interfacial tension of the two phases of the solvent system (see Martin, 1949).

water content can be measured. In the author's experience this method is not altogether satisfactory with the solvent systems commonly used for steroids, probably because most of them do not swell cellulose as much as the aqueous stationary phases used by Campbell. Fairly satisfactory columns are achieved equally well by packing the swollen cellulose under gravity and then exerting strong steady pressure on the top of the column after a filter-paper disk has been placed there (e.g. Simpson *et al.*, 1954). Martin's technique of packing can also be used but the optimum size of hole is smaller than with Celite and a separate set of packing disks is needed.

It is best to use precision-bore tubing for chromatographic columns and with Martin's technique it is essential to select tubing that is of good quality and as even a bore as possible by testing with a packer before making the column. It is decidedly irritating to fit a tap and possibly a sintered disk to a piece of tubing and then find that the packer fits neatly at the top of the tube but sticks fast half way down. A good fit with the packer at its circumference is essential for even packing; the clearance should not be much greater than one-quarter to one-half the diameter of the holes with which the disk is perforated. To avoid asymmetry the column should be rotated between the packing of each small zone of powder.

(c) Design and Operation of Columns

Several features of column chromatography are common to both adsorption and partition methods; others are conveniently considered together in order to emphasize points of contrast between the two methods.

It is common to use straight glass tubes with stopcocks for both types of column. Difficulty is sometimes found in keeping stopcocks in good condition since grease cannot be used for badly fitting taps. Since it is also common to use a head of pressure to maintain a convenient rate of flow through the column it is worth considering the use of swan-neck columns (Axelrod, 1954) in which a stopcock is unnecessary (Fig. 3.6). Slight spillage from the tip during packing can be dealt with by tying a conical flask loosely to the spout of the column, or by resting the spout over a large boiling tube or measuring cylinder.

The dimensions of columns vary widely. The length: diameter ratio is best kept in the range 10: 1–40: 1 and is usually lower for adsorption columns which may be operated with ratios as low as 2: 1, or as "beds" rather than "columns" with ratios of 1: 4, when displacement conditions are desired. The solute: adsorbent ratio is also very variable for adsorption columns, and rather more limited with partition columns since displacement conditions are unlikely to be useful with the steroids and have not been attempted. For preparative work and simple mixtures adsorption columns are usually run with a solute: adsorbent ratio in the range of 1: 20–1: 50. For very complicated mixtures and for routine methods in which it is essential to secure the elution of each compound at a reproducible point in the elution sequence much lower ratios are used with adsorption columns, usually in the range 1: 200–1: 2000. With partition columns the largest ratio that is practicable for most problems is about

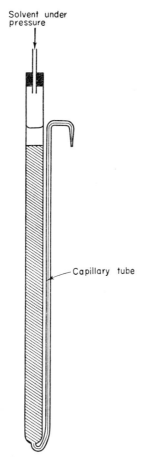

Fig. 3.6. Swan-necked column.

1: 500 (weight of substance: weight of Celite plus stationary phase) and for complicated mixtures, and the estimation of small amounts of steroid in the presence of much impurity, ratios of 1: 10,000 must be used. Using aqueous stationary phases on Celite, Hegedüs et al. (1953) found that slightly larger ratios could be used for separating plant glycosides on the preparative scale. These authors give a very useful table for the dimensions and running conditions of Celite: water columns. A typical large column had a stationary phase of 3 kg Celite: water (1: 1 by weight) measuring 115×6.5 cm, a maximal flow of 1·2 ml/min, and could handle 6·0 g of mixed glycosides. In the author's experience neutral steroids of the C_{19} and C_{21} series, when run in aqueous methanol systems on Celite columns, need a rather smaller solute: Celite ratio, but can be run successfully at rather larger rates of flow.

With both types of column it is best to test the performance of a new type or size of column with dyes, in order to detect serious channelling or cracks in the column, and to find the optimum flow-rate (Butt *et al.*, 1951). A dye moving with an R_F in the range of 0·5–0·7 is run on the column with a very fast flow-rate so that the band is obviously longer than it should be in normal running. The retention volume of the peak and the total volume between head and tail of the dye are recorded. The procedure is repeated reducing the flow-rate by two- to four-fold steps. The bandwidth will fall to a minimum value and the highest flow-rate giving this value is adopted (Fig. 3.7). While this is a useful rough method it is sometimes necessary to proceed further and make a final run at the supposed optimum flow-rate and map out the profile of the effluent dye concentration by taking accurate small fractions. The symmetry and dimensions of the peak (p. 81) can then be used to calculate the number of theoretical plates in the column. Good Celite partition columns should achieve about 15–20 theoretical plates per cm with aqueous methanol systems and diameters of 1·0–2·5 cm (Tait and Tait, 1960). The rough order of dimensions required for typical problems is given in Table 3.1. Hard and fast rules are not very useful since so many factors are involved and it is besides always sensible, and not usually inconvenient, to err on the side of safety— that is in the direction of low solute: column weight ratios, and low flow-rates. For instance the separation of three steroids such as progesterone and its 11-oxo-and 11β-hydroxy-derivatives in an extract containing large amounts of cholesterol would need a large partition column to avoid displacement effects between progesterone and cholesterol if aqueous methanol systems were used. A much smaller column would suffice if acetic acid systems were used because of the much lower R_F value of progesterone that can be obtained with a given non-polar mobile phase. A still smaller column would suffice if more polar steroids with very low R_F values had to be separated from cholesterol and from each other, such as a mixture of cortisone and cortisol.

In most cases, the dyes commonly used as markers on columns (see p. 181) behave similarly to the steroids in partition systems in terms of diffusibility, and hence in the number of theoretical plates achieved by the column under given conditions of temperature, solvents, and flow. In some cases, however, a given dye or steroid is much worse or better than a standard dye used to assess the column. In that case, the performance of

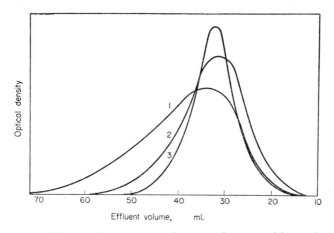

Fig. 3.7. Finding the optimum flow-rate for a partition column.

1, 2, 3 represent the elution curves for a dye of R_F 0·5–0·8 using successively slower rates of flow. For practical purposes the optimum has been reached when the peak approaches complete symmetry as with 3 in the figure. (Only approximately to scale.)

the column must be reassessed using a more comparable dye, or by measuring the elution profile of a standard steroid. This divergence of behaviour most often occurs with acid or alkaline solvent systems in which particular dyes with unfavourable ionizing groups streak or form complexes in certain ranges of pH and not others.

Some of the difficulties attending the adsorption chromatography of steroids have already been dealt with (p. 38). A good account of artifact formation is given in Lederer and Lederer (1957) and in Neher (1958a, 1959). In general, little difficulty should be experienced with 17-oxo-steroids and other robust compounds. Reproducibility may be difficult with different batches of all types of adsorbent, and careful standardization with a set of dyes should be done routinely when starting a new batch. The papers of Reichstein, Dobriner, Lieberman and their colleagues

Table 3.1. Examples of Partition Columns

Reference	Butt et al. (1951)	Hegedüs (1953) Hegedüs et al. (1953)		Simpson et al. (1954)	Simpson et al. (1954)	Simpson et al. (1954)	Tait and Tait (1960)	Kritchevsky (1954)	Wilson et al. (1958)
Purpose	Estimation of progesterone & corticoids	Strophanthus glycosides, etc.		Adrenal extract	Adrenal extract fractions	Steroids and acetates	Urinary steroids	Urinary steroids	Urinary steroids
Supporting material	Kieselguhr	Kieselguhr		Kieselguhr	Cellulose	Kieselguhr	Kieselguhr	Silica gel etc.	Aluminium silicate
Solvent system	Hydrocarbon/ aq. MeOH or Et(OH)$_2$	Benzene → BuOH/water		Petrol → benzene/water	Hydrocarbon/ aq. MeOH	Hydrocarbon/ aq. MeOH	Hydrocarbon/ aq. MeOH	Hydrocarbons and CH$_2$Cl$_2$/EtOH	Hydrocarbons and CHCl$_3$/aq. EtOH
Length (cm)	6·0	50	65	115	~115	26	30	60	~40	30
Diameter (cm)	0·6	2·6	4·5	6·5	~6·5	2·6	1·0	1·0		1·5
L/D ratio	10	20	15	18	~18	10	30	60		20
Weight of stationary phase+ support (g)	1·5	200	700	3000	~2800	~80	~20	40	~10 ~52	~15
Load possible	~1–2 mg	0·4 g	1·4 g	6 g	27 g	320 mg		~20 mg	~20 mg ~100 mg	—
Flow rate (ml/hr)		4	18	70	100	8		12	12 6	36
Stationary phase/ support		1:1 (wt.)	1:1 (wt.)	1:1 (wt.)	1:1 (wt.)	1:2 (v/w)	1:2 (v/w)	1:2 (v/w)	1:4 (v/w) 2:5 (v/w)	2:1 (v/w)

Approximate figures are inferred from weights of *individual* fractions given in the reference quoted: it may be possible to exceed them. These columns are of general application and not confined to the purposes exemplified in the table.

should be consulted for details, and with a new or unfamiliar family of compounds one should always be on the alert for artifact formation, particularly with alumina. The very considerable heat of adsorption of silica gel with polar solvents should be remembered and the addition of polar solvents to sequences of elution should be made by a gentle gradient or by small steps rather than large ones. Charcoal is rarely used for adsorption of steroids, particularly since Meyer's (1953) demonstration of serious artifact formation with steroid α-ketols. However, the importance of considering all the factors involved is illustrated by this adsorbent since Savard (1960) found no artifact formation when other solvents, particularly pyridine, were used for elution.

Gradient elution has become steadily more popular for both adsorption and partition columns (Alm *et al.*, 1952: Lakshmanan and Lieberman, 1954; Morris and Williams, 1955; Kellie and Wade, 1957). Many types of apparatus are in use (Donaldson, Tulane and Marshall, 1952; Heftmann and Johnson, 1954), but the device of Morris and Williams (1955) is particularly neat in that the gradient curve is a result of the slopes and mutual orientation of the two solvent vessels and mixing is achieved by small turbulence chambers rather than by stirrers (Fig. 3.8). On the other hand, a given pair of vessels always gives the same gradient curvature. In order to combine the neatness and reproducibility of this method with the greater versatility of the methods using independent vessels with stirrers, a modification of Morris's device is suggested in Fig. 3.9: here cylindrical vessels are used and the geometrical variations are provided by inserting solid cores of various shapes. It is far less expensive and difficult to make a set of such cores than to make a large number of linked non-cylindrical vessels to provide different gradients.

For large columns the front-straightening method of Claesson is worthy of attention although it has not apparently been applied to steroid chromatography as yet (Claesson, 1947).

Various types of fraction-collector are used for dividing up the eluate from columns. These are based on set volumes provided by siphons; set volumes provided by a drop-counter; or set time intervals provided by a clock with relays. For very large fractions a ball-valve device has been devised by Philip and Schenk (1957). Haines (1952) described a complicated but useful machine which only changes the sampling vessels when a given ultraviolet absorbing zone has been completely eluted: other recordable properties can also be used but most of these would not be easily applicable to the common run of steroids. Many of the commercial models involving electrical mechanisms are unreliable or very variable in quality from model to model. In particular, most of the siphons supplied with them are unsuitable for light solvents. Two faults are common.

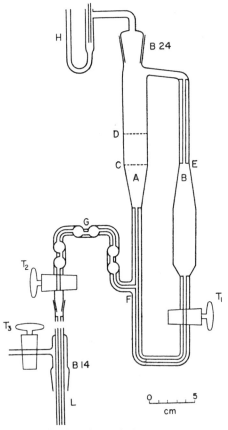

Fig. 3.8. Gradient elution apparatus.
(Morris and Williams, 1955).

(a) H—manometer giving raised pressure in vessels A and B.

G—turbulence bulbs in capillary tubing. As the first solvent falls in A from D to C the amount of the second solvent (in B) delivered to the column at L is negligible. As A falls below C, the concentration of the second solvent discharged from B rises progressively till finally almost pure B is delivered to the column.

Reproduced by kind permission of Professor C. J. O. R. Morris and Messrs. J. and A. Churchill Ltd.

TECHNIQUES AND APPARATUS

Firstly, the siphons often have delivery stems with too large a bore so that light solvents form a broken column which leaves drops in the stem after delivery; or they go on delivering dribbles of solvent from the broken column in the stem which pass into the next tube after the main body of the fraction has been collected. The second common fault is that the rocker arm on which the siphon moves, thus actuating the relay which

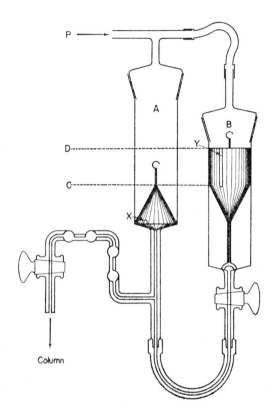

(b) A more versatile variant of (a) using cylindrical vessels and solid fillers. The arrangement shown will give nearly the same solvent gradient as (a). Any desired gradient can be obtained simply by changing the solid fillers to other shapes and by raising or lowering B relative to A. For most solvent systems, stainless steel, Teflon, polypropylene, or nylon fillers can be used.

X and Y are grooves to allow the passage of solvent.

turns the turntable, is both insensitive with solvents of low specific gravity and difficult to adjust; and it also tends to oscillate at the critical point, rather than rocking decisively into the "down" position when a few drops of solvent more than the balancing volume have been collected. In some collectors the period of oscillation is disastrously close to the lag time of

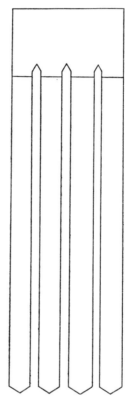

Fig. 3.9. Curtain pattern for multiple-strip paper chromatogram (after Zaffaroni *et al.*).

the relay and turntable: in this case, the siphon descends after a few drops enough to actuate the relay and the slight jolt delivered by the abrupt rotation of the turntable jerks the rocker arm back so that it actuates the relay again as it descends a second time. A nicely resonating system obtains until the siphon is sufficiently full not to be jerked back by the motion of the turntable, and ten or twenty tubes may be passed over between each fraction.

These faults are overcome by the gravity-operated models designed by James, Martin and Randall (1951) which are extremely simple and reliable and do not rely on electrical mechanisms. There is, of course, no reason why an electrically operated machine should not embody the mechanical properties of the collectors designed by James *et al.* (1951). This is yet another example of the slightly greater practical difficulty of using the solvent systems commonly employed in steroid work than those employed in other fields. Where possible it is helpful for instance to substitute dense solvents such as decalin, tetralin, and halogenated hydrocarbons for lighter ones which may have been used for paper chromatography; this tends to overcome the mechanical faults and usually the surface tensions and viscosities are great enough to make it unnecessary to use siphons with specially narrow outlet tubes. Precautions must also be taken to see that volatile eluants do not evaporate too rapidly or too variably from the siphon; another fault seen with mobile solvents if no precautions are taken is the formation of a ring of deposit on the outside of the tip of the column due to solvent creeping round on to the outside and evaporating. These faults which rarely arise with dense and involatile solvents can be avoided by drawing out one side of the outlet to form a pointed gutter and enclosing the tip and the top of the siphon in a glass envelope or funnel to prevent evaporation.

Continuous recording of column effluents has been used widely for radioactive (Deal *et al.*, 1956), light-absorbing (Stein and Moore, 1948; Stark *et al.*, 1957; Spackman *et al.*, 1958), and titratable substances (Howard and Martin, 1950; Liberti, 1953) among others but is not easily applicable to many classes of steroid. Capacitance or conductivity methods should be feasible with some classes of steroid for monitoring columns used on the preparative scale (see Lederer and Lederer, 1957, p. 18) and the elegant thermistor bridge of Sternberg and Carson (1959) should be worth adapting to steroid work.

(d) Multiple-Column Chromatography

Vestergaard (1960a) has described a linear fraction-collector which allows the collection of 20 fractions from each of six columns running simultaneously. This is simply a rack with six rows of 20 tubes which runs down an inclined plane; it is started and stopped by a ratchet actuated by a solenoid which is controlled by a timer. This apparatus was used in conjunction with six alumina columns to carry out analyses of urinary 17-oxosteroids (Vestergaard, 1960b). It was found that for small columns the gradient elution device of Lakshmanan and Lieberman (1954) was not sufficiently accurate; a syringe continuous infusion apparatus was used instead to introduce the second solvent of the eluant mixture. The reproducibility of the method and the agreement between the pattern of eluted steroids from the six columns appears to be excellent, and the method as a whole is a masterpiece of ingenuity. For a complete demonstration of the peaks of six typical 17-oxosteroids, 85 fractions were collected, taking about 28 hours for the whole run.

3. Paper Chromatography

(a) Preparation of Paper and Sheets

Several surveys have been made of the properties of different types of filter paper, largely with the paper chromatography of amino acids in view. Much of the information is also useful to those working on steroids if sensible allowances are made for the different physical properties of the solvent systems commonly used in the two fields (Kowkabani and Cassidy, 1950). In general, the apparent pH of most filter papers is quite strongly acid but some have been treated with basic polymers to raise the wet-strength of the paper. These may be useful in retarding acidic contaminants of neutral steroids (p. 213) but probably should be used with caution for acidic steroids. Small amounts of strong acids and bases are retained very strongly by filter paper and this should be remembered when considering washing procedures.

The ether-soluble matter in good filter paper can be as high as 0·05% by weight (Balston and Talbot, 1952). Since ether is a poor desorber for paper much larger amounts are probably eluted by alcohols: some of this material is precipitated by diluting ethanol eluates of filter paper with benzene or hexane (E. B. Romanoff, personal communication) and some of its properties have been described by the author (Bush, 1954a). A typical zone of a chromatogram containing 50 μg steroid will cover about 300 mg of paper: ethanol or ethyl acetate–methanol may be expected to elute up to 300 μg of non-specific material together with the steroid if the paper is unwashed. To some extent, however, this material will be carried with the solvent front (p. 69) so that the amount of non-specific material eluted from zones of chromatograms is usually less than this in practice, although still far from negligible.

With large-scale screening work it is worth while getting special sizes and shapes of filter paper stamped out commercially (cf. Neher, 1958a). Unless rectangular sheets or strips are used there is always the problem of irregularity of solvent flow. The arrangement whereby several strips hang from a common curtain of paper which hangs in the trough is convenient but usually gives rise to a larger velocity of the mobile phase at the edges of the strips than at their centres, resulting in bands which are concave downwards on the finished chromatogram. This can be avoided by folding the curtain into the trough in such a way that only the separated parts of the strips emerge from the solvent. Otherwise it is important to cut the slots in the original sheet as narrow as possible (slot: strip width ratio 0·1: 0·15) and the author prefers to avoid this arrangement altogether by using rectangular strips mounted on a special light frame (Fig. 3.15). The suitability of any particular shape and size of sheet or strip under given conditions is best tested by running some dyes and observing their behaviour during the run; if they are deposited as thin straight "bars" on the origin, edging or coning will be shown up by the coloured bands becoming concave downwards or upwards respectively.

The most important aspect of the preparation of paper for chromatography is its chemical cleanliness. There are two sorts of problem involved: firstly the question of getting the paper cleaner than it is on arrival from the supplier; secondly keeping it clean once it arrives or after it has been washed. For many purposes we have found that Whatman papers need no extra cleaning, but it must be confessed that the problem is far from solved; many different procedures have been used for washing filter paper, none of which can be claimed to be more than partly successful, and a slow decomposition of paper occurs which is not fully understood and at present not preventable.

Paper from the makers is stored in the packets provided, care being taken to close the packets or store them in a more or less dust-proof box after withdrawing sheets for use. Paper which has been cleaned by any of the steps given below is stored in packs of 100–200 sheets either between clean sheets of glass in a dust-proof box, or wrapped in aluminium foil. For 2·5 or 5·0 cm wide strips we have found it useful to pack them in lengths of channel-section aluminium with the open side of the channel closed with aluminium foil secured with sticking plaster; the strips are then kept clean but are easily removed singly with forceps without exposing the others (Fig. 3.10).

Papers supplied from the makers are in our experience not sufficiently clean for the best results with the following procedures:

(a) Elution of zones for ultraviolet spectroscopy;
(b) Elution of zones for ultraviolet spectroscopy in concentrated H_2SO_4;
(c) Elution of material for infrared spectroscopy;
(d) Elution of material for several absorptiometric estimation methods such as those using alkaline blue tetrazolium or *m*-dinitrobenzene.

Satisfactory results can usually be obtained by various washing procedures followed by careful storage of the washed paper, and preferably early use. However, there is at present no absolutely satisfactory method of washing filter paper and it depends entirely upon the quality of the results desired and the class of compounds under investigation how far one needs to go.

A good method in our experience is to wash the papers chromatographically in packs of 100–200. This is much more economically done if work requiring washed paper is done with strips of paper not more than 10 cm wide. They can be washed chromatographically in a direction *perpendicular* to the length of the strip with relatively small volumes of solvent in the simple apparatus shown in Fig. 3.11 (cf. Isherwood and Hanes, 1953; Neher, 1958a, p. 66). This apparatus avoids the need for a specially machined trough and can be used for widely varying numbers of strips. For a pack of 200 strips of Whatman No. 2 paper, 22 in. × 2 in., the following sequence is satisfactory:

1000 ml ethanol–ammonia (sp. gr. 0·880) (5: 1 by vol.),
500 ml water,
1000 ml 2N-acetic acid in water,
500 ml 50% aqueous ethanol (by vol.),
1000 ml ether. (For purity of solvents see Appendix I.)

Vogt (1960) found that a wash with light petroleum was needed to remove oils but in our experience this is achieved by the final ether wash in the above sequence. Lewis (1956, 1957) uses a 24hr run in 2N-ethanolic NaOH followed by water till alkali-free, and then 3 hr with ethanol. This removes most of the materials giving high blank values in quantitative work.

The paper washed in this way is sometimes improved for elution of material for spectroscopy in H_2SO_4 by extraction with methanol–benzene (1: 1 by vol) or methanol in a Soxhlet extractor for 6–18 hr. Neher (1958a) finds Soxhlet extraction with chloroform and methanol–benzene satisfactory. The steroid glucuronosides can be run on unwashed paper without

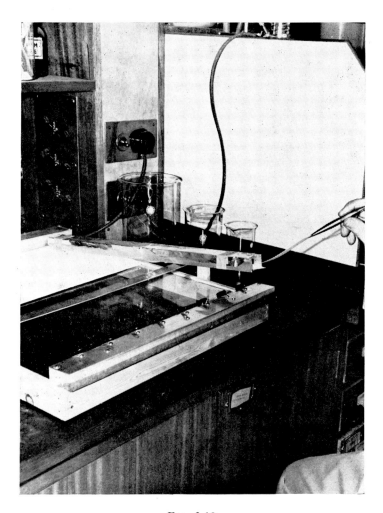

Fig. 3.10

(see p. 159)

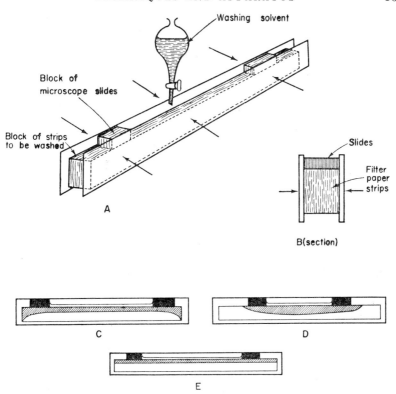

Fig. 3.11. Cross-washing of paper strips. The strips ($2 \times 22\frac{1}{2}$ in.) are laid in packs of about 200 (Whatman No. 2) on a plate-glass sheet (4×24 in.) and a second glass placed on top. The block is compressed by springs (arrows) in a frame in the position shown and two packs of microscope slides laid on top to form a trough. Gaps between the slides and side-plates are filled with $\frac{1}{2} \times 2$ in. strips of Whatman 3MM paper. The solvent is allowed to drip on to the strips fast enough to form a fluid layer $0 \cdot 3 - 0 \cdot 8$ cm deep between the two packs of microscope slides. The microscope slides are moved along the top edge of the block of paper strips until the leakage past them just produces a straight solvent front at the ends of the block of paper strips. The block and its supporting frame are mounted on a trough with suitable drainage, in the fume cupboard. C: microscope slides too close to ends. D: slides too far from ends. E: correct position of slides. Shaded area represents solvent area in early stages of washing. With aqueous solvents the microscope slides are best replaced by filter paper packs which match the swelling of the main block of paper strips.

any tailing or other artifacts being noticeable; untoward experiences with steroid conjugates, however, should lead the investigator to try the washing methods of Hanes and Isherwood (1949) and Isherwood and Hanes (1953) designed for avoiding artifacts with organic phosphates, carbohydrates, and organic acids.

For quantitative chromatography using the scanning method, reasonable results can be obtained with unwashed Whatman No. 2 paper (Bush and Willoughby, 1957; Bush, 1960a). The background is, however, improved by using the chromatographic washing procedure given above; with phosphomolybdic acid or alkaline m-dinitrobenzene this improvement is considerable with some batches of paper.

(b) Impregnation of Paper with Stationary Phases

The original introduction of filter paper as a supporting material for partition systems made use of the ability of cellulose to swell and imbibe solvents from a vapour phase (Consden *et al.*, 1944, 1947). This technique is the one of choice in that it probably secures the most even distribution of the stationary phase over the paper. However, it was soon realized that solvents which were relatively involatile might be useful in partition systems, and that there was no reason why filter paper should not be impregnated directly with them, rather than by using the inordinately long period of equilibration with the vapour phase that would be needed for the orthodox procedure. The first use of this method was probably the impregnation of filter paper with non-polar solvents such as petroleum jelly in order to achieve "reversed-phase" partition chromatography (e.g. Winteringham *et al.*, 1950). Later the technique was extended to polar solvents such as formamide and propylene glycol by Zaffaroni and his colleagues and such methods were used widely for the steroid hormones, and their metabolites, and also for steroids of vegetable origin. Other similar techniques were evolved which used chemically modified paper such as acetylated paper (Kostir and Slavik, 1949; Ritter and Hartel, 1958), paper esterified with stearato-chromyl chloride (Kritchevsky and Kirk, 1952b) and paper treated with silicone (Kritchevsky and Tiselius, 1951), or rubber latex (Boldingh, 1948). Adsorption chromatography on paper impregnated with alumina (Datta, Overell and Stack-Dunne, 1949; Bush, 1950; Shull, Sardinas and Nubel, 1952) or with silicic acid (Kirchner and Keller, 1950) has also been used for non-polar substances, including steroids, but will not be discussed here since the method is rarely, if ever, any better than an equivalent partition method.

The outstanding features of impregnation methods are, first the possibility of obtaining a larger ratio of the cross-sections stationary/mobile phase than by the orthodox technique; second the possibility of avoiding a period of equilibration longer than a few minutes; and finally the difficulty of obtaining really reproducible results, unless a rigidly controlled technique of impregnation is used, because of variation in the amount of stationary phase taken up. In the author's view there is little, if any, advantage in the more complicated methods of chemical modification of filter paper since they are unnecessary for securing stable systems with polar solvents, and a useful range of non-polar solvents will give stable reversed-phase systems on untreated paper if certain precautions are observed (Winteringham et al., 1950; Mills and Werner, 1952, 1955; R. P. Martin, 1957).

To obtain a stationary phase with an involatile polar solvent the usual and well-tried technique is to dip the paper through a solution of the desired solvent in a suitable volatile solvent such as methanol or acetone. The paper is lightly blotted with clean filter paper, allowed to dry at room temperature for a few minutes, and is then ready for use. The impregnating solution usually contains from 10 to 25% by volume of the involatile solvent in the volatile carrier. The older method of dipping the paper through the undiluted involatile solvent and then squeezing out the excess with a mangle or press has now fallen into disuse. If concentrations of more than 25% of the involatile solvent are used to impregnate the paper, the mobile phase will run very slowly; this, however, may not be a nuisance if a system of large capacity is required. This technique is very simple and with care reasonably reproducible R_F values can be obtained. With certain solvents such as ethylene glycol and some of the cellosolves a humid atmosphere causes trouble. With the phenyl cellosolve system (Neher and Wettstein, 1952) it seems that exposure of the paper to a damp atmosphere *before* impregnation leads to serious variations in behaviour of the system, while exposure after impregnation has little effect (Loomeijer and Luinge, 1958). With solvents such as formamide and propylene glycol evaporation is so slow in still air that there is plenty of time for preparing the chromatograms without losing more than traces of the stationary phase. The only disadvantage with such solvents is the difficulty of removing them completely when the chromatogram has been run (see p. 66).

Certain polar solvents are more volatile than the glycols and can be used either in the orthodox manner, or as impregnating solvents (Sjövall, 1954; Bush, 1960b). Concentrated aqueous acetic acid (70–90% by vol.) is extremely useful in this respect (Liberman *et al.*, 1951) but care must be taken with such solvents to see that as little time as possible elapses between impregnating the paper and hanging it in the chromatography tank, and that this interval is standardized. The best technique in such cases is the modified impregnating method described below. The great advantage of this group of solvents is that they are easily removed when the chromatogram has been run. As with the other impregnating solvents, they are applied to the paper as a 10–25% solution by volume in a volatile carrier which is then allowed to evaporate. Thus, for example, to obtain a stationary phase of 85% acetic acid in water, the paper is dipped through a mixture of 85 ml acetic acid, 15 ml water, and 560 ml ether or methylene chloride.

Sjövall's (1954) technique of impregnation with 70% acetic acid followed by drying at 100° "till just dry" seems to be far less reproducible than the use of a 15–20% (by vol.) solution of the 70% aqueous acetic acid in ether, followed by lightly drying at room temperature for 3–5 min. The latter technique only needs 10–20 min equilibration in the tank before running the chromatogram, while Sjövall (1954) needed equilibration for 16 hours with 70% acetic acid systems with or without previous impregnation of the paper.

Stable non-polar stationary phases can be obtained on *untreated* dry filter paper with certain relatively involatile solvents. Petroleum jelly, various high-boiling paraffins, tetralin and decalin seem to be the most suitable (e.g. Winteringham *et al.*, 1950); Mills and Werner, 1952; R. P. Martin, 1957; Inouye *et al.*, 1955; this book, p. 129). The method of impregnation is exactly the same as with polar solvents except that in the author's experience the carrier solvent and the concentration of the solvent in the carrier are more critical than with the latter. These solvents give systems which are stable in the temperature range 15–38 °C and which do not seem to be more sensitive to changes of temperature than "straight-phase" systems in this range of temperatures. The optimum concentration of these solvents in the carrier (volatile) solvent is in the range 15–20% by vol. With concentrations of 10% the system is very apt to break down and the stationary phase is then displaced completely from the cellulose. With 25% impregnating solutions the polar mobile phase usually runs excessively slowly and may even fail to enter the paper in some cases. It is usual to use a non-polar solvent such as hexane or chloroform as the carrier of the impregnating medium. In several instances the author has found that attempts to use a volatile carrier containing small quantities of polar solvents such as alcohols has led to an unstable system, while no such trouble had been encountered with dry, completely non-polar carriers such as hexane or methylene chloride.

If these precautions are observed the running of reversed-phase systems on untreated paper is as simple as any other method using impregnated paper. I shall not, therefore, consider the more complicated methods using chemically modified paper in detail since they appear to be unnecessary, except when there is good reason to suppose that better separations will be achieved with a more volatile non-polar solvent as stationary phase, e.g. benzene or chloroform. In such cases the use of siliconed, esterified, or rubberized paper is necessary and may be worth while: it is unlikely, however, that many situations will arise in which such volatile solvents will be any improvement on, say, tetralin (see Chapter 2). Again the volatility of the stationary phase may affect the details of the technique. Thus petroleum jelly allows the operator plenty of time to prepare his chromatograms but is a nuisance to get rid of after running them. "Odourless kerosene" (Mills and Werner, 1952) is almost ideal in being slow to evaporate in still air but easily removed in a warm draught (R. P. Martin, 1957). Tetralin is excellent but needs slightly more careful control of the time interval between impregnation and hanging the paper in the tank. The stationary phase may also be applied to the paper by spraying, with or without a volatile carrier (e.g. Inouye *et al.*, 1955), but in the author's experience this is not an easily reproducible procedure. In order to avoid "thin spots", for instance (which may lead to the complete breakdown of the system), heavy spraying is needed to be on the safe side. Papers impregnated with stationary phase pose several problems in the preparation of the chromatogram when impregnation is carried out by the above techniques. These have been described on p. 66. In particular, the use of the "running up" technique in the deposition of extracts is impossible or at least very unreliable. A simple method of impregnation was therefore devised to overcome these difficulties. It has the advantage that since it is performed *after* the chromatogram has been prepared (i.e. the extracts and standards have all been deposited and "run up" to the origin, etc.) the time between impregnation and hanging in the tank is much more surely controlled, and can be short enough to use a far wider range of more volatile solvents than is possible with the usual procedure for impregnation. The paper is marked and extracts are deposited and run up exactly as for a chromatogram run in the orthodox manner (vapour-phase equilibration) with volatile solvents. The dry paper is then dipped smoothly into the impregnating medium so that the latter advances through the paper as a straight front (Fig. 3.12) thus removing small bubbles of air. The movement of the paper is stopped when the origin line of the chromatogram is within 0·3–0·5 cm of the surface of the medium, and the front is allowed to

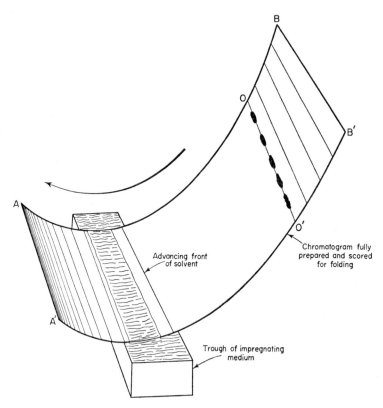

Fig. 3.12. Modified technique for impregnating paper with a stationary phase.

The fully prepared chromatogram is dipped at AA′ and drawn through the impregnating medium (10–20% stationary phase +90–80% volatile carrier) until surface of solvent is within 0·3–0·5 cm of OO′. The solvent front is allowed to advance to 0·1–0·2 cm beyond OO′ by capillary penetration and the sheet is then rapidly withdrawn and the carrier dried off by waving the sheet about. It is then dipped at BB′ in the reverse direction and the same procedure followed when the front approaches OO′. Thin strips are more easily done since they can be dipped vertically into the medium using a large measuring cylinder.

advance to this line by capillary forces (i.e. as if it were an ascending mobile phase). When the solvent mixture has reached this line the paper is instantly removed and waved in the air to evaporate the volatile carrier. The paper is now inverted and again dipped into the medium, this time from the opposite end; again the front is allowed to seep slowly into the last 5–8 cm of paper until it reaches and just "penetrates" the origin line, after which it is removed and rapidly dried by waving it about. With a little practice it is quite easy to get a good straight front which reaches the whole length of the origin line at the same time to within a second or less. It is essential that the approach from the opposite direction should be allowed to run "into" the origin line and overlap the original excursion of the impregnating medium by 1–2 mm: otherwise there is the risk that material will be adsorbed by bare areas of dry cellulose at the origin. The impregnation of strips by this method is easily mastered using a large measuring cylinder: sheets take a little more practice since economy forces one to use a trough and the sheet must be dipped obliquely through the medium which is more difficult to control. This method has given extremely reproducible results in the author's laboratory and is believed to possess numerous advantages over the earlier methods.

The useful properties of acetic acid as a stationary phase have already been discussed in Chapter 2. An attractive feature of this solvent is that it is easily used by both the orthodox and impregnating methods for obtaining a stationary phase. Thus for rapid sub-fractionation the impregnation method is used, obtaining a system of large capacity and avoiding the need for more than 15 min equilibration. For final fractionations and the determination of accurate R_M values the orthodox method is used in which slightly more reproducible results are obtained.

Systems based on the more viscous hydrocarbons such as decalin and tetralin similarly have advantages for some separations (Chapter 2) and additional practical advantages. Thus the system tetralin– or decalin– 80–90% aqueous methanol can be used in "straight" or reversed-phase manner in the same tank; and, with two troughs, at the same time. Similarly, tetralin– or decalin– 85–90% aqueous acetic acid can be used in "straight" or reversed-phase manner; and can be used in the orthodox fashion or by the impregnation technique, with either phase as the stationary one.

Aqueous methanol systems do not lend themselves easily to the use of the impregnation method because of their great volatility, although useful results have been obtained on occasions (Burstein, personal communication). However, it is possible that a modified procedure may be useful in shortening the period of equilibration. Thus the paper can be "wetted" by immersing in 85% acetone, drying off the organic solvent quickly, and then hanging in the tanks. Dowlatabadi (personal communication) finds that after this pre-treatment the equilibration period can be shortened to 30 min with aqueous methanol systems. Two factors probably operate in this method. First the *direct* contact with (liquid) water probably swells and loosens the cellulose fibres more rapidly than does exposure to the final 50–85% methanol vapour phase in the tank (see p. 71). Second the chromatogram probably contains a "stationary phase" of something like 10–20% acetone when first introduced into the tank; the acetone and some water will evaporate and facilitate equilibration of the paper with the vapour phase by setting up convection currents in the tank (cf. Hanes and Isherwood, 1949).

(c) *No-Touch Technique for Preparing Paper Chromatograms*

Apart from the techniques described above for cleaning paper and ensuring low volumetric error, the following special techniques have been used routinely for the last five years, both for quantitative work and for the identification of compounds by sequential degradation. They form essentially a "no-touch" method analogous to that used in orthopaedic surgery; certain features are specially designed to ensure the clean and smooth handling of the large number of strips needed for quantitative paper chromatography.

The paper is handled entirely with chromium or stainless steel forceps of the ordinary straight dissecting type and is laid only on glass or filter paper when not being handled. Special forceps (Fig. 3.13) are used for marking and labelling the strips, and also for bending them. These instruments are washed before use with ethyl acetate and methanol; they are easily made from standard aluminium alloy strip.

The apparatus shown in Fig. 3.14 is used for "running up". This enables large numbers of strips to be "run up" with small volumes of solvent and with no danger of them touching each other. The gap at the top is adjustable to take any convenient width of strip. Finally the strips are mounted in glass and aluminium frames of the type shown in Fig. 3.15. The frames are light and have two screws projecting at the sides, at the top of the vertical members: these act as trunnions and enable the frame to hang from the complementary framework in the tank, thus allowing large numbers of strips to be placed in the tank smoothly and quickly. The top ends of the strips are then turned down with forceps into

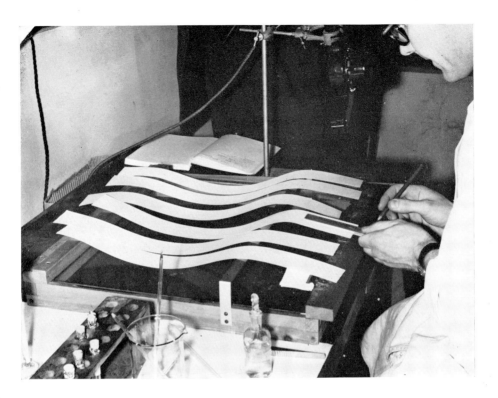

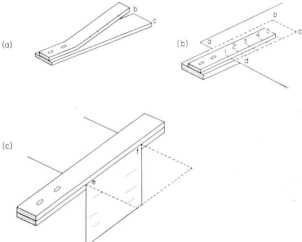

Fig. 3.13. Marking and bending forceps. (a) The marking forceps in open position. (b) Being used to mark a sheet of paper. The numbers show the standard marks for the folds for the trough (2,3), the fold over the antisyphoning rod (4), and the origin of the chromatogram (5). Note that bc is well beyond the upper blade to allow firm code letters to be written on the paper. (c) The bending forceps in use. The fold ef is made by firm pressure with the outer surface of a pair of large dissecting forceps. For cleanliness the folds can be lightly scored rather than pencilled.

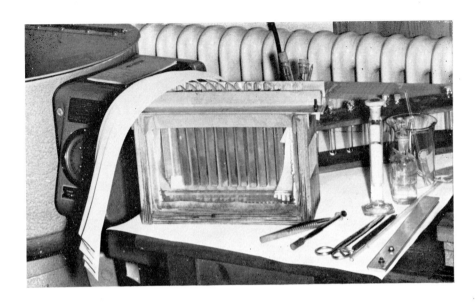

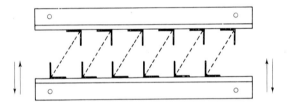

Fig. 3.14. Running-up apparatus for strips: A. General view showing strips in place and the extent of the strip-angle guides. The strip emerges for about 2 cm to reach the trough of solvent. The supporting frames are adjustable so as to take different widths of strip. B. Plan diagram showing arrangement of strip-angle guides, and paper strips in place (dashed lines).

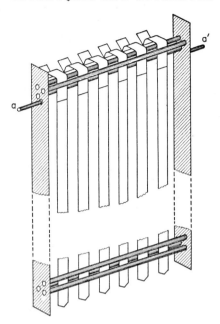

Fig. 3.15. Frame for mounting strips: Made of light aluminium and six glass rods. The two side members are rigidly connected by a stainless steel or aluminium rod at their bottom ends. These have been omitted to make arrangement of the strips and glass rods clear.
The pins (a, a') engage with notches cut in the supporting frame inside the chromatography tank.

the troughs and secured by a glass rod. Since the weight of the strips is carried by the frame they can be adjusted very easily to lie regularly and square to the edge of the trough.

After running the chromatograms they are removed from the tank (on the frame) and placed in a fume cupboard to dry. They are only touched by hand on removal from the frame and then only at points outside the zone of interest. The strips are inserted and removed from the frames in a direction that avoids scraping the main surface of the strip over the glass rods. These frames are easily made and it is hard to overemphasize their value in making the handling of large numbers of strips a clean, rapid and easy matter. Sheets of paper are easily mounted in the same way if the tanks are set up for such frames.

(d) Operation of Chromatography Tanks

The best results are obtained by having well-sealed lids and introducing the mobile phase by a pipette or funnel through holes in the lids which are stoppered or covered with a microscope slide and starch-glycerol paste. Each tank should be sealed with weights and an elastic material at the junction of lid and tank. Adhesive medical tape of the zinc oxide type is excellent since it is very slightly porous. The pressure of the vapour phase in the tanks is greater than atmospheric in the period immediately after making up the solvent system, or after removing the lid to insert or remove chromatograms. A slightly porous seal allows a slow leak of the vapour until equilibrium at atmospheric pressure is reached. If the lids are sealed with a paste or jelly and no weights, the excess pressure with these volatile solvents usually lifts the lid abruptly and the vapour is discharged so violently that more outside air is introduced before the lid falls again: it takes much longer to establish true equilibrium if this process continues. With impregnated-paper systems starch paste was found satisfactory by Burton et al. (1951a). The problem is more difficult with the volatile solvent systems; Vaseline or Plasticine are rapidly dissolved by the non-polar components, while starch paste is dehydrated by the alcohol component. The latter trouble has been overcome by using 50% glycerol for making a starch paste (Edwards, in Smith (1950)). For the last five years we have used zinc oxide self-adhesive plaster in two or three layers with excellent results; as long as solvents are not spilled on the tape the zinc oxide adhesive lasts for at least six weeks. Haslewood (1954) uses several layers of polythene and rubber.

Good equilibration is secured by lining the tanks with wads of 4–6 sheets of Whatman No. 4 paper which dip in the solvent covering the bottom of the tanks and extend to within 1 cm of the lids. To prevent these sagging to touch chromatograms it is best to press these papers against the walls of the tank with sheets of expanded aluminium alloy (an open mesh). These sheets take up the more polar of the two phases if both are present on the bottom of the tank. If the less polar phase is the lighter of the two, equilibration with this phase occurs via its upper surface and is adequate. If, however, it is the heavier of the two phases then it can only come to equilibrium with the vapour phase via the upper more polar layer of liquid. In some cases this is insufficient in practice (e.g. the systems based on ethylene dichloride for steroid conjugates), and then two sets of equilibrating sheets are needed with a separate trough, or set of small beakers, for

FIG. 3.15a. Photograph of frame shown in Fig. 3.15.

Fig. 3.15b. A set of frames with paper strips ready for insertion in a tank.

TECHNIQUES AND APPARATUS   171

the heavier mobile phase (i.e. as described originally for volatile systems; Bush, 1952). In all cases the polar phase seeps up the main sheets and equilibrates with the vapour phase from their surface.

Edwards (in Smith, 1958) recommends siliconed side-papers in the tanks. This, if it works, should soak up the non-polar phase preferentially: this would be excellent for systems in which the non-polar phase was denser than the polar phase, but not the best for the majority of systems.

Various frames for holding the troughs and glass rods are in use. Stainless steel or aluminium are best for neutral systems but have a very limited life in acid or alkaline ones. Wood soaks up organic solvents and hinders equilibration if it is bulky; it also needs prolonged drying out when the tank is used for a different solvent system. Tufnol, a bonded plastic, has been found ideal by Dr C. L. Cope (personal communication) for all neutral systems; it is easily shaped and machined but fairly expensive. The frames should be rigid, and tank and frames should be levelled individually in a way that is permanent. The tanks should be spaced about 3 in. apart to allow good circulation of air if a convecting system is in use, and covered with corrugated cardboard. This keeps the interiors dark, prevents sudden changes of temperature when people enter and leave the room, and stops radiation causing hot-spots on the internal framework or troughs.

The theoretical virtues of different solvent systems, running speeds, and so on have already been dealt with. Within certain limitations, however, numerous variations of technique are possible to meet different needs and situations. The volatile solvent systems are believed to be rather more adaptable in this sense than the impregnated-paper systems, but the latter are almost certainly capable of more variations than hitherto published in detail (e.g. Schindler and Reichstein, 1951; Hegedüs, 1953). As an illustration of what can be done, Table 3.2 lists the conditions needed to secure a good separation of six important urinary 17-oxosteroids, in all of which the relative positions are nearly identical. Also in this table are the conditions which will secure good separation of the important urinary triad tetrahydrocortisol, allotetrahydrocortisol, and tetrahydrocortisone. In general, the longer slower runs give the sharpest zones and best separations but reasonably good resolution is obtained with the fastest methods given in the table.

These are merely examples to show that the systems defined in published papers are largely convenient members of an almost infinitely variable series. It is almost certain that similar variations (largely confined to the

Table 3.2. Various Conditions which Secure Approximately Similar Chromatograms of Some Urinary Steroids

	Approximate distance moved (cm)						
Tetrahydrocortisol	9.0	25.5 °C No. 2 paper 15 hr B/50	20 °C No. 2 paper 24 hr B/50	20 °C No. 2 paper 9 hr TE11/50	25.5 °C No. 3MM paper 7 hr B/50		20 °C No. 4 paper 8 hr B/50
Allotetrahydrocortisol	12.5						
Tetrahydrocortisone	16.8						
11β-Hydroxyaetiocholanolone	17	25.5 °C No. 2 paper 12 hr LT21/85	20 °C No. 2 paper 16 hr LB21/80	20 °C No. 2 paper 13 hr LT11/70	20 °C No. 3MM paper 6 hr LT11/70	25.5 °C No. 2 paper 40 hr DX21/85	*22-25 °C No. 1 paper 72–96 hr Lig/PG
11β-Hydroxyandrosterone	22						
11-Oxoaetiocholanolone	32						
Dehydroepiandrosterone	19	20 °C No. 2 paper 18 hr L/96	25.5 °C No. 2 paper 13 hr L/96	25.5 °C No. 2 paper 16 hr LD/11/85	25.5 °C No. 2 paper 36 hr D/95	25.5 °C No. 2 paper 72 hr L/85	
Aetiocholanolone	24						
Androsterone	32						

Solvents as in glossary: L = petrol   Lig = ligroin   T = toluene   B = benzene
                          D = decalin   X = xylene   E = ethylene dichloride
                          M = methylene chloride   PG = propylene glycol

/Figure = % MeOH in water v/v., equal vol. to mobile phase
Figure/ = ratio of the components of mobile phase v/v.

These values are drawn from individual experiments in routine work and are approximate: however, they are a reasonable guide for adjusting methods and material to suit individual and local requirements. In all cases the strips were approximately 40 cm long.

* Savard, 1954.

composition of the mobile phase, however) are possible with the impregnated-paper systems. Only a few problems of separation require unique methods for their solution and are considered in more detail in Chapters 2, 4, and 6.

(e) Ways of Running Paper Chromatograms

The only essential feature of a chromatogram is a counter-current operation in an extended finely divided medium. The fact that paper presents two dimensions in which to operate, is flexible, and works mainly by capillary penetration rather than gravity or applied pressure, affords many possible ways of using it for chromatography. It is by now no novelty (see for example Martin, quoted in Lederer and Lederer, 1957, p. 134) that paper can be run upwards, downwards, sideways, horizontally (Cox and Finkelstein, 1957) radially, and even spirally (Schwarz, 1953). These methods all present special technical points of advantage or difficulty which are usually due to differences in the relative importance of the various factors affecting the achievement of equilibrium. Since difficulties of separation are commonly met with in the steroid field, these being best overcome by using overrun chromatograms with solvent systems giving low R_F values (Jermyn and Isherwood, 1949), it is usual to work by the "descending" method. Trial runs of experimental systems however are more conveniently run upwards in measuring cylinders or jam jars.

For similar reasons it is usual to sub-fractionate steroid mixtures and re-run the fractions on unidimensional chromatograms rather than use the two-dimensional method (see, however, p. 212).

The velocity of the mobile phase diminishes greatly when the solvent front reaches the bottom of a sheet of paper. This reduction in velocity is greatly reduced by clipping a "receiver" pad of filter paper to the bottom of the sheet (Hanes and Isherwood, 1949). This is inconvenient with thin strips and in our experience the most reproducible results are obtained by cutting the bottom of the strips to a blunt point.

It is often very convenient to avoid the need for sub-fractionation by "multiple chromatography" in *one* direction. This can be done with the same or different solvent systems (Jeanes *et al.*, 1951) and can be combined with subsequent development in other directions (Short, 1960—"to-and-fro'matography"!). The principal 17-oxosteroids of human urine are very conveniently handled in this way. The urine extract is overrun by a factor of 2–3 with a group-A solvent system and the strip then dried for 10–15 min at room temperature. It is then re-run (after equilibration) in a group-B system (single-length). Androsterone, aetiocholanolone, and dehydroepiandrosterone are compressed to three sharply defined bands near the front; while 11β-hydroxyaetiocholanolone, 11-oxoaetiocholanolone, and 11β-hydroxyandrosterone are well separated over the middle third of the strip. Short (1960) has described other uses for this type of procedure but on the whole it is not as good as the more usual subfractionation methods (e.g., Neher, 1958a, p. 63; Neher, Meystre and Wettstein, 1959; Chapter 6) for quantitative work.

"Double-length" tanks (Reineke, 1956; T. F. Gallagher, personal communication) are valuable but need more trouble taken to secure equilibration. There is little broadening of bands due to diffusion, and difficult separations are greatly improved. Particularly notable in this respect are the 180-cm long strips described by Reichstein's group for separating 15–20 cardiac aglycones and glycosides at a time (Schindler and Reichstein, 1951; Schröter et al., 1958). They seem to be eminently suited to the steroid field and could be used more frequently with profit.

(f) Temperature Control

For best results good temperature control is needed. This is not always emphasized in the literature and it may seem that some authors make an unnecessary fuss over this factor. The reason for this apparent discrepancy is almost always explained by the prevailing local conditions of the authors concerned. Many American workers make little or no mention of temperature control using impregnated-paper systems, and several workers using volatile systems soon discovered that the raised temperatures used originally by the author were unnecessary. Most Americans, however, are blessed with central heating and a more or less stable continental climate, ambient room temperatures being usually 24–28 °C, while the author began work in a north-west corner room in Cambridge, whose temperature fluctuated between 5 and 18 °C in winter and between 10 and 23 °C in autumn and spring.

Columns are readily jacketed with insulating material, or with fluid jackets under thermostatic control. Celite columns need more careful protection than gel or cellulose columns, and, as with tanks for paper chromatograms, care must be taken to avoid hot-spots caused by radiation from heating elements of any kind, or from tungsten lamps. Slightly variable but perfectly practical results can be obtained with paper chromatograms if the tanks are in a reasonably warm room (16–22 °C) and wrapped on all sides, top, and bottom, with corrugated cardboard, particularly if the slightly modified systems of Bush (1954a) are used. Particular care, however, is needed to avoid concealed pipes or other sources of conducted or radiated heat. The best sign of trouble due to uneven distribution of temperature is usually condensation of solvent on the walls and lid of a tank or, if the top is hotter than the bottom, evaporation from the troughs; the former is the commoner fault.

If elevated temperatures are to be used it is best to consult a standard textbook on heating and ventilation before designing the room or box to be used. The investigator will not go far wrong, however, if the design incorporates three principal features: first really rapid convection with a fan; secondly heaters with a low heat capacity; and thirdly the avoidance of radiation directly to the tanks. The fan must run all the time and *not* be switched off when the heaters are switched off by the thermostat. A room with thick walls and preferably away from an exposed corner of the building is desirable; concealed hot water pipes must be avoided, or the wall or floor over them covered with corrugated cardboard surfaced with aluminium foil to prevent radiation. The design of Edward's box (Smith, 1958) appears to follow the original faulty design of the author's, in which the heaters were directly beneath the tanks with no radiation shields.

Surprisingly crude apparatus will work if these points are observed. For example a room of approximately 600 ft^3 (18m^3) was adequately controlled by one small laboratory hot-plate set permanently at "medium" heat, and a second one, set at "low", controlled by an air thermostat; both hot-plates were placed in the draught of an oscillating office fan, about 18 in. from the fan, the whole assembly being about 36 in. from the floor. Our present assembly consists of the convector shown in Fig. 3.16 used in a room of about 400 ft^3 ($\sim$11m^3). Heating is provided by an oil-filled electric radiator controlled by a Simmerstat and six 500W tungsten lamps controlled by a thermostat. Convection is obtained by a 9 in. Ventaxia fan set at maximum speed. This fan has now run continuously for over two years without stopping and without any trouble developing. The room is provided with an exhaust fan and a set of louvres which are used occasionally to change the air when very toxic solvents are in use.

Temperature control is undoubtedly more important in the steroid field than most others because the solvent systems commonly used are more volatile, and their phase composition more sensitive to changes of temperature.

(g) General Features of Methods of Detection

The large books on chromatography cover this field in detail and many ideas for new methods can be gleaned from them. An excellent review of

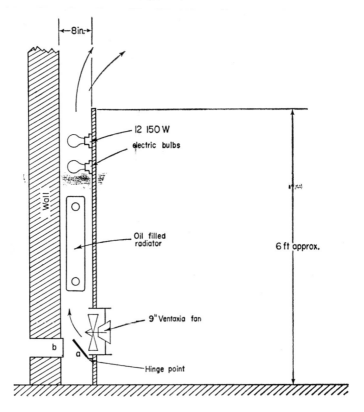

Fig. 3.16. Convector for chromatography room. (Section). This is constructed of Dexion strip and asbestos and attached to the wall of the room with Rawlplugs. Large arrows show the direction of airflow.

The deflector, a, can be lowered to allow the fan to be reversed and suck fresh air in through the louvres at b which are normally kept shut.

these methods is given by Neher (1958a, p. 54). Fussy or excessively sophisticated techniques should be avoided in general unless they are specifically required for a particular problem. For instance it is difficult to control the use of hot solutions of colour reagents and they are a nuisance in other ways: on the other hand Axelrod's method of detecting 17-hydroxy-steroids of the C_{19} series by oxidizing with a quick dip through chromic acid in boiling acetic acid is undoubtedly a valuable procedure, as are his other tests for differentiating side chains of the C_{21} series (Axelrod, 1955). The technical details described below are only given because failure to observe them may give indifferent results or complete failure.

For the last five years all colour reagents in our laboratory have been applied to paper chromatograms by dipping the paper through, or rather over, the reagent using a watch glass supported on a cork ring. Neher (1958a) and Smith (1958) also favour this method. While this method was adopted in the first place for quantitative estimation by the scanning method we also find it superior to spraying for qualitative work. If the reagent is dissolved in a good solvent for the steroids under examination, it is important to ensure that the paper is wetted from the lower surface only. This is easily done by using a large watch glass and sliding the strip down on to the surface of the reagent along the watch glass, e.g. a 15cm glass with the reagent covering the middle 7 cm or so (Fig. 3.17).

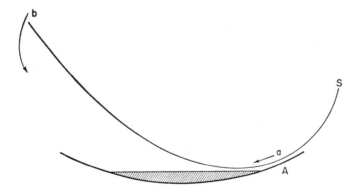

FIG. 3.17. Dipping a strip in a watch-glass. A large glass is used and the strip S is bent so as to contact the dry part of the glass at A. The point of contact, a, is then pushed forward gently until the liquid is in contact with its lower surface, by lowering S at b while compressing the strip towards a. The strip is then pulled smoothly over the reagent in the direction b→A.

In order to control the amount of reagent taken up by the paper, a lamp is placed on the far side of the glass so that when the paper is thoroughly soaked the light is reflected from the sheen on the top surface of the paper.

The paper is passed over the reagent at a speed, and with a length in contact with the surface, giving the desired degree of sheen on the part of the strip just losing contact with the surface of the reagent. The length in contact is a useful factor to control since the speed of passage cannot be less than the speed of capillary absorption of the solvent by paper; otherwise the zones will be partly eluted into neighbouring areas by the entering solvent.

With reagents in volatile solvents it is important to avoid draughts and to change the reagent as soon as appreciable evaporation has occurred (e.g. with alkaline *m*-dinitrobenzene in ethanol). This difficulty is the one disadvantage of dipping compared with spraying.

It is a common practice to cut thin strips out of wide sheets of filter paper that have been run as preparative chromatograms and use these strips to locate steroids with colour reagents or similar methods. In many cases, really accurate location is required and the following details become important. The zones on large sheets used for preparative work are rarely absolutely straight (Fig. 3.18) and apart from slight irregularities there are often lines of greater than average concentration resulting in thickened bands or some degree of edging. If the steroid to be isolated does not absorb in the ultraviolet it is useful to pencil in some of the irregularities by tracing a few of the fluorescent zones usually found, in a brief exposure to ultraviolet light (a 366-mμ source is usually the best, and least destructive to most steroids). The same zones are pencilled in on the thin strips before treating them with reagent and used to obtain an accurate relationship between the fluorescent impurities and the steroid to be eluted. Finally the test strips must be dried in such a way that their pattern is not distorted by shrinkage or movement of the coloured or fluorescent zones due to uneven drying. This is not likely to happen unless aqueous reagents are used or a prolonged period of drying is involved. If the strips are in danger of this type of distortion they must be dried horizontally on glass plates with their ends secured with spring paper-clips. The effect of bulk movement of reagent during vertical drying is easily seen by carrying out the combined tetrazolium–NaOH fluorescence test for adrenal steroids containing both an α-ketol and a Δ^4-3-ketone group, and marking the formazan spot with pinholes while wet. If the strip is hung vertically for half an hour before heating and drying, the fluorescent spot will be found well below the pinholes marking the formazan spot.

When trying out potential methods of detection, it is valuable to have a rule that chromatograms are always examined in ultraviolet light whether or not a successful colour reaction has been obtained. The NaOH fluorescence reaction was discovered in this way during an experiment to see if the conditions of the Zimmermann reaction with alkaline m-dinitrobenzene

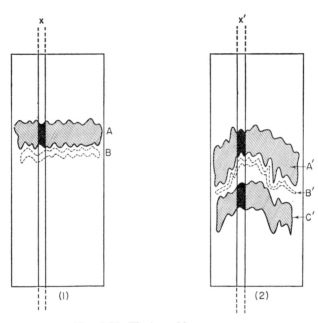

FIG. 3.18. Elution of large zones.

1. The single strip x gives only a moderately good indication of the real position of the large zone A. A fluorescent or absorbing zone B, if present, can give useful information on the small irregularities of the zone A and lead to less wastage on elution of A.

2. Seriously distorted large zones of A' and B' very poorly indicated by a test strip x' but fairly accurately defined by a fluorescent band B' correlated with the colour reaction on the test strip x'. This sort of error can also be dealt with by the excision of several test strips well spaced across the sheet (e.g. Simpson et al., 1954, Fig. 1).

could be altered to give a good colour reaction with Δ^4-3-ketosteroids. Only a faint colour was obtained with the usual reagent in alcohol but, on examination in ultraviolet light, faint yellow fluorescent spots were observed. It was easy to demonstrate that the alkali rather than the m-dinitrobenzene was responsible for this fluorescence reaction; the latter reagent, in fact, quenches the fluorescence very strongly and if the ultraviolet lamp had not been allowed time to warm up to give full intensity it is probable that the original weak fluorescence would have been missed.

(h) Special features of Methods of Detection

A wide variety of methods of varying specificity are described in the large textbooks on chromatography. Some of the most useful reagents in steroid work are summarized in Appendix III together with optimum or satisfactory conditions.

(i) The Use of Dyes and Markers

While R_F values are reasonably constant under controlled conditions, the rate of flow of solvent varies with different batches of paper, especially if the extracts being run vary greatly in weight and composition (e.g., Reineke, 1956). These effects are of little moment when single length runs are used but become important if reproducible results with overrun chromatograms are desired. It is also useful to have a check on the behaviour of columns, which vary with tightness of packing and sometimes other factors. For this purpose it is useful to have coloured substances with known R_F values to act as markers. Neher (1954) has used several commercial fat-soluble dyes for this purpose, especially for following the progress of chromatograms of adrenal steroids. The author has investigated dyes suitable for markers on chromatograms of 17-oxosteroids and conjugates; one of these turned out to be identical with Neher's F9. Many of the others were made in a search for compounds with useful R_F values. Since this work is largely unpublished these dyes are given in Table 3.3 with their chromatographic properties. The dye III2 is useful for separating the $C_{19}O_2$ from the $C_{19}O_3$ 17-oxosteroids of urine; in the system light petroleum–toluene–methanol–water (67:33:85:15 by vol.) it overlaps the tail of dehydroepiandrosterone (DHA) in a single length run. The urinary fraction A (Bush and Willoughby, 1957) is obtained by cutting the strip at the tail of dye III1 (or F11). On the routine final chromatogram of this fraction in light petroleum–methanol–water (100:96:4 by vol.) the dye has a very much lower R_F value than DHA, probably owing to the absence of an aromatic component in this system, and does not interfere with the isolation or estimation of DHA itself. The division between fractions B and C on Bush and Willoughby's pre-fractionation of urine is conveniently marked by the middle of the dye F9.

For the steroid conjugates of human urine methyl orange is very useful for marking the division between the steroid sulphates and the least polar

Table 3.3. R_M Values of Marker Dyes (for Neutral Steroids)

Dye	Colour	L/96	LT21/85	D/A90	B/50	CHCl$_3$/F	T/PG	B/F	DX21/A90
4′-Diethanolamino-2′-methyl-4-nitroazobenzene (F5)	Red	—	—	—	—	—	—	—	—
1,4-Diaminoanthraquinone (F9)*	Violet	~1·5	1·3	~2·0	−0·60	−0·75	0·68	0·03	1·1
4-Amino-1-methylaminoanthraquinone (F11)	Blue	0·95	0·19	−0·98	−0·55	−0·62	0·42	−0·12	0·53
1,4,5,8-Tetra-aminoanthraquinone (F14)	Blue	—	—	—	—	—	—	—	—
1-(2′-Methoxyphenyl)-azonaphthol (F63)	Red	−0·65	~1·6	−0·45	0·50	0·22	1·3	0·90	−0·83
4-Hydroxyphenylazobenzene (I1)	Orange	0·58	—						
4-Hydroxy-2-methyl-phenylazobenzene (I2)	Orange	0·33	—						
4′-Hydroxy-4-nitroazobenzene (III1)	Orange	0·95	0·26						
4-Hydroxy-2′-methyl-4-nitroazobenzene (III2)	Orange	0·83	0·08						
4′-Hydroxy-4,2′-dimethyl-2-nitroazobenzene (II2)	Red	0·75	−0·03						
4′-Hydroxy-4-methyl-2-nitroazobenzene (III1)	Red	0·95	0·19						
2′,4′-Dihydroxy-4-methyl-3-nitroazobenzene (II4)	Red?	1·1	—						
1′-Amino-4-nitroazonaphthalene (III3)		0·62							
4-Hydroxy-2′-methoxy-2-methylazobenzene (VI2)		0·73							
4-Hydroxy-2,2′-dimethylazobenzene (VI5)		0·47							
4-Hydroxy-2,2′,5′-trimethylazobenzene		0·67							
4,2-Hydroxy-4′-methoxyazobenzene (VII4)		0·96	−0·15						
4-Hydroxy-2,4′-dimethylazobenzene (VIII2)		0·36	1·1						
4-Hydroxy-2,2′-dimethylazobenzene (IX2)		1·6							

Sources: Dyes with F numbers—Neher *et al.* (1959), columns 5–8.
Remainder—D. J. Short and the author (unpublished).
Solvent systems—see Glossary p. xvii. The first four and last columns are the author's (unpublished) data.
All figures for 22–27 °C.

* "Lacquer violet" (I.C.I.).

steroid glucuronosides in alkaline solvent systems, especially butanol–ammonia (Bush, 1957). Methyl red runs with androsterone and aetiocholanolone glucuronosides in butanol–ammonia but with much lower relative mobility in the acid systems used for the final resolution of these glucuronosides.

Neher (1959b) marks a pencilled circle on the chromatogram before running and instructs the nightwatchman to take the strip out when a given dye has just entered the circle. We have used dyes largely as markers for the rather crude extracts of urine which are roughly fractionated before the final runs used for isolating or estimating individual steroids. It is precisely with such extracts that internal markers are needed as a check on variations in flow of solvent, edging, slight displacement, or other deviations from ideal behaviour. They have also been used to place cuts in fractions from columns used for preliminary fractionation of crude extracts; the same deviations from ideal behaviour can occur here as with paper, and the use of dyes also controls variations due to differences in packing which are difficult to avoid when a large number of columns have to be prepared in a short space of time. For accurate cuts of small fractions, calculations must be done in terms of R_M values (p. 75): for broad fractions with a large margin of safety this is not necessary.

(j) Elution of Material from Paper

Many divergent opinions are to be found on this important operation. Numerous workers have found it impossible to obtain quantitative recovery of steroids from paper chromatograms, and some have implied that others would find it equally impossible. It is the author's belief that with one exception we know the causes of such failures and that these difficulties are by no means insuperable. It seems most likely that usually at the root of such failures is the assumption that the steroid to be eluted lies in an area smaller than is actually the case (Bush 1960a; see also p. 223). It is probable that more or less irreversible adsorption occurs if the paper is dried too vigorously or allowed to stay dry for too long before attempting elution (cf. amino acids); another factor is probably the dryness of the eluting solvent. Adsorption sufficient to cause material to remain at the origin of chromatograms or to prevent complete elution can be caused by drying the area of deposition (or of a zone after running) with a draught of *warm* air rather than a cold draught. Chromatograms should never be dried with warm air before running; nor at any stage if elution is contemplated. We have found the following techniques reliable and quantitative.

The area to be eluted is marked off lightly in pencil and is made *three times* the length of the "spot" that would be visible with a method of detection sensitive to 1–2 $\mu g/cm^2$. This requirement entails accurate location of zones and separations rather better than would suffice for qualitative work or "scanning". With dirty extracts, or when other factors cause slight variations in mobility, the substance must either be a coloured derivative, or visible in ultraviolet light, or located by other techniques. Where possible, it is best to run duplicate samples and use one chromatogram of each pair to locate the steroid with a suitable colour or fluorescence reaction; this can be made extremely accurate if the desired substance is close to some coloured, fluorescent or ultraviolet-absorbing impurity, or to a suitably chosen marker dye. Another technique is to add a dye to the extract with a known mobility relative to the substance desired. It should be emphasized that accurate location is the most important step of all.

Most steroids are readily eluted from paper by simply soaking the paper in a suitable solvent (Zaffaroni, 1953; Vogt, 1955) with or without agitation. Chromatographic elution (Consden *et al.*, 1947), however, is superior in needing a much smaller volume of solvent for complete elution; a combination of the two is also very useful. The solvents to be used are given in Table 3.4. We have used two techniques for various purposes.

In the first (Fig. 3.19) the rectangle of paper is "run up" as in the original preparation of the chromatogram until the material is concentrated at the top edge (small areas); or in the middle by successive runs from each end (large areas). The only modification is that the solvent is run on to the edge very gradually at first, by standing the paper on a disk of filter paper soaked in the solvent, in order to recover material on the extreme ends of the rectangle. Several segments of paper can be "run up" in a beaker covered with a watch-glass. The area containing the concentrated material (about 3 mm wide) is now cut out with a margin of 5 mm each side and chopped into bits about 3×10 mm with scissors. These bits are placed in a small test-tube and covered with about 2 ml of solvent, some of which is used to rinse the scissors and forceps which have been used in the final manipulations. The paper is then extracted at 45 °C for 1 hr. The solvent is decanted off, and the paper residue is washed with several lots of changing solvents to ensure good mixing by convection (e.g. paper extracted with ethyl acetate–methanol is washed in turn with small volumes of methanol, ethyl acetate, and again methanol). When necessary (e.g. spectrophotometry or treatment with H_2SO_4) the combined extracts are filtered through sintered glass (porosity 3–4).

Table 3.4. SOLVENTS FOR ELUTING AND "RUNNING UP" STEROIDS

Type of Steroid	Solvent mixture (v/v)	
Sterols and esters, very low polarity; and non-polar, neutral steroids (e.g. androstenedione)	Benzene–MeOH CH_2Cl_2–MeOH $EtCl_2$–MeOH	3–9 : 1 3–4 : 1 3–4 : 1
Neutral steroids, medium polarity (e.g. cortisone)	EtOAc–MeOH CH_2Cl_2–MeOH	3 : 1 2 : 1
Neutral steroids, very polar (e.g. cortol)	EtOAc–MeOH MeOH–H_2O	1–2 : 1 9 : 1
Neutral steroids, polarity range progesterone–cortisol in presence of fat: effectively "reversed-phase" (p. 144)	MeOH–H_2O EtOH–H_2O	85 : 15–90 : 10 80 : 20
Steroid conjugates (acid) and steroid carboxylic acids	EtOAc–MeOH–HOAc–H_2O, MeOH–2N-NH_4OH	8 : 8 : 1 : 1 19 : 1

These solvents are all well tested except the first, but many variants are quite suitable for the majority of steroids. In a few cases, special solvents or mixtures may be necessary, e.g. very insoluble steroids, steroid acetates (see p. 186): but these will need individual attention as they arise.

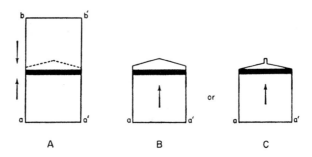

FIG. 3.19. Eluting material from paper segments. A. The material is concentrated by running up in both directions from aa′ and bb′. B, C. The segment is cut for chromatographic elution with solvent starting at aa′, and the segment mounted in the inverted position. In C a small tongue (1 × 3 mm) speeds up elution and reduces the volume needed.

In the second method the material is "run up" as above and the paper cut to a blunt point just beyond the concentrated zone (Fig. 3.19). The material is then eluted chromatographically by mounting the pointed rectangle in a simple apparatus of which many variations have been described (Consden *et al.*, 1947; Haines, 1952; Moore and Boylen, 1953; Gregory, 1955). For some purposes filtration is necessary which requires some dilution with extra solvent. In our experience, a blunt point is essential for elution in a small volume. The effect of a sharp point is shown in Fig. 3.20.

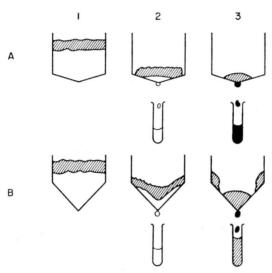

Fig. 3.20. Elution with sharp and blunt points. With sharp points material is often trapped at the edges as in B_3. Very large volumes are needed to elute this trapped material. The diagram is a sketch of an actual experiment with a dye.

Another method is to sew the tip of the paper itself into the origin of a fresh strip for further chromatography (Boggs, 1952; Oertel, 1954; Gregory, 1955).

Orthodox Soxhlet extraction is of course possible but results in much more "dirt" if the paper is not absolutely clean (i.e. nearly always!) and many more particles of filter paper. Scrupulous cleanliness and careful filtration are essential if material eluted in any of these ways is to be suitable for infrared spectroscopy or spectroscopy in concentrated H_2SO_4. Dr W. Mason (Rochester, New York, personal communication) reports that the best way to clean up such eluates for infrared spectroscopy is to take up the extract in water and extract the steroid with ether; this seems a very valuable wrinkle, and probably applicable to other forms of spectroscopy.

Some workers (personal communications) have suggested that about 5% water in the solvent is essential for eluting neutral steroids quantitatively. We have not found this necessary (except for conjugates); atmospheric moisture in England and the circumstance that elution is always carried out within an hour of removing the strips from the tanks, or after storage in the cold-room, seem to ensure sufficient wetness.

Dr R. Peterson, New York, reports that complete elution of steroid acetates is possible from chromatograms run with formamide as stationary phase, but that only 70–80% recovery is possible from chromatograms run with aqueous methanol as stationary phase. Since an adequate area was eluted in these experiments it must be presumed that quasi-irreversible adsorption was the cause. It would not be surprising, however, if formamide itself, or some other compound, added to the eluting solvent in small quantities was found to succeed in eluting this adsorbed residue. This could then be removed by the ether–water partition method. Dr Peterson has had no difficulty in the quantitative elution of various free steroids from paper, and elution has formed the basis of methods used by many workers (e.g. Zaffaroni, 1953; Bondy, Abelson, Scheun, Tseu and Upton, 1957; Lewis, 1957).

Miss M. L. Helmreich (personal communication) found that an earlier elution apparatus devised by the author (Fig. 3.21) always gave dirty eluates if paper was used as the solvent feed; carefully washed paper gave clean eluates for a short time only. This trouble was overcome by using glass wool, instead of filter paper, for the solvent feed. Glass "paper" is now available for this (Whatman).

(k) Preparative Paper Chromatography

Paper chromatography started as a purely analytical method but the convenience of filter paper as a ready-made and evenly porous material soon led to attempts to use thick sheets, or piles, of filter paper for large-scale separations. The first method was the "chromatopile" of Mitchell and Haskins (1949) in which disks of filter paper were piled on top of one another and then compressed between perforated steel plates. The sample was spread over several disks which were then placed at the top of the pile, and the mobile phase introduced with a solvent "spreader" at the top.

After passage of the mobile phase the column was broken up and the individual disks extracted and their contents checked by an analytical paper chromatogram. This technique was applied by Hassall and S. L. Martin (1951) to the separation of plant glycosides.

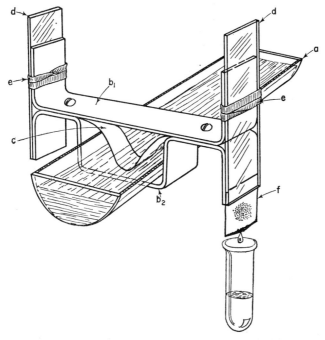

FIG. 3.21.

(See p. 186.)

Methods using blocks or rolls of filter paper run in the direction of the original paper sheets have found rather greater use, the most successful being the "chromatoblock" of von Arx and Neher (1956). The arrangement is shown in Fig. 3.22 and is usually used with propylene glycol or formamide by the impregnation method. By this method it is possible to run as much as 1·5 g of material with about 10 mg on each sheet in the block. Simpler methods use large numbers of separate sheets with the material spread as a line on the origin across a sheet of 20–40 cm wide. Alternatively, single thick sheets (e.g. Eaton and Dikeman, No. 320, used by Frierson, Thomason and Raaen (1954)) can be used and have a large capacity. Whatman papers Nos. 7 and 17 are very useful for this purpose in the author's experience. Von Arx and Neher (1956) have described an extremely useful form of trough for running many separate sheets under

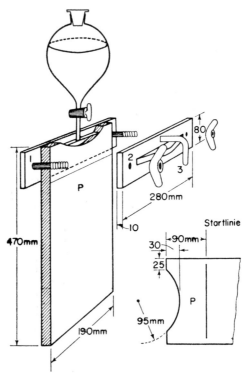

Fig. 3.22. The Chromatoblock (von Arx and Neher, 1956). 1, 2—stainless steel clamping plates; 3—overflow; P—pack of filter paper sheets. Reproduced by kind permission of Dr R. Neher and the Editorial Board of *Helvetica Chimica Acta*.

similar conditions (Fig. 3.23). This consists of a shallow tray perforated with numerous slits having their edges raised to the level of the surrounding wall of the tray. The paper sheets pass down through the slits and are folded over their edges to dip into the solvent in the tray. In this way the solvent level and hence the rate of running is kept the same for all the sheets. A horizontal arrangement was used by Guendet *et al.* (1953) for very thick sheets: at least 20mg xylose/cm width could be run satisfactorily.

The rolled paper column of Svensen *et al.* (1955) is available commercially (LKB Produkter, Stockholm, Sweden) but at present is limited

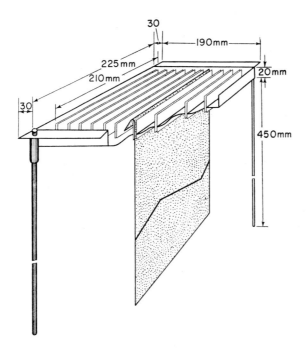

Fig. 3.23. Multi-slit trough (von Arx and Neher, 1956). A constant level of mobile phase is maintained for each sheet of paper.
Reproduced by kind permission of Dr R. Neher and the Editorial Board of *Helvetica Chemica Acta*.

in range because of its polythene construction. Unfortunately, the paper roll must be wound on its central spindle by a special machine if even running is to be obtained.

Despite the ingenuity and attractiveness of these techniques they have numerous snags and need a fair amount of skill to get good results. On the whole, the author prefers to set up a suitable partition column with Celite or silica gel whenever there is too much material for more than one or two thick sheets of paper. To avoid using very large columns, the mixture can be sub-fractionated on a medium-sized column and the sub-fractions handled separately on small or medium-sized columns. This is quickly done without the need for complicated calculations if gradient elution is used and the sub-fractions defined by a suitable set of marker dyes which have been tried out with a small fraction of the mixture on an analytical paper chromatogram.

Among the troubles besetting preparative methods using paper sheets, the following need mention. Firstly there is the difficulty of mounting very thick sheets which cannot be bent to hang in an ordinary trough without cracking. One method is to sew a thinner sheet to the thick one at the top: this can be bent to fold into a trough in the usual way and also acts as a brake on the usually rather rapid flow of the mobile phase with thick sheets of filter paper (Mueller, 1950).

Another problem is the excessive flow of solvent at the edges of thick sheets. Frierson *et al.* (1954) wrap the edges of thick sheets in platinum foil to prevent this, and it can also be overcome by careful cutting of the top of the sheet to give a tapered end. In arrangements such as that of von Arx and Neher (1956) there is also the danger of excessive solvent flow *between* the sheets making up the block: some methods get over this by compressing the whole block between large glass or steel plates. The results of these authors however suggest that this trouble can be overcome by careful choice of dimensions and other conditions. Finally all these methods need the construction of accurately machined parts in stainless steel or other more or less expensive materials, and, last but not least, they use a large amount of paper which is usually not recoverable at the end of the separation.

With very thick sheets of paper it is difficult to get an even distribution (i.e. across the thickness of the paper) of the material at the origin by the usual methods of depositing the material on the surface of the paper, even if material is deposited on both sides of the sheet. This can be seen easily with solutions of dyes. While this is partly overcome by depositing in a large enough volume of solution to "fill" the whole thickness of the paper and "running up", there is still a tendency for the material to be concentrated at the surfaces of the sheet as the "running up" solution evaporates. Surprisingly these thick sheets seem to need little or no increase in the equilibration time over that used with thinner sheets. They can also be used by the impregnation technique and in this case do need a longer time for the volatile carrier to evaporate. Here again there tends to be displacement of the stationary phase towards the surface. This is avoided if the carrier is as volatile and as non-polar as possible; e.g. methylene chloride or ether for impregnating with stationary phases of 70–90% acetic acid rather than methanol; and acetone for propylene glycol rather than methanol. Maldistribution of the stationary phase also appears to be corrected automatically if a short period of equilibration is allowed before starting the run; e.g. 60 min instead of the 10–20 min usual with thinner impregnated paper.

4. Preparation of Extracts Suitable for Chromatography

(a) General

Some adsorbents such as aluminium silicate (Kraybill, Thornton and Eldridge, 1940; Thornton, Kraybill and Broome, 1941) will adsorb steroids or other material from a crude oil or aqueous concentrate, but extracts of organic material usually require purification before they are suitable for chromatography. Partition methods require more care in this respect than adsorption methods. Even with the latter, the author wonders whether some adsorption methods are not operating in practice as quasi-partition systems, due to the uptake of water or other impurities to form an effective bulk liquid or gel phase. This seems particularly possible with silica gel, some silicates, and some preparations of alumina. The zone in which such a separate bulk phase exists probably only extends a short way down the column, but it seems probable that mixtures of hydrocarbons or chloroform with methanol or ethanol on silica gel transform the column into a partition system of the type introduced by Katzenellenbogen *et al.* (1952) due to the uptake of the alcohol by the gel. The same is true for water in the extract or solvents used with silica gel, not to mention incomplete drying of the gel itself.

There is little evidence in the literature of trouble with displacing agents in adsorption chromatography of steroids. This is probably because it has been usual to operate at large adsorbent: adsorbate ratios with the "liquid" chromatogram technique of Reichstein, and because displacement or carrier-displacement chromatography has not been used for steroids.

Partition chromatography on the other hand is mainly sensitive to substances which alter the composition of the phases, and in paper chromatography to impurities catalysing oxidation because of the large surface over which the solutes are exposed to the atmosphere. With amino acids and carbohydrates the main impurities giving trouble are inorganic salts. With steroids it is more usually lipid material, but salts are also troublesome with the conjugated steroids and glycosides if present in the starting materials. Since the type of impurity depends more upon the starting material than on the type of steroid, specific problems will be grouped under the different sources from which steroids are commonly extracted.

(b) Animal Tissues

Soft tissues are easily extracted by one of the classical methods used for adrenal glands, testes, ovaries, or corpora lutea (Pfiffner, Vars and Taylor, 1934; Pfiffner, Wintersteiner and Vars, 1935; Grollman, 1937; Kendall, 1937; Reichstein, 1938; Gallagher and Koch, 1934; David, 1935; Allen, 1930, 1932; Neher, 1958b). For non-polar steroids such as progesterone or sterols and their esters the usual aqueous ethanol (or methanol)/hexane (or light petroleum) partition to remove glycerides gives poor yields unless numerous extractions and back-washings are used. Good yields are obtained, however, by cooling a solution of the extract in 70% aqueous methanol to $-15\,°C$ (Butt *et al.*, 1951) which precipitates glycerides and leaves small concentrations of steroid in solution. We have found a preliminary chromatogram on a silica gel column useful (see below). Alternatively a reversed phase chromatogram (Morris and Williams, 1955) on siliconed Celite, or with filter paper impregnated with high-boiling paraffin (R. P. Martin, 1957) can be used.

In unpublished experiments on whale adrenal glands with M. Brachi, the author found that excellent extraction of steroids was obtained by macerating the chilled glands in ethyl acetate in a Waring Blendor using the low speed to avoid emulsification. Extraction and maceration were carried out with five volumes of water-saturated solvent for 5 min. After decantation, the extraction was repeated twice, the final extract being collected by filtration through Hyflo Supercel on Whatman No. 1 filter paper, under suction. Yields were the same as with orthodox extraction with dehydrating solvents and the extract contained much less triglyceride, phospholipid, and water soluble material. It is important to use *water-saturated* ethyl acetate and to use the low speed on the macerator to avoid emulsification. This method has also been applied successfully to muscle, kidney, liver, heart, spleen, lungs and brain. Yields of cortisol and corticosterone were the same as with acetone or ethanol extraction and the extracts were far easier to purify. It should be emphasized however that it is not yet known whether this method is successful with other types of steroid, nor whether it extracts all such steroids in tissues; it is only claimed to be the equal of the more conventional methods and far more convenient in terms of ease of purification.

Edwards (in Smith, 1958) suggests a similar method using chloroform. In our experience, this is a poor extractant of the very polar corticoids (cf. Burstein, 1956) and much more prone to form emulsions.

(c) Blood and Plasma

Numerous methods have been described for this material (Table 3.5)

Table 3.5. Typical Methods for Extracting Neutral Steroids from Plasma

Nelson and Samuels (1952)	Dilute with equal vol. water. Extract with $CHCl_3$–ether (1 : 4 by vol.), evaporate. Partition 70% EtOH/hexane. Evaporate EtOH.
Bush (1952)	(a) 4 vol. MeOH: filter, evaporate. Acetone: $MgCl_2$ precipitation of phospholipids. Partition light petroleum/60% MeOH: partial evaporation. Extract with EtOAc. (b) Dilute with 1 vol. H_2O. Extract 1 ×6 vol. $CHCl_3$. Acetone, etc. as in (a).
Bush (1953)	(c) As (b) but using EtOAc and omitting acetone-precipitation step.
Morris and Williams (1953)	3 vol. EtOH: partial evaporation. Extract ethyl acetate: evaporate. Dissolve 70% MeOH or 20% EtOH and freeze out fat. Wash with CCl_4 at 50% EtOH concentration. Evaporate ethanol. Reversed-phase Celite column to remove traces of fat, etc.
Bush and Sandberg (1953)	Extract 3 ×3 vol. EtOAc: evaporate. Silica gel column to de-fat. Elute corticoids, etc. with EtOAc–MeOH.
Levy and Kushinsky (1954)	Various large-scale methods—dialysis, charcoal adsorption, use of isopropyl acetate, etc.

and the problem varies considerably with the type of steroid under consideration. In view of earlier publications (Bush, 1952) it seems important to emphasize that methods based on ethanol or acetone extraction are inefficient for the extraction of free neutral steroids because equally good yields are obtained with other methods that yield far purer extracts. An important new technical advance discovered simultaneously by several workers is the addition of sodium hydroxide to plasma before extraction with an immiscible solvent. This reduces very considerably the amount of triglyceride and other undesired lipid that is extracted while giving full recovery of neutral steroid hormones such as progesterone, androstenedione (Short, 1957) and cortisol, and corticosterone (Bondy *et al.*, 1957; Bondy and Upton, 1957; Somerville and Deshpande, 1958). Chloroform or methylene chloride extracts of plasma are usually cleaner than ethyl acetate extracts whether or not sodium hydroxide is used, but depending upon the lipid content of the sample; thus Silber (personal communication) finds such extracts are sufficiently clean to allow the estimation of corticosterone and cortisol by fluorimetry in concentrated sulphuric acid without any chromatographic step, and obtains results in fair agreement with methods using chromatographic isolation. The addition of NaOH appears to stabilize plasma lipoproteins which undergo an irreversible change (Dr L. Vallee, London, personal communication). Some of the newer methods suitable for extracting small volumes (5–10 ml) of plasma are given in Table 3.6. These methods all give extracts which are suitable for paper chromatography without the need for paraffin/aquous alcohol partition, or silica gel chromatography, as preliminary purification steps. The extraction of large amounts of lipid however cannot be prevented by NaOH if the plasma is too old, and we are in the habit of extracting plasma within a few hours of collection. Certain post-prandial samples also give rise to large amounts of non-polar lipid whether or not NaOH has been added before their extraction. Such samples need a preliminary purification before chromatography.

Fresh plasma contains all the cortisol and cortisosterone found in blood by specific means (Bibile, 1953) (cf. radioactivity after injected [4-^{14}C]-cortisol, Peterson *et al.*, 1955; Migeon *et al.*, 1956) so that separation of plasma by centrifugation is a valuable step. The extraction of whole blood yields a far less satisfactory extract by all these methods. In some recent experiments in which the author used whole blood very low recoveries of added sub-microgram amounts of steroid were obtained (30–60%), suggesting that poor extraction or actual destruction of cortisol was

Table 3.6. NEWER METHODS FOR EXTRACTING NEUTRAL STEROIDS FROM BLOOD PLASMA

Bondy and Upton (1957) Bondy et al. (1957)	Add 1/20 vol. N-NaOH. Extract 3 × 1 vol. CHCl$_3$. No washing of extract.
Short (1957)	Make plasma to 0·45% (w/v) NaOH with 0·5 N or 1·0 N-NaOH. Extract 1 × 1 vol. ether (progesterone, etc.). No wash. Partition hexane/70% MeOH.
Lewis (1957)	Extract 2 × 4 vol. EtOAc. Wash N-NaOH, 2% HOAc, H$_2$O. Reversed-phase "run up" on siliconed end of paper strip.
Peterson et al. (1957)	Extract 1 × 5 vol. CH$_2$Cl$_2$. Wash dilute NaOH, HOAc, H$_2$O.
Peterson (1957)	Extract 1 × 2·5 vol. CCl$_4$–CH$_2$Cl$_2$ (1 : 1 by vol.). Wash N/100-NaOH, N/50-HOAc, H$_2$O (extracts corticosterone almost selectively).
Bush and Mahesh (1959a)	Extract 2 × 3 or 1 × 6 vol. ether–EtOAc (2 : 1 by vol.). Wash NaOH, H$_2$O. Add HOAc (trace). Reversed-phase "run up" on untreated paper.

The advantages of EtOAc were recognized by early workers on adrenal extracts and rediscovered for blood and urine by Bush (1953), Meyer (1953) and Burstein (1956).

CH$_2$Cl$_2$ (and similar solvents) is both adequate for extracting cortisol and gives rather less unwanted material of a polar nature: *without* NaOH treatment of the plasma, however, it extracts *more* non-polar material than EtOAc.

occurring. This problem needs further investigation with sub-microgram amounts of steroid: such difficulties have not been found with larger quantities of steroid.

In the author's opinion the more complicated methods of extraction by means of charcoal adsorption or dialysis (Hechter, 1949; Zaffaroni, 1953; Grollman, Firor and Grollman, 1935; Levy and Kushinsky, 1954) have nothing to recommend them now that it has been shown by large numbers of workers that clean extracts of plasma can be obtained in a fraction of the time by simple extraction with immiscible solvents.

The extraction of oestrogens from blood or plasma is accomplished by various means (e.g. Brown, 1955) but the problem is complicated by the fact that some oestrogen at least is "bound" in a way that resists direct extraction by organic solvents (e.g. Szego, 1956). It is, however, by no means clear whether this "bound" oestrogen is "free" or "conjugated" and, until its nature is firmly established, it seems unwise to prescribe a rigid routine.

The extraction of steroid conjugates from plasma is also still in its infancy and little guidance can be given until further work has been done. Unpublished work by Miss Muriel Gale and the author has shown that recovery of DHA sulphate by direct extraction of plasma with n-butanol is incomplete, even when saturation with salts and acid conditions are used. This conjugate is completely extracted, however, by adding four volumes of ethanol to the plasma, centrifuging, and re-extracting the residue with ethanol. Eik-Nes and Oertel (1958); Oertel and Eik-Nes, (1959) have published similar results although they suggest that a major part of their material is DHA phosphate. The major part of conjugated etiocholanolone however is not extracted by this procedure and further work is needed. Ethanol extracts are readily purified and concentrated by the procedure given in Table 3.7. This procedure recovers all the discoverable DHA sulphate of plasma but has not been tested with other conjugates.

Bile acids present an even more difficult problem (Wootton and Osborn, 1959) but their method gives reproducible though low recoveries.

It should be emphasized in conclusion that these methods have been considered largely from the point of view of their success or otherwise in providing an extract suitable for chromatography, particularly on paper. Detailed consideration of the yields and types of steroid obtained is beyond the scope of this book and it should be realized that, with the possible exception of some steroid hormones, much work is needed before the significance of the steroids extracted by these methods is fully understood.

(d) Sweat, Faeces, Urine

Sweat can be collected by washing the skin with water or acetone, the latter also extracting much sebum and surface lipid, or by swabbing with tissue and squeezing out the latter. Little work has been done on the

TABLE 5.1. EXTRACTION AND PURIFICATION OF CONJUGATES FROM PLASMA (HUMAN)

250 ml plasma
†Add 700 ml EtOH. Stir.
Centrifuge ×10 min.

Supernatant

Evaporate at 45 °C in vacuum to 150 ml

Leave at 0–2 °C for 16 hr. Centrifuge briefly or filter off separated fat, etc.

*A Pass through 20 × 1-cm column of 15-ml bed volume "Decolorite" resin (OH⁻ or Cl⁻ form) at 2–3 ml/min. Wash column with 40 ml H₂O. Elute conjugates with 250 ml MeOH–conc. NH₄OH(0·880 sp. gr.) (9:1 by vol.) Evaporate 10 min at room temperature in vacuum, then complete at 40–45 °C.

Fractionate on 15-cm wide sheet Whatman No. 3MM in butanol/2N . NH₄OH 22–25·5 °C. Equilibrate 5 hr: run 11 hr (×1).

Sediment
Add 6 vol. EtOH and stir well
Centrifuge as above

Supernatant

Sediment
Reject

*B Half-saturate with (NH₄)₂SO₄. Make to 2N-NH₄OH Extract 3 × 300 ml EtOAc. Wash extract 1 × 50 ml 2N-NH₄OH. Filter through Whatman No. 2 paper. Evaporate 40–45 °C in vacuum.

Extract
Sulphates

Aqueous residue
Glucuronosides, etc.

Miss M. Gale and author (unpublished); see Eik-Nes and Oertel (1958) for similar method. Method B: gives relatively pure sulphate fraction free of glucuronosides. See also Burstein and Lieberman (1958).   † A plasma–ethanol ratio of 1:4 is better.
* Method A: gives crude conjugate fraction.

steroids of sweat but aqueous skin washes or swabbings yield easily purified extracts if ether, methylene chloride, or ethyl acetate is used as extractant (Lewis and Thorn, 1955; Nakaga, 1958). Acetone washes would require the removal of much sebaceous lipid by one of the methods used for plasma extracts.

Faeces are usually extracted by one of the methods suitable for soft tissues. In the author's limited experience the extracts are more tedious to purify than urine or blood extracts but present no more difficulty than soft tissue extracts. Silica gel chromatography is an especially useful preliminary step.

Urine presents a field too large for complete discussion so that only a few typical problems will be considered (see e.g. Lieberman and Teich, 1953). A detailed system of examining urine will be suggested in Chapter 6.

Free steroids are readily extracted by any of the methods for plasma. S. Burstein (personal communication) reports that exceptionally clean extracts are obtained by chloroform or ethyl acetate extraction if the urine is collected in a bottle containing enough sodium chloride to provide a saturated solution of the salt. For small volumes (10–50 ml.) the author extracts urine collected under toluene with one-fifth volume of hexane or light petroleum to remove toluene and non-polar lipid and then once with five volumes of water-saturated ether–ethyl acetate (2: 1 by vol.). This is washed once with one-twentieth volume N-NaOH and twice with saturated NaCl. The extract is then dried with about 2 g anhydrous sodium sulphate, filtered, and two drops of glacial acetic acid are added before evaporation under reduced pressure. If the non-polar steroids such as progesterone or 2α-methyladrenosterone are to be recovered the toluene preservative and hexane wash must be conserved and worked up separately. Silica gel or reversed-phase chromatography may be necessary as preliminaries to paper chromatography.

Extracts prepared by continuous ether extraction at pH 1–2 usually need considerable preliminary purification before chromatography on paper, but are pure enough for adsorption chromatography or partition column chromatography if suitably washed and dried. Apparatus for emulsion-free extraction is described by Cohen (1950) and Hill (1952).

Extracts of urine which have been treated with one of the usual preparations of glucuronidase are easily purified for chromatography. In our experience the extract needs three washes with one-twentieth its volume of N-NaOH, and two of water or saturated NaCl for the best results (Bush and Willoughby, 1957). These conditions need some explanation since lower concentrations of alkali are more commonly used. It was found that an important factor in the successful removal of impurities (presumably phenolic or giving rise to quinones) was oxidation during the NaOH washing, indicated by the very strong colour of the aqueous layer. If dithionite ("hydrosulphite") was added to the alkali, the wash layer remained colourless and the residue remaining after evaporation of the extract was far greater in amount than after washing with N-NaOH alone. In our experience the extract from 1/20–1/10 of a 24 hr specimen can be run directly on the full width of a 5×40 cm strip of Whatman No. 3MM paper (Bush and Willoughby, 1957) but occasionally urines are encountered which give extracts which need a 10 cm wide strip for good results.

Some degree of flexibility is desirable if a method of extraction is to give the greatest efficiency with a particular problem. The method of Bush and Willoughby (1957) for instance is designed to give an extract clean enough for preliminary separation into useful groups while obtaining reasonable and reproducible recoveries of steroids as different as androsterone and tetrahydrocortisol. Recovery of cortolone, however, is low (55%) and if it was desirable to study this and other more polar steroids in greater detail it would be better to substitute ethyl acetate for ether–ethyl acetate (Burstein, 1956b). This would give a dirtier extract but this could then be overcome by adding a silica gel chromatogram or a partition between benzene and water to the method. Similarly, a routine method concerned only with 17-oxosteroids could use ether or benzene as extractant and obtain a cleaner extract initially (Lieberman et al., 1958; Kellie and Wade, 1957).

Many workers still find Girard's reagent useful for obtaining a clean ketonic fraction from urine. While this is obviously useful for isolation work, it is not a method which is convenient or even reliable for most estimation methods. Many steroids undergo serious alterations during the hydrolysis of their Girard derivatives; for instance 16α-hydroxyoestrone rearranges to 16-oxoestradiol, and Zaffaroni (1953) found that the orthodox procedure gave rise to nine products when applied to deoxycorticosterone. Complete hydrolysis of the derivatives is sometimes difficult to achieve without employing undesirably vigorous conditions (see Brooks (1958) for a careful investigation of this). In the author's experience, this procedure can always be circumvented by appropriate chromatographic procedures, and has not been necessary for methods of microestimation of steroids in blood and urine (see Chapter 6). If this procedure is unavoidable, then the hydrolysis of the condensation products with neutral formaldehyde appears to offer considerable advantages over the older procedure (Teitelbaum, 1958). On the other hand, Lindner (1960) found that hydrolysis of the Girard hydrazones of 17-oxosteroids was far from complete: while this procedure has a certain usefulness for the selective extraction of some types of oxosteroids (particularly Δ^4-3-ketones) it cannot be used at present with confidence for isolating oxosteroids of unknown constitution.

CHAPTER 4

QUANTITATIVE CHROMATOGRAPHY: COLORIMETRIC AND RADIOISOTOPE TECHNIQUES

1. Introduction

ALMOST all methods of quantitative estimation involve a compromise in the effort expended on two parts of the process, namely a specific way of obtaining a measurable property of the substance to be determined, and a specific way of separating that substance from others interfering with its determination. The two extremes of this compromise are seen in the flame photometric estimation of sodium in plasma, which involves no separation process at all but uses a very specific means of eliciting a physical property of sodium; and in the estimation of hydrocarbons in petroleum fractions by gas–liquid chromatography in which a highly selective method of separating the desired fractions is coupled with one of several methods of determination (e.g. gas density balance or ionization detector; James and Martin, 1954; Martin and James, 1956; Lovelock, 1958) which are almost totally non-specific. These two basic parts of a quantitative determination involve a compromise because, for a given degree of specificity, the effort spent on one leads to less effort being necessary on the other. Both, however, are influenced by other important aspects of such methods, namely the quantity of the substance in question and the material on which the estimation is carried out. Finally the choice of method is influenced considerably by the accuracy and specificity required, the number of estimations to be done, and the time and facilities available.

The quantitative estimation of amino acids in order to determine the composition of a protein, for instance, requires a very high degree of accuracy, and such estimations are done relatively infrequently. In such a case, laborious and sophisticated techniques are essential and therefore reasonable (Chibnall, 1954). The detection of changes in the urinary excretion of metabolites of adrenal steroids in human beings during stimulation with adrenocorticotrophic hormone requires a simpler technique since the normal response involves five- to ten-fold increases in excretion. On the other hand the estimation of aldosterone needs a much more complicated method because, although significant changes in excretion are equally large, the amounts are very small and numerous substances interfere with its estimation. The disappearance of the Δ^4-3-ketone group of steroids incubated with dilute liver homogenates can be studied by simply extracting the medium with an organic solvent and following the change in ultraviolet absorption at about 240 mμ: on the other hand the disappearance of this group in steroids incubated with sewage would need a far more specific method. It is important, therefore, to describe the limitations of quantitative methods and, if possible, to give some idea of the type of compounds likely to interfere with them; others can then judge more easily whether the method will suit their purposes with the materials they wish to study.

Until recently, quantitative methods for estimating steroids in small amounts have been unsatisfactory, particularly in blood and urine; the considerable technical problems involved with these important fluids have tended to lead many workers to develop rather complicated techniques in the search for an "ideal" method free from all snags and interference from other compounds. It seems doubtful if such specialized solutions to the problem are always absolutely necessary (they sometimes probably are) and this chapter will deal with what is believed to be a useful general approach to the problem, rather than with a detailed review of the methods that have been proposed. Steroid workers who become depressed in what sometimes may seem an endless search for an "ideal" quantitative method can rapidly recover their spirits by reading the analytical section of *Chemical Abstracts*: although they probably dimly remember estimating copper or iron at school they will find that others are apparently still dissatisfied and writing busily on their ingenious efforts to improve what they consider to be a dismal situation.

The main feature of chromatographic methods of estimation is of course that the effort needed to obtain a highly selective means of separating the substance it is desired to estimate is enormously reduced. However, since the problems to be tackled mostly involve microestimations and often involve a material such as urine in which many other steroids are present, not to mention an extremely variable number and quantity of other compounds such as metabolites of drugs and foods, it is unusual to find a problem in which one can rely entirely on the chromatographic step to provide all the specificity desired. Similarly, although numerous methods of determining steroids by fluorimetry or absorptiometry are known, few if any of these are so specific as to remove all doubts about them. Indeed all such methods for steroids are specific for groups of steroids rather than individual compounds, and all of them are capable of estimating, and hence of being interfered with, by classes of organic substances other than steroids. This problem "will always be with us", so it seems best to try and outline a general approach to it rather than expect that a rigorously controlled and specialized technique must always be developed to solve each problem as it comes along.

It will be taken as agreed that quantitative methods will depend upon a chromatographic step for a large part of their specificity with most problems in steroid biochemistry. Enthusiasm for this technique should not, however, blind us to the great possibilities of chemical specificity, especially in situations where circumstantial knowledge is available to the effect that interference from other substances is likely to be small, or that measurement of a group of steroids rather than of individual members is all that is needed. Outstanding examples of the use of chemical specificity are the colorimetric estimation of corticosterone and cortisol in plasma by the method of Peterson *et al.* (1957) and the spectrophotometric methods for various groups of 17-hydroxy-C_{21} steroids in urine developed by Norymberski and his group (Norymberski, 1952, 1953; Appleby, Gibson, Norymberski and Stubbs, 1954; Appleby and Norymberski, 1955; Norymberski and Stubbs, 1954, 1956a, b). The chief aim of this chapter, however, will be to decide how chromatographic methods can be used to obtain the maximum possible specificity of the separation steps needed for the microestimation of steroids, and chemical specificity will be treated as subsidiary to this.

2. General Procedure

The main problem to be solved is independent of the actual chromatographic technique employed and is more a matter of careful design of the method as a whole. This is that, however good one's technique, many problems of importance deal with materials that may contain substances that interfere with the final method of determination, and which are not completely separated from the steroid under study in one or several solvent systems. This sort of interference can often be shown to be insignificant because of the nature of the material studied, or because the steroid concerned is in large concentration; and it is of course always reduced by adopting techniques giving high resolving power and chemical specificity.

The number of practical situations in which it is serious, however, is large, and the number of conceivable interfering substances enormous, so that the general procedure for quantitative chromatographic estimations should attempt to minimize this type of interference. The fact that we shall often find it possible to take short cuts in no way reduces the value of such a procedure as a general approach to new problems; it is easy enough to take a short cut or make a simplifying modification, but it is often difficult to avoid getting hopelessly bogged down if a piecemeal empirical attack on a difficult problem fails to work after several attempts with varying conditions.

The suggested solution to this general problem is a simple one and one that has already been used in several quantitative methods for steroids. It is that two chromatographic steps should always be used and designed so as to give the maximum difference in the relative contributions of the different substituents to the R_M value of the substance to be measured: furthermore that in general the best way of achieving this is to modify the substance chemically between the two chromatographic steps in the procedure, thus using ΔR_{Mr} effects as well as ΔR_{Ms} effects (Chapter 2). In most cases the greatest specificity will be achieved by using two solvent systems in which the relative ΔR_{Mg} values of different types of substituent are as different as possible. In situations where highly specific colour reagents exist and the compounds under study are present in relatively large concentrations one or both of these features may be dropped if they are found to be inessential. Before going into the details of such methods it is worth considering how certain published methods fit this approach.

The methods of Brown (1955) for oestrogens in blood and urine, and of Klopper, Michie and Brown (1955) for pregnanediol in the same two fluids, both illustrate the principle of two chromatographic steps with a modifying step between them, with one or two interesting differences of detail. In Brown's method careful use of solvents and control of pH in the extraction and washing steps leads to two fairly pure phenolic fractions, one containing oestrone and oestradiol, and the other oestriol. In this method, the modifying step results in such a large change in solubility properties that orthodox partition between solvents achieves a considerable degree of purification and effectively substitutes for a chromatographic separation step: the method used is the methylation of the phenolic hydroxyl group of the oestrogens. In addition, alkaline peroxide is used to remove impurities (see below). It is worth noting that methylation changes the most polar type of hydroxyl group in the molecule into a non-polar ether group leaving the less polar secondary hydroxyl groups unchanged. Only some alcohols and phenols will give non-polar derivatives with this reagent under these conditions; acetylation would not distinguish between alcoholic and phenolic hydroxyl groups and would besides give the less easily separated polyacetates of oestradiol and oestriol. The final step in Brown's method is the separation of the oestrogen ethers by elution from a column of alumina.

In the method of Klopper et al. (1955) a less specific modifying step, namely acetylation, has to be used since pregananediol offers only secondary hydroxyl groups for chemical manipulation. This step therefore has to be both preceded and succeeded by chromatographic steps, in this case separation on alumina columns. An additional feature of this method, as in Brown's (above), is the inclusion of a special type of modifying step, namely one designed to modify undesirable interfering substances while leaving the substance to be measured unchanged. In this case it was found that such material was removed by oxidizing the extract with alkaline permanganate before the first chromatographic step.

These two methods are not only elegant and successful but valuable examples of the argument of this chapter. Thus it is not accidental that the colour reactions used for oestrogens and pregnanediol are both very easily interfered with by numerous unknown substances in natural extracts; further, both methods aim to measure very small quantities of steroid in the presence of large amounts of unwanted materials. These two factors cause the specificity of each method to depend very largely upon the efficiency of the separation steps employed, and hence justify the elaboration of this part of the method.

Both the previous examples use adsorption chromatography but it is probable that equally efficient methods could rapidly be developed using partition chromatography if similar modifying steps were employed (see p. 315). An example of these principles using partition chromatography is the method of Peterson et al. (1957b, 1959) for aldosterone in blood and urine. Similar principles underlie the method of Ayres et al. (1957a), Ayres, Pearlman, Tait and Tait (1958), but Peterson's method is more instructive since it also illustrates the way such principles can be adjusted to suit the powerful method of labelling with two isotopes. The extract of blood or urine is acetylated with tritium-labelled acetic anhydride and a known amount of aldosterone [*carboxy*-^{14}C] diacetate added at the end of the acetylation. The acetates are then run on a paper chromatogram for 14–18 hr after 1 hour's equilibration; adrenosterone is run as marker, to avoid using the rather rare aldosterone diacetate itself, and is detected with short-wave ultraviolet light (2537 Å). The zone is eluted and run on a second chromatogram in which the diacetate runs at the same rate as 11β-hydroxyandrostenedione, which is used as the marker (p. 180). After elution and evaporation the material is oxidized for 10 min with chromic acid in acetic acid (Mattox *et al.*, 1956). The oxidized product (the lactone monoacetate, see Fig. 5.10) is then run again in the first solvent system with 11α-hydroxyprogesterone as marker, for 24–36 hr. The eluate from this chromatogram is then dissolved in a phosphor and the tritium and carbon-14 measured in the Packard Tricarb Scintillation Counter. (This instrument has a particle energy discriminator which enables the two isotopes to be counted simultaneously; a less satisfactory alternative would be to use a combination of counters suitable for measuring the isotopes separately.)

The first two solvent systems are chosen to give a definite change in the type of solvent–solute interaction involved (see p. 118); this is shown by the fact that the desired substance moves with very different (chromatographically speaking) marker substances in the two systems. The modifying step is the highly characteristic production of the lactone-monoacetate by oxidizing the diacetate of aldosterone with chromic acid. In order to be able to use an internal radioactive control the equally useful modifying step of acetylation could not be used between two chromatographic steps because labelled aldosterone was not yet available and the diacetate made with ^{14}C-labelled acetic anhydride had to be used.

Once again it is found that a highly selective means of separation is needed when the material to be measured is present in very small quantities, and when the method of estimation is relatively non-specific. In this case, for instance, the method of labelling with radioactive acetic anhydride (Avivi, Simpson, Tait and Whitehead, 1954) is a method of general application to all esterifiable substances, and it or similar agents such as those based on *p*-iodobenzenesulphonic acid (Keston, Udenfriend and Levy, 1947, 1950; Fresco and Warner, 1955; Leegwater, 1956) have been used for other steroids and for amino compounds. Another important feature of the method of Peterson *et al.* is the use of long runs in solvent systems giving low R_F values. In the second chromatogram, only part of the aldosterone diacetate is cut out and eluted. This is a general feature of double-isotope methods (or any other type of method using an internal control) since there is no need to recover more material than is required by the sensitivity of the method of measurement.

Other methods for aldosterone may be briefly compared with the foregoing one. Methods of the type of Neher and Wettstein (1956a) and Moolenaar (1957) use no modifying reaction step and seem to be unsatisfactory in principle; they are often unsatisfactory in practice (see e.g. Nowaczinsky, Steyermark, Koiw, Genest and Jones, 1956). Methods of the type of Ayres *et al.* (1957) use one or two partition chromatograms followed by acetylation and chromatography of the acetate on columns, paper, or on both. The final estimation is made by spectrophotometric or fluorimetric methods according to the quantity available. The biggest differences in relative R_F values of aldosterone and its usual companion substances seem to be obtained by using first an acetic acid system and then one of the t-butanol systems. However it is not certain that two chromatograms before acetylation are always really necessary since much published work appears to have taken little or no advantage of the use of overrun chromatograms. These and other matters will be considered in more detail in Chapter 6.

With the 17-oxosteroids of urine most published methods use a far simpler type of separation and rely on the specificity of the Zimmermann reaction to a considerable extent. Some methods with alumina columns confirm the identification of each peak by using part of each for an infrared absorption spectrum. The common 17-oxosteroids of human urine, however, make up by far the greater proportion of all urinary substances giving a typical colour with the Zimmermann reaction (cf. other species in which β-ionones are the larger fraction of such substances (Klyne and Wright, 1957)) so that a good chromatographic system should give a reliable identification of each substance, and infrared absorption spectra seem unnecessary for the majority of routine purposes. A more elaborate scheme of separation would, however, seem desirable for special tasks such as the estimation of very small amounts of 17-oxosteroids in the urine of patients who have been gonadectomized and adrenalectomized, or hypophysectomized.

Furthest from the ideal chromatographic method come those methods which rely almost entirely upon the specificity of a spectrophotometric or fluorimetric determination, and in which the chromatographic separation is poor. Typical of such methods is that of Nelson and Samuels (1952) for 17-hydroxycorticosteroids, (i.e. with a dihydroxyacetone sidechain), in blood and its modification for urine (Glenn and Nelson, 1953). In this method the neutral extract is put on a column of Florisil and the column washed with chloroform. The desired fraction is then eluted with chloroform–methanol and the steroids estimated with the Porter–Silber reaction. The first eluate contains most of the common 17-oxosteroids and some deoxycorticosterone; the second, however, contains almost the whole gamut of possible corticosteroids (Bush, 1954b). Nevertheless the method has been of enormous usefulness because of its simplicity and the fact that except for low values in blood there is little interference from other substances giving this colour reaction. The extension of the method to other classes of steroid in the second eluate using the reduction of tetrazolium salts or ultraviolet absorption (Weichselbaum, Margraf and Elman, 1953) is however of very doubtful value since the specificity of these spectrophotometric methods is low, and numerous materials interfering with such methods exist in blood (Bush and Sandberg, 1953).

Finally we pass without any really sharp dividing line on to those methods mentioned previously which do not depend upon a chromatographic step at all. These are outside the field covered by this book but they are useful in showing that chromatographic methods simply represent one extreme of the compromise mentioned at the beginning of this chapter; namely an extremely powerful means of increasing the specificity of a method of estimation by concentrating on the selectivity of the separation steps of such methods. It is not always essential to use such an approach.

An example of a successful but rigid technique is that of Aitken and Preedy (1956, 1957) for oestrogens. Here a very closely controlled gradient-elution partition chromatogram on a Celite column is used. Numerous variations of this solvent system and of the gradient failed to separate certain oestrogens from common components of urine extracts which interfered with the fluorometric method of determination that was employed. While this method is obviously an elegant one for human urine, it is typical of such specialized solutions of a particular problem in that it may be inapplicable to other species of body fluids, and is in danger of being found useless for urines containing metabolites of certain drugs. Each new application of such a method may need an extensive period of development and modification.

3. Chromatographic Selectivity

The problem can be considered in two parts. Firstly the two solvent systems used should give the largest possible difference in the relative ΔR_{Mg} values of the different groups found in the steroids. Secondly they should give the largest possible differences in the ΔR_{Mg} values of groups found in non-steroidal substances. Certain groups of course are found in both steroids and other classes of substances, but we have seen in previous chapters that there are good grounds for supposing that they will tend to show similar behaviour in all classes of compounds. For this type of substituent group, therefore, we have only to look after the steroids and "the rest will take care of themselves".

The commonest groups with which we are concerned are, of course, ketone and hydroxyl groups in the steroids, and there exist important classes of steroids in which we also have to consider phenol, carboxyl, lactone, ester, and conjugated sugar or acid groups. It was seen in Chapter 2 that although a few sets of steroids usually exist whose relative R_M values in one system are reversed in another, the biggest general difference was found when comparing the systems based on t-butanol or acetic acid with the rest. It was also suggested that systems with the highest concentrations of t-butanol gave the most striking differences. Since this property of t-butanol systems seems to be due largely to a fall in the ΔR_{Mg} value of all hydroxyl groups relative to those for ketone groups, and since most steroids of interest contain both types of group it seems likely that confusions between neutral steroids will in general be reduced to a minimum if one of the two solvent systems of our suggested general method is based on a t-butanol or acetic acid system. The other should be based on methanol, formamide, or propylene glycol. Similar but smaller general differences are found with systems based on dioxan (Peterson et al., 1958) and nitromethane, but less is known about them at present.

It will be a matter for investigation as to the order in which the two systems will be used in a particular case. If, as is usual, the modifying step yields derivatives which are less polar than the original steroid the second system will have to be capable of giving low R_F values with the derivatives and this may be easier with one type of system than another. However, it should not be forgotten that almost all of the available systems can be used in reversed-phase fashion or have reversed-phase equivalents in which the solvent–solute interactions are similar (p. 164).

Again the type of reaction used as the modifying step should be taken into account when deciding the order and types of solvent systems to be used. Suppose for instance that the modifying step is acetylation and it is desired to estimate individual reducing steroids (20,21-ketols) in urine with the minimum of confusion of various isomers. Since the ΔR_{Mg} values of epimeric acetoxyl groups generally differ less than those of the equivalent epimeric hydroxyl groups, and the latter differences are largest with aqueous methanol and propylene glycol systems it follows that such systems should be used for the first chromatographic step. After acetylation it will then be appropriate to use an acetic acid or t-butanol system: the ΔR_{Mg} values of ketone groups will remain largely unchanged in the latter system, but those of unesterified hydroxyl groups will be very much lower than in the first system. Both types of hydroxyl group (unesterifiable and esterifiable) will then have been given large ΔR_{Mr} or ΔR_{Ms} values by the combination of the acetylation and the change of solvent system, in contrast to the ketone groups which will have similar ΔR_{Mg} values in the two chromatographic steps. If on the other hand ketalation was chosen as the modifying reaction for another class of steroids the order of using the two solvent systems would be determined only by the need to use one which gave low R_F values for the rather non-polar ketals in the final chromatographic step.

This problem becomes more complicated with some modifying reactions, for example oxidation with sodium bismuthate. In this case the ΔR_{Mr} values for the oxidation of different types of side chain are very different, but nuclear substituents are unaffected by this step. In a simple case (e.g. pregnanetriol, p. 340), where only one nuclear substituent is present at position 3 and the 17-ketone group is used for the final determination, no change in the type of solvent system is needed because of the great chemical specificity of the Zimmermann reaction coupled with bismuthate oxidation, and the equally great specificity of the ΔR_{Mg} differences for single functional groups. In more complicated cases, however, it would be desirable to use two solvent systems of different type (Fig. 4.1).

Strictly speaking, however, the maximum efficiency would only be obtained by using a separate second modifying step for the nuclear substituents followed by a third chromatogram, and it is important to note that in the more complicated cases the combination of two modifying steps between two chromatograms is inviting trouble. Thus, for instance, it can be calculated that βP-3α,6β,17α,20α-ol and αP-3α,6α,17α,21-ol-20-one would probably be incompletely separated in aqueous methanol systems although easily distinguishable by colour reactions (the former gives a colour with the periodate-Zimmermann reaction, the latter with blue tetrazolium). On the other hand the related 17-ketones produced by oxidation with bismuthate would probably be separated by about 0·3–0·5 R_M units in aqueous methanol systems. The diacetates of these 17-ketones would, however, be difficult to separate in most of the available systems.

If, therefore, acetylation were combined with bismuthate oxidation between the two chromatographic steps the estimation would confuse the two derivatives of the original substances on the final chromatogram: if the

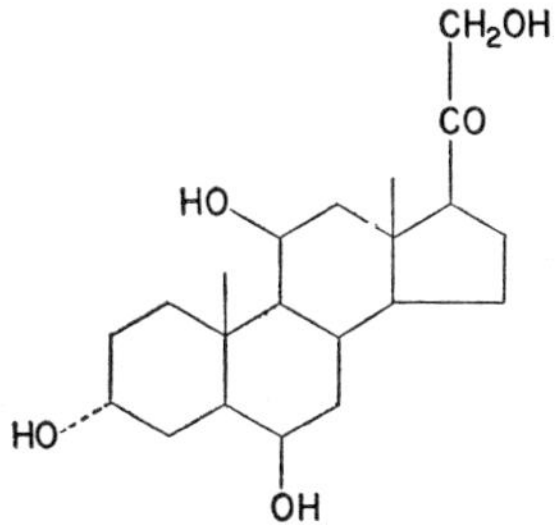

Fig. 4.1. Obtaining ΔR_M effects for maximum number of substituents.

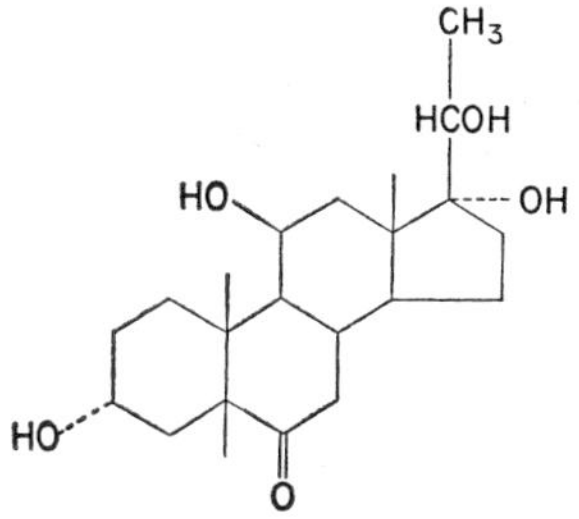

I. (a) Modifying reaction acetylation: ΔR_{Mr} of acetylating 3,6, and 21-hydroxyl groups. (b) Change from methanol to t-butanol system after acetylation: 11β-hydroxyl, negative ΔR_{Ms}; 20-ketone, zero or small negative ΔR_{Ms}.

II (a) Modifying reaction oxidation with $NaBiO_3$: ΔR_{Mr} of 17α,20α-diol → 17-ketone. (b) Change to t-butanol system after oxidation: negative ΔR_{Ms} of 3- and 11-hydroxyl groups; negative or zero ΔR_{Ms} of 6-ketone group. (c) If necessary, carry out second modifying step of acetylation: ΔR_{Mr} of acetylation of 3-hydroxyl.

acetylation were deferred until after the second chromatogram the acetates would still have the same R_M value on a third chromatogram but they would have been eluted from different zones of the second chromatogram thus allowing an individual estimation of the two substances (Fig. 4.2).

The greatest selectivity will be obtained if each substance to be measured is eluted separately from the first chromatogram. With few exceptions the combination of the two types of solvent system will achieve

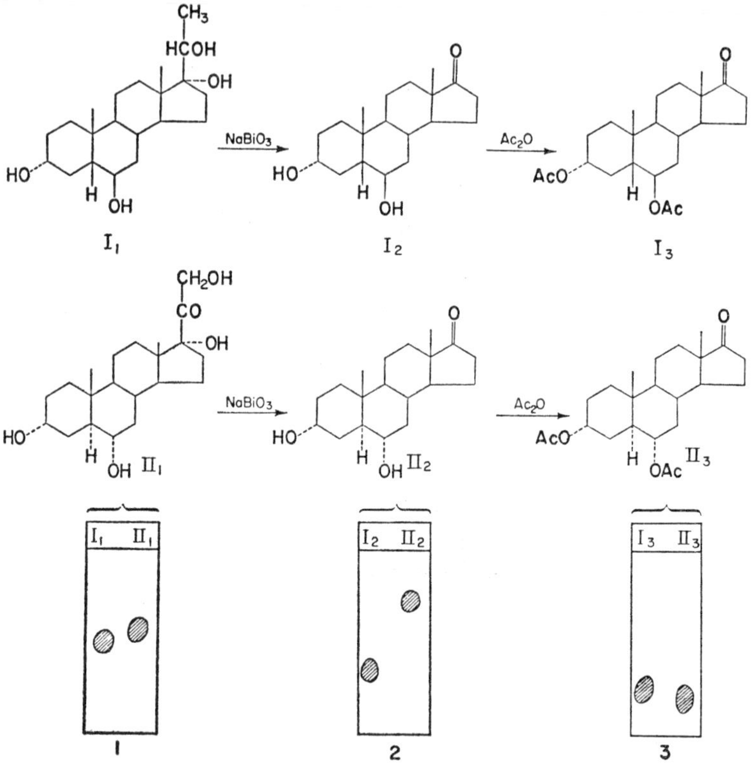

Fig. 4.2. Confusion caused by combining two modifying steps. The three chromatograms show the probable positions on an over-run B/50-type system (1) B-group system (2) and a single-length A-group system (3) of the original two substances and their derivatives. Note excellent separation of first pair of derivatives, little or no separation of final pair (see text p. 210).

no increase in selectivity of separation if wide cuts are taken from the first chromatogram; the differences in relative ΔR_{Mg} values of different substituents in the different solvent systems available are too small in most cases. This raises the question as to whether two-dimensional chromatography is likely to be useful for quantitative work with steroids.

This seems to be unlikely in the author's estimation because of the difficulty of "spreading" of the spots which makes quantitative elution even more difficult than usual: two-dimensional "scanning" is costly in time and apparatus. However, this difficulty would be absent in methods using doubly labelled steroids since quantitative elution is not required: with the "atypical" systems now available two-dimensional chromatography is perhaps worth considering.

It is probable that two solvent systems of different types such as those suggested will give moderate to large differences in the relative ΔR_{Mg} values of all neutral polar substituents in whatever class of organic substances. The possibility of confusion between different steroids, and between any steroid and a non-steroidal contaminant, will be reduced to a minimum by adopting such a sequence of solvent systems. The remaining problem with non-steroidal materials is concerned with basic and acidic groups. Since the majority of these can form hydrogen bonds either as donors or acceptors there will probably be little difference in the relative ΔR_M values of such groups in the two types of system dealt with so far. One cannot rely on orthodox solvent partition such as is used in the preparation of "neutral" extracts to remove all contaminants containing basic and acid groups; such compounds may be present in enormous quantities in the original material that was extracted, or be carried into the organic solvent phase as complexes, or colloidal micelles, because of low dissociation constants. Caffeine (Beatty, 1949; Abelson and Borchards, 1957) is a common contaminant with chromatographic properties similar to corticosteroids in neutral systems, and pentobarbitone has an R_F value close to that of dehydroepiandrosterone in light petroleum–aqueous methanol (Short, personal communication). Numerous antibiotics, including chloramphenicol, have R_F values similar to those of steroids in neutral systems. The streptovaricin complex and various halogenated derivatives of chloramphenicol were separated routinely with aqueous methanol systems (Sokolski, Eilers and Simonoff, 1957; Folkertsma, Sokolski and Snyder, 1957; C. G. Smith, 1958). There are numerous situations in steroid biochemistry in which such possible contaminants may be met, particularly in the examination of human urine and other biological material.

Most of the steroids of interest are stable to formic or acetic acid, and to ammonia, on paper chromatograms, with the the exception of C_{21} α-ketols which are destroyed in ammoniacal systems, and it is relatively easy to devise modifications of the usual neutral systems for steroids which incorporate such reagents and provide R_F values for the neutral steroids similar to those obtained with the original solvent systems (Bush, 1960b). Such modifications have not been tried out extensively but on the general argument of this chapter would seem to be valuable as a means of improving the selectivity of separations. It has been seen that, as with other ionizing compounds, such changes in pH are of great use with the acid steroid conjugates (p. 32); their use with the neutral steroids on the other hand has another purpose: it is designed to produce large ΔR_{MS} values of possible ionizing contaminants.

For this purpose the acetic acid systems would seem to be excellent in providing both typical ΔR_{Mg} values for ketone relative to hydroxyl groups and high and low ΔR_{Mg} values for basic and acidic groups respectively. Similarly the addition of ammonia to aqueous methanol systems causes little change in the ΔR_{Mg} values of steroid hydroxyl groups, and only very small positive changes in the ΔR_{Mg} values of 3-ketone groups (unpublished experiments): on the other hand very large ΔR_{Ms} values of acids and bases will be produced in the opposite direction to those of the acid systems (e.g. Reichl, 1955; 1956).

4. Modifying Steps

With the neutral steroids the ΔR_{Ms} values that are obtainable by changing solvent systems are much smaller than with acids and bases; furthermore they are more often related to the total number of ketone and hydroxyl groups in the molecule than to specific changes in the ΔR_{Mg} values of individual substituents. With certain steroids the changes in relative ΔR_{Mg} values of the various substituents cancel out so that the ΔR_{Ms} values of some pairs of steroids are negligible on going from methanol to equivalent t-butanol systems. This difficulty is reduced very considerably if a modifying reaction is used which depends upon the number and type of substituents in the molecule and which introduces a group causing a large ΔR_{Mr} value. For quantitative work such groups should either not interfere with the method of determination or be easily removable. In some cases the groups can in fact be the means of determination, as in the case of chromophores or isotopically labelled groups.

Acetylation is one such method (p. 283). It is usually easy to find conditions in which esterification is complete, and the acetoxyl group rarely interferes with the usual methods of determination. Thus the acetates of 20,21-ketols are easily estimated with blue tetrazolium, and the acetates of the common 17-oxosteroids can be estimated as usual with alkaline m-dinitrobenzene. The 20-acetates of 17,20-diols are not oxidized by periodate or bismuthate for estimation as the related 17-oxosteroids, but are readily saponified if this method of estimation were to be used. The availability of tritium-labelled acetic anhydride makes this a method of great potential value since the [^{3}H]acetoxyl group both enables the determination of less than microgram quantities of material and enhances the selectivity of the separation procedure.

The great value of acetylation from the point of view of selectivity of the separation steps is that not only does the resulting ΔR_{Mr} value afford a ready means of separating compounds with different numbers of esterifiable groups, but it also gives different ΔR_{Mr} values for hydroxyl groups in different positions and conformations. Thus a steroid 20,21-ketol containing, say, one additional axial esterifiable hydroxyl group and one unsaturated ketone group might be confused on the first chromatogram with one containing one additional equatorial hydroxyl group and one saturated ketone group—because of the larger ΔR_{Mg} value of the conjugated ketone group (p. 290). Although both gave diacetates (x- and 21-) the former would show a smaller ΔR_{Mr} on acetylation than the latter and the two esters would almost certainly be completely separated on the second chromatogram.

In certain cases it may be better to form the ethylene ketals (Reineke, 1956) producing a negative ΔR_{Mr} depending upon the number and position of reactive ketone groups in the molecule. Isotopic estimation has not yet been carried out with such derivatives but should be possible using tritium-labelled ethylene glycol or ^{35}S-labelled ethylene dithiol to give the thioketals. In general this would seem less useful than acetylation since the ΔR_{Mr} values on ketalation are less sensitive to differences in the position of the ketone groups (apart from complete hindrance of the reaction): also ketone groups are often the basis of spectrophotometric methods of determination, and the ketal group is less easily removed quantitatively than the acetoxyl group. Numerous condensing agents are useful for producing chromophores with ketones. The thiosemicarbazones are useful in this way (Pearlman and Cercero, 1953; Bush, 1953b) since they are readily separated on paper chromatograms and are more polar than the parent steroid ketones. Dinitrophenylhydrazones are readily separated on alumina (Reich, Sanfilippo and Crane, 1952) and steroid dinitrophenylhydrazones and p-nitrophenylhydrazones have been separated well on paper chromatograms (unpublished experiments): the p-nitrophenylhydrazones give better results since they have lower R_F values. The derivatives of 17-oxosteroids formed with alkaline m-dinitrobenzene have been used for identification by Kellie and Smith (1956) and run well on paper chromatograms. If the procedure could be made to give quantitative yields this would form the basis of a good quantitative method for 17-oxosteroids since not only is a characteristic ΔR_{Mr} produced but the group introduced is potentially chromogenic, needing only treatment with alkali to produce the characteristic colour of the Zimmermann reaction, a treatment which should be easily controllable and give very low background colour. Furthermore the original treatment with the reagent yields the usual coloured derivatives thus allowing accurate elution from the first chromatogram.

More specific modifying steps can be used in certain cases such as the oxidation of aldosterone diacetate to its lactone monoacetate (Peterson et al., 1958). A very useful one is the use of periodate or bismuthate to degrade various classes of 17-hydroxycorticosteroids and their metabolites to 17-ketones which are then easily estimated with alkaline m-dinitrobenzene after a second chromatogram. This allows of several useful permutations following very much the lines of Norymberski's subsequent modifications of the original method (p. 343). A very good example is the estimation of pregnanetriol (βP-3α,17α,20α-ol) which is difficult to separate from 11β-hydroxyaetiocholanolone. Although it can be detected fairly specifically (Neher, 1958a, p. 58) by its fluorescence with trichloroacetic acid (De Courcy, 1956) it is more convenient to use bismuthate or periodate followed by the Zimmermann reaction for quantitative determination. After pre-fractionation by the method of Bush and Willoughby (1957), a part of the fraction containing the 11-oxygenated 17-oxosteroids (fraction 2) is oxidized with bismuthate and the products run in system A. The aetiocholanolone derived from the oxidation of pregnanetriol is then estimated with alkaline m-dinitrobenzene by scanning or elution.

The specificity achieved by combining Norymberski's method with chromatographic separations is very great because of the very considerable differences in ΔR_{Mr} of the reactions with different types of side chain (see Chapter 5, p. 292). The direct oxidation with bismuthate is usually quite sufficient since the ΔR_{Mr} value is more than enough to distinguish between the different sort of side chain involved; for certain purposes the chemical specificity of Norymberski's sequences involving reduction with borohydride can also be used with profit, or more specific oxidation to 17-ketones obtained with periodate. Thus although a variety of steroids can give aetiocholanolone on oxidation with bismuthate their ΔR_{Mr} values are widely different from that of pregnanetriol. For routine purposes the whole of fraction B is used in this method and this would include the 20β-epimer in the estimation (but not the isomers epimeric at C-3 and C-5); the estimation can be made completely specific for the 20α-epimer when necessary by taking for the oxidation a cut from fraction 2 that includes only the (11β-hydroxyaetiocholanolone + pregnanetriol) zone.

In certain cases it is more convenient to use a modifying step aimed at possible contaminants than one for the class of substances to be measured (cf. Brown, 1955; Klopper et al., 1955). For the estimation of progesterone or other steroids lacking esterifiable hydroxyl groups the ketone groups afford the easiest methods of determination and chromophoric derivatives may be inconvenient (e.g. dinitrophenylhydrazones or thiosemicarbazones). In this case it would be best to use a methanol or propylene glycol system followed by an acetic acid system after acetylation: the two systems give almost equal differences between ΔR_{Ms} values of acetoxyl and ketone groups and those of hydroxyl and ketone groups. The change of systems gives large ΔR_{Ms} values of certain hydroxyl and ketone groups, and the order in which they were used would be immaterial, except for the need to obtain low R_F values in the final system.

With double-isotopic labelling methods it is possible to elute the substance to be measured incompletely by cutting out only part of the zone (p. 322). With the use of two types of solvent system the position of the zone that is eluted relative to the centre of the zone on the chromatogram may with profit be reversed on the second chromatogram in certain cases. Thus the nearest 3-hydroxylic contaminant of a 3-ketonic substance to be measured will lie in front (i.e. higher R_F) on a t-butanol or acetic acid system: the area cut out should therefore include the tail of the ketone zone but exclude the head. On the second chromatogram (methanol type) any residual contaminant of the ketone will now lie behind it, and the area cut out should therefore exclude the "tail" of the zone. This procedure may be useful if the type of contaminant is known: for example the measurement of cortisol and tetrahydrocortisol using bismuthate as the modifying step. These two will overlap on a t-butanol chromatogram with the head of the former over the tail of the latter: the related 17-oxosteroids will overlap in the opposite sense in an aqueous methanol system. (There are of course easier ways of estimating this pair but this method may be needed in order to deal simultaneously with other types of compound.) In general, however, partial elution of zones should be symmetrical with the centre of the zone on both chromatograms to obtain the greatest selectivity of the separation steps.

In several important cases the ΔR_{Mr} of the modifying step is far more valuable than a change of type of solvent system. Thus for instance if one wanted to estimate tetrahydrocortisol and closely related metabolites in urine using acetylation as the modifying step, (e.g. in the presence of prednisolone which runs with allotetrahydrocortisol in methanol and propylene glycol systems) the published t-butanol systems would only give a poor resolution of these compounds from tetrahydrocortisone, while the methanol systems give a moderately good separation of the 11-ketones from the 11β-hydroxysteroids after formation of the diacetates of this type of compound. In this case there is good evidence that there is little or no contamination of this class of compounds with non-steroidal reducing substances, so that the general rule of employing two types of solvent system is of less importance than the empirical knowledge that resolution of all the above free steroids is excellent with methanol systems, with the exception of the confusion of allotetrahydrocortisol with prednisolone (see p. 344). Since the latter confusion is easily resolved by an acetylation step it was found best to use two methanol systems and rely mainly on the modifying step for selectivity of separation (Bush and Mahesh, 1959b).

5. Chromatographic Technique in Quantitative Methods

It is probably impracticable for most workers to use the double-isotope method for more than one or two compounds at a time so that most techniques aiming to estimate more than one or two compounds should in general be designed to give the maximum specificity of separation, and a quantitative recovery of each compound to be estimated. In other words we want to find the most convenient way of obtaining zones of a chromatogram which, taking account of the specificity of the method of determination, each contain one and only one of the substances it is desired to estimate; i.e. 98% pure or better (see p. 81). "Zone" is understood to apply to either areas on a column or strip of paper, or fractions of the eluate from columns or strips of paper.

It was seen in Chapter 2 that a given structural change causes a larger linear separation on a single-length run when the solvent system gives R_F

values in the range 0·25–0·75; the best separations are however achieved by using the R_F range 0·05–0·25 (Jermyn and Isherwood, 1949) and overrunning the system because this method increases the number of theoretical transfers undergone by the solutes. Tait and Tait (1960) have given a valuable quantitative treatment of this problem together with an experimental comparison of the relative efficiency of columns and paper strips. As they point out the usual mode of operating columns is the equivalent of a very considerably overrun paper strip so that the number of theoretical plates passed through by each solute is very large. As they state, the average column method is far superior to paper strips run with solvent systems giving R_F values in the range 0·5 for the solutes under examination. Their implication that paper chromatograms are "normally operated" with this R_F range for quantitative estimations is, however, incorrect. Bush and Willoughby (1957) pointed out that the best quantitative results were only obtained by using overrun chromatograms with solvent systems giving low R_F values; for the major urinary steroid metabolites systems are used giving R_F values in the range 0·1–0·35 and are overrun by a factor of 2–3 on 40 cm strips (Chapter 6). Under such conditions Tait and Tait (1960) have shown that the height of one theoretical plate on Whatman No. 2 paper with volatile solvent systems is of the order of 0·26 mm, while for well-packed Celite columns with the same solvent systems it is 0·75 mm when the solutes are present in amounts less than 50 μg per 2·5 cm width of paper. The superiority of paper breaks down rapidly however as the amount rises above this figure while columns of 60×1–3 cm show no decrease in efficiency (i.e. increase in height of theoretical plate) up to at least 1·0 mg of solute. The Taits suggest that the ideal method is probably to use a column technique followed by paper chromatography under optimum conditions; the column is used first because with its large capacity it is less likely to be upset by the impurities present in the early stages of fractionating natural extracts.

The Taits' paper showed conclusively that with amounts of less than 50 μg there is no intrinsic inferiority of paper chromatography *vis-à-vis* columns (cf. e.g. Morris, 1955) and underlined the importance of using low R_F values. In problems where the chromatography of a bulky and very impure extract cannot be avoided a column technique or even the counter-current apparatus must be used. It seems unlikely that many problems of this type exist, however, and the great saving in time and space gained by using paper techniques makes it well worth seeing whether column techniques are really unavoidable whenever they appear to be so at first sight. The use of thick paper, blocks of paper sheets, or wide sheets of paper (Neher, 1958a, p. 65) is often sufficient to circumvent the use of columns. Often the amount of extract that needs to be used can be reduced by increasing the sensitivity of the method of determination; this of course is the basis of the method of Peterson *et al.* (1958) which only requires 10–20 ml of human urine. Again an extract which exceeds the capacity of a reasonable width of a thick paper using volatile solvent systems can often be handled satisfactorily with one of the impregnated-paper methods (Reineke, 1956; Neher, 1958).

In the author's view it is rare that any sacrifice of accuracy is an essential feature of paper as against column techniques, as long as the extra precautions given in Chapter 3 are taken, and optimum conditions are achieved. Probably one of the most carefully developed quantitative column methods is that of Stein and Moore (1948) for amino acids using starch columns. The error of the method for 100 μg amounts of amino acids was of the order of $\pm 3\%$ with synthetic mixtures of nineteen compounds. The error of the scanning methods for reducing steroids (Bush and Willoughby, 1957) and 17-oxosteroids has been of the order of $\pm 4\%$ for the former and $\pm 7\%$ for the latter when used under optimum conditions with synthetic mixtures: it is already known that the errors of planimetry and various manual procedures can be reduced very considerably by adopting relatively simple automatic apparatus. The statement that column methods are "always the best... unless a very large number of samples have to be analyzed" (Tait and Tait, 1960) seems to be questionable, although a similar conclusion was reached by Chibnall (1954) when considering methods for amino acids. However, the controversy is a genuine one and the reader must decide for himself which side he takes in the battle between the "columnists" and the "paperhangers"!

It is, however, generally agreed that methods using paper strips are more productive when large numbers of compounds and large numbers of samples have to be handled. A rough comparison of the logistics of the two types of method has been given previously (Bush, 1960a). Columns of any type have to be packed carefully and need automatic fraction collectors or automatic recording instruments (e.g. James *et al.*, 1951; Sternberg and Carson, 1959). The majority of steroids do not lend themselves to automatic spectrophotometric methods, nor can they be separated by the very rapid gas–liquid methods.[1] Column methods therefore involve the

---

[1] Added in proof: see however p. 328.

determination of a large number of fractional eluates if more than a few compounds are to be estimated, and to be productive when numerous samples have to be analysed a lot of space is required for the necessary number of fraction collectors. The manpower needed for the analysis of the fractions and the preparation of columns, and for the maintenance and running of the apparatus is far larger than that for paper techniques of comparable productivity. Since many problems of quantitative estimation in the steroid field involve the measurement of large numbers of compounds in large numbers of samples it is probably not unreasonable to suggest that paper chromatography is likely to be the method of choice for many workers.

These considerations are reduced in force but not eliminated by the introduction of automatic gradient-elution methods in which several columns are run simultaneously (e.g. Vestergaard, 1960a, b). Using a linear fraction-collector, the space needed to run six columns at a time is by no means excessive although it would still suffice for the running of three or four large chromatography tanks in which up to 120 paper strips of 2·5 cm width could be run at the same time. Again the time taken to separate the six major 17-oxosteroids by this method is about 28 hr and for six urine samples, about 520 fractions have to be treated with Zimmermann's reagent and their optical density measured. It seems likely that this task would take up most of the working time of two people during the two days over which such a set of columns could conveniently be run. The same analysis using paper strips would take 20 hr for equilibration, running, and drying the strips (N.B. all waiting time free for other tasks) and 30–40 min for scanning with automatic apparatus (see p. 243). In fact it would be uneconomical as a routine to make less than 12 analyses at a time with the automatic apparatus, and this additional load would only add 20 min to the time taken over scanning and reading the results. The column method could of course be modified to reduce the working time on the Zimmermann estimations by pooling many of the fractions; this, however, would reduce very considerably the value of the information obtained and run a serious risk of mis-identification of peaks in pathological urines.

There is no doubt that the best method of operating columns for quantitative work with steroids is that of the "liquid chromatogram" (Reichstein and Shoppee, 1949). The fractions are presented in the form of solutions of equal volume to which it is easy to apply any method of determination. There is no need to discuss the general technique of handling such fractions. There are, however, a variety of ways in which quantitative determination can be attempted with paper chromatograms and these need some discussion. The first method is to determine the position of the desired steroid, cut out the appropriate area, and elute the steroid; the resulting solution is treated as if it were a fraction from a column. The second method is to carry out a colour reaction on the paper strip and then elute the coloured material for spectrophotometry (e.g. Hoffman and Staudinger, 1951; Schwarz, 1953); this method has the advantage that slight variations in the position of the compounds do not prevent the accurate elution of each zone. The third method is to measure the fluorescence, radioactivity, or absorption of light of the material on the strip of paper itself, either with or without a chemical treatment of the strip to produce light absorption or fluorescence. This method is usually called the "scanning" method.

It is clear that all these methods with their many variations have their technical difficulties (Chapter 3). Elution methods need very wide separations to allow quantitative recovery of pure "zones" and very accurate location of each zone (Fig. 4.3). This difficulty is reduced when the colour reaction is carried out on the paper before elution or when the substances are detectable by reason of a chromophore. The last two methods, however, both involve the difficulties of securing a quantitative reaction in the paper strip, and the scanning method in addition is subject to all the difficulties of obtaining satisfactory agreement with Beer's law when carrying out densitometry on the paper strips. With a small number of compounds to be estimated it is probably best to use one of the two elution techniques since no special apparatus is needed, and the first method is usually the easier of the two since it avoids the difficulties of securing the right conditions for getting quantitative chemical reactions in filter paper. Most quantitative methods using paper strips in fact use one of these two methods. The only sources of error outside the method of determination itself are variations in the "dirt" from the paper affecting the background absorption or fluorescence, and the accuracy with which the required zones are located and eluted. The necessary techniques have been discussed in Chapter 3.

6. Direct Scanning of Paper Chromatograms

The scanning method (Block, 1948) is considered usually to be greatly inferior to the elution methods; in most fields the "error" is usually said to be of the order of $\pm 10\%$ of the determined value. However, it has been suggested that this method may well be the one of choice for many problems, particularly where a large number of compounds have to be estimated (Bush, 1960a). Thus the method has the advantages that very slight degrees of contamination with compounds different from those to be measured can be detected by doubling of the peaks of the scanning record, or by inflections; the rather time-consuming step of elution is avoided; the separations need not be so large since the separate contributions of overlapping peaks on the scanning record can be assessed by a simple geometrical construc-

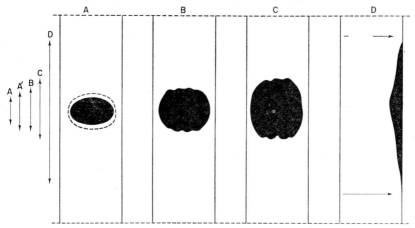

TABLE 4.3. Real and apparent distribution of zones.

Five identical chromatograms of 30 μg cortisol were run (single-length, toluene–methanol–water, 4: 3: 1 by volume) and treated to colour or fluorescence reactions of varying sensitivity. A, spot obtained with triphenyl tetrazolium chloride. A', as seen with ultraviolet light (2537 Å). B, with "blue tetrazolium". C, fluorescence with NaOH. D, the spot was run to the edge with ethyl-acetate–methanol and then examined by the NaOH-fluorescence reaction. (Reproduced by permission of the Editors of *The Journal of Endocrinology*).

tion (Bush and Willoughby, 1957); very accurate correlation of absorption or fluorescence with radioactivity can be obtained for different parts of each peak, thus giving a close check on identity when isotopic methods are used (Tait and Tait, 1960); with care there is rather less difficulty with variation of the "paper blank". The method is also capable of being made semi-automatic (unpublished experiments). In view of these advantages and the potential productivity of the method it was decided to devote a fair amount of effort to see if its error could be reduced. This was relatively easily achieved for the estimation of reducing steroids with blue tetrazolium (Bush and Willoughby, 1957) and with slightly more difficulty a satisfactory method for 17-oxosteroids was developed using the Zimmermann reaction (Bush, 1960a; Bush and Mahesh, 1959a). Ayres *et al.* (1957a; Ayres, Simpson and Tait (1957b)) developed a very satisfactory method for measuring Δ^4-3-oxosteroids with the NaOH fluorescence reaction using direct fluorimetry of the paper; similar methods were developed with slight modifications by Gowenlock (1959) and Brooks (1960). The literature on other classes of substance suggests that fluorimetric methods are more easily applied directly to spots on paper than absorption methods, at any rate when the latter use ultraviolet absorption. Very much smoother scanning records are obtained with a much more regular background from the paper (Brown and Marsh, 1953).

The techniques required for the scanning method will be described here; the general techniques needed for quantitative paper chromatography have been described in Chapter 3.

(a) Treatment of Paper Strips with Reagents

The aim is to obtain a stoichiometric reaction with the substances on the paper and a constant optical density of the blank areas of the strip. Surprisingly trivial differences in technique can make all the difference between success and failure. In most cases the main problem is to get enough reagent on the paper to give a stoichiometric reaction with the high concentrations of material at the centre of each zone occupied by the steroids, (see e.g. Cope, 1960). It is difficult to cope with zones containing more than 50–80 μg of steroid on a 2·5 cm wide strip; it is not difficult to obtain a linear calibration curve over the useful working range 0–50 μg per steroid on such a strip. With the Zimmermann reaction and blue tetrazolium accurate results can be obtained in the range 5–60 μg on a 2·5 cm strip, and useful results (95% confidence limits approximately $\pm 20\%$) down to 1·0 μg. With blue tetrazolium accurate results can be obtained for the range 0·5–15 μg on 1·0 cm wide strips but the optical arrangements are slightly more difficult to maintain.

In our experience the even deposition of the rather concentrated reagents required is only secured by dipping the strips over the surface of the reagent in a watch-glass. The best technique with blue tetrazolium was found to be a double dip through the reagent; in this case the time allowed for the paper to swell after the first dip was extremely critical and the second dip had to be made in the same direction and at the same speed as the first so that the interval between the two dips was the same for each zone of the paper (Bush and Willoughby, 1957). Sufficient accuracy was achieved by using an interval timer (small divisions = 20 sec) to maintain a uniform speed of immersion (Fig. 4.4). Other aspects of this technique are described in Chapters 3 and 6 (pp. 177, 341).

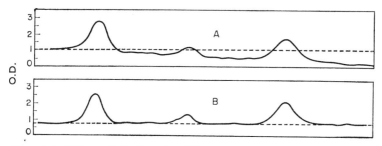

Fig. 4.4. Effect of dipping in different directions. Reducing steroids treated with alkaline "blue tetrazolium".

A. Dipped twice with a to-and-fro (pendular) movement: first dip started at left-hand end of strip.

B. Two dips in same direction: i.e. both dips made from the left-hand end of the strip.

With blue tetrazolium (Mader and Buck, 1952) the main factor determining the error is the quality of the reagent (Bush and Gale, 1958; see also Burtner, Bahn and Longley, 1957, on neotetrazolium). With impure samples, the depth of colour is not simply or constantly proportional to the amount of reducing agent. Apparent proportionality may be observed under one set of conditions but the factor of proportionality may change capriciously—even for instance with a change in the solvent system—as much as twofold. These are chemical, not just optical effects, and thus this type of error will also affect methods such as those of Hoffmann and Staudinger (1951) and Cope and Hurlock (1954) in which the colour is developed on the strip and then eluted (Bush, 1960a).

With the Zimmermann reagent the main difficulty is that the alcoholic reagent does not soften the strip as it is dipped; this makes it much more difficult to dip the strip evenly and avoid immersing it totally in the reagent. Slight unevenness of the speed of dipping also leads to distortion of the resulting bands of colour because the steroids are very soluble in the reagent and are eluted from their original positions if the rate of movement of the strip across the surface of the reagent ever becomes less than the rate of capillary flow of the reagent through the paper. Too thick a film of excess reagent on the paper is also a cause of distortion since it "rolls" about the strip during the preliminary drying period. The manual difficulties of dipping with this reagent are greatly reduced by using aqueous alcohol for the reagent but such reagents commonly give a much lower sensitivity and usually a higher and less even background.

These difficulties are avoided by using a mechanical "dipper" of the type shown in Fig. 4.5. The paper strip is seized by two pairs of stainless steel bands which grip it by about 2 mm of each edge, and is guided on to the bands by a pair of grooved plates. The reagent is fed on to the strip by a stainless steel or plastic roller, the amount being controlled by the relative speeds of the steel bands and the surface of the roller. This unit is the first part of a machine capable of the automatic treatment and scanning of paper strips (p. 243) but can easily be used alone when evenness of dipping is the main requirement for reproducible results.

(b) *Solvents and Reagents Suitable for Scanning Methods*

Some of the complications introduced by impure reagents or alcoholic vehicles have already been discussed. In the author's experience, zones containing 5–50 μg of steroid on a 2·5 cm strip need much more concentrated reagents than usual if linear calibration curves are to be obtained. This becomes reasonable when one considers the very small volume of reagent taken up by the paper and the relatively high concentrations of steroid at the centre of each zone. It can be calculated from the figures of Tait and Tait (1960) that the surface concentration of a steroid with an R_F of 0·5, at the peak of a zone containing 50 μg, will be of the order of 30 μg/cm^2 on a single-length chromatogram using a strip 40 × 2·5 cm; with steroids of lower R_F value it will be very much larger. On the other hand the amount of an aqueous reagent taken up by one dip is of the order of 0·02 ml/cm^2 with Whatman No. 2 paper, and rather less with alcoholic reagents. For 50 μg of a typical 17-oxosteroid the maximum surface concentration to be handled is thus about 0·1 μmole/cm^2 and for a uni-equivalent reagent one equivalent of reagent will be made available if its concentration in the vehicle is about 5mM. When the Zimmermann reaction is carried out in solution it is usual to have a twenty- to forty-fold (molar) excess of *m*-dinitrobenzene in the reagent solution in order to secure complete reaction: the concentration of this reagent needed to carry out the reaction on filter paper under conditions comparable to the usual method for absorptiometry in cuvettes is therefore of the order of 0·2 M, or 3·36% (w/v).

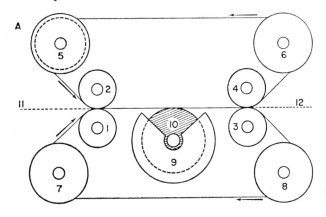

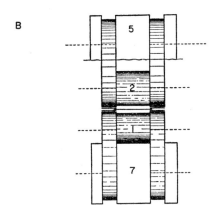

Fig. 4.5. Mechanical reagent applicator.

Stainless-steel or PTFE tapes run continuously over the raised faces of the rollers 1, 2, 3 and 4 and the grooved pulleys 5, 6, 7, 8. The drive is to roller 1, 2, 3 or 4. 1 and 3 are in fixed bearings, 2 and 4 press down on 1 and 3. 9 is the reagent trough fed from a small constant-level reservoir. 10 is a roller of adjustable speed of rotation feeding reagent as a film on to the paper strip. The paper is fed in on grooved guides at 11 and meshes with the free edges of the tapes. It emerges at 12. For aqueous reagents (which weaken the paper considerably) the bottom tape is taken over a third roller to the right of 12 before returning via pulley 8.

A, elevation. B, section through and around rollers 1 and 2 and pulleys 5 and 7.

The usual reagent for paper chromatograms has a final concentration of about 1% *m*-dinitrobenzene (w/v) or only about one-quarter the molar concentration that is used in the cuvette method. This is effectively reduced still further by sublimation of the reagent during the heating process, and probably in addition by adsorption and occlusion of some of the reagent by the cellulose fibres. The latter factor is of unknown magnitude but several features of colour reagents on filter paper suggest that it can be large. In practice a more concentrated reagent gives trouble with the background so that it is better in this case to use the usual mixture (Bush and Mahesh, 1959a) and take considerable care in standardizing the conditions under which the reagent is used.

With aqueous reagents it seems probable that the trick used with alkaline blue tetrazolium (Bush and Willoughby, 1957) may be generally useful in getting the maximum amount of reagent on to the paper by dipping. This consists in using two consecutive dips in a moderately concentrated solution of the reagent rather than one dip in a more concentrated solution. It seems clear that full swelling of the partly dry cellulose fibres takes at least 30–40 sec at room temperature (22 °C) and takes place without obliterating the interstitial spaces of the paper strip (see p. 3); in the first 20–30 sec after each zone is dipped the cellulose fibres swell and suck the solution out of the interstitial spaces so that more solution is taken up by capillary attraction when the strip is dipped for a second time. With alkaline solutions it is also possible that the fibres are softened and partly broken down thus opening up new channels for the absorption of solution by capillary attraction. This process can be seen by watching the reflection of a bright light from the damp surface of the strip immediately after it has been dipped; over a period of about 30 sec the sheen of the film of fluid on the surface (and continuous with the interstitial fluid) disappears and is replaced by the matt surface of the damp paper fibres. Evaporation is not the cause of this since the same phenomenon is not seen with ethanolic solutions (in still air) which are far more volatile but cause very little swelling of the fibres.

Another important feature of reagents used for producing reactions in filter paper is the capacity of the solvent to dissolve the steroid. On lightly dried paper the steroid is partly in solution in the residual cellulose gel and partly adsorbed on the unswollen and crystalline parts of the cellulose fibres; it must therefore be extracted if the reagent is applied in a water-immiscible vehicle, and also desorbed whatever the nature of the vehicle. It seems probable that solvents which mix with water are most likely to carry this out effectively, and that solvents which swell the fibres are likely to be most efficient. Water is thus the most suitable vehicle if it can be used, or an aqueous mixture which will leave a considerable amount of water in the paper as the volatile organic component evaporates. Reactions which take more than a few minutes for completion are best done either with a very involatile solvent or by keeping the damp strip in a chamber saturated with the vapour of the solvent to prevent evaporation and allow time for the reaction to be completed. These factors have already been found to be important with the amino acids. Ninhydrin applied in acetone or butanol solution and allowed to evaporate during heating gives only an incomplete reaction; the residual amino acid can be eluted from the zone and full colour developed by a further application of the ninhydrin reagent to the eluate (Morris, 1960). A more complete reaction is obtained, in the case of electrophoresis strips, by very careful control of the reaction by keeping the paper in an atmosphere saturated with the reagent vehicle, (Morris, personal communication). Block (1950) uses an oven saturated with water for the ninhydrin reaction.

(c) Treatment of Strips after the Reagent has been Applied

With rapid reactions in volatile solvents there is little difficulty in removing the solvent without harming the strip or altering the coloured or fluorescent zones obtained by the reaction. With prolonged reactions conditions must be arranged to avoid evaporation of the vehicle until the reaction is complete; to avoid the introduction of contamination in the shape of dust, metal, fumes, and sometimes atmospheric carbon dioxide; to prevent distortion or breakage of the damp paper; and to avoid migration of the vehicle through the paper due to gravity or uneven evaporation. Bush and Willoughby (1957) hung strips treated with alkaline blue tetrazolium horizontally between glass rods in a still atmosphere to avoid the worst of these effects, but slight increases in the background colour in the middle of the strips were sometimes observed if the strips had been slightly overcharged with reagent and sagged too much between the rods. This can be overcome by drying the strips on glass plates but sometimes this causes the zones to become "dappled" due to uneven contact with the glass leading to droplet accumulation and uneven drying; the zones are broken up by patches of lighter colour about 2mm in diameter. The ideal arrangement is to support the paper strips by a grille of glass strips mounted on their edges and supporting the strips by their outer 2 mm (Fig. 4.6).

Some reagents are unstable or give a high background on paper if they are allowed to dry; the coloured or fluorescent products may also be unstable under these conditions. With alkaline blue tetrazolium in a largely aqueous vehicle all these difficulties are encountered, mainly because of the high final concentration of alkali on the dried paper. The high background colour can be partly attributed to hydrolysis of the cellulose with the formation of glucose which reduces the reagent in strong alkali, and can be partly overcome by reducing the time needed for drying by using a powerful draught of cold air. For the best results, however, the alkali must be neutralized (cf. neutralization with acetic acid in the cuvette

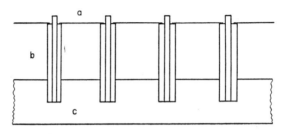

FIG. 4.6. Supporting strips for even drying. A section through part of the apparatus is shown. (a) section through strips; (b) glass plates in triplets; (c) baseboard.

method (Morris and Williams, 1955)), and this is conveniently done with sulphur dioxide introduced into a glass tank from a siphon. This has, in addition, a slight bleaching action on the paper. It is possible that washing the strips with dilute acid would give an even lower background but this has not been tried because of the risk of distortion or breakage of the paper (cf. use of water, Touchstone and Chien Tien Su, 1955); although the blue formazan that is measured is extremely insoluble in water there is also the risk of losing some of the formazan by mechanical detachment of the precipitated material or of particles of paper containing the precipitate. Care must be taken to ensure that the whole length of the strip is saturated with sulphur dioxide. For ideal results the strips must be dried off horizontally but they are already partly dried at this stage of the procedure so that little trouble is caused if they are dried quickly by hanging them vertically in the draught of a powerful fume cupboard.

Some coloured products are unstable because they are transient intermediates in complicated reactions, because of reversibility with changes of pH, or for other reasons. An example is the Zimmermann reaction, although the factors responsible for fading are not completely known. It is known that the complex formed with 17-oxosteroids is fairly stable in neutral conditions (Kellie and Smith, 1956). When trying to get this reaction to work quantitatively in the scanning method it was found that fading was extremely rapid and prevented the accurate use of scanning by the manually operated Evans Electroselenium Scanner, which is rather slow to use. It was noticed, however, that chromatograms stored between the leaves of an experimental notebook retained most of their colour for 12–16 hr; these strips were also stiff and dry while strips left in the open air became limp and slightly damp. It was found possible to stop serious fading for long enough to carry out accurate scanning by storing the treated strips, in batches of five, between two blank strips of Whatman No. 3MM paper treated with the reagent and dried by heating immediately before this use. These strips act as a protective blanket preventing the absorption of water and carbon dioxide from the atmosphere. It is probable that they also reduce the rate of sublimation of m-dinitrobenzene, which is appreciable at room temperature, but the importance of this factor in the usual fading of the colour is not known. The change of colour due to instability of the coloured complexes at high pH is too slow to interfere with the method for 17-oxosteroids and certain 20-oxosteroids, but appears to be more rapid with some 3-oxosteroids. It might be difficult to get good results with 3-oxosteroids unless the conditions were altered or more rigidly controlled.

The application of a second liquid reagent is quite feasible for qualitative work (e.g. Axelrod, 1955; Bush, 1955; Romanoff and Hunt, 1955) but has so far defeated attempts to make it work quantitatively (Bush, 1960a). The difficulty is that for reasons given above the first reagent usually has to be rather concentrated, and it has not been found possible to wash out the excess without losing some of the steroid. The capillary absorption of a second reagent therefore tends to be irregular and inefficient. Regular calibration curves have been obtained with the periodate-Zimmermann reaction sequence but they have not been linear. This problem is probably not insoluble but at the moment it seems better to avoid such methods or else in suitable cases (as with the periodate or bismuthate-Zimmermann sequence (see p. 343)) to use the first reagent in solution as the modifying step in the overall method of estimation. Gases are easily applied as second reagents as in the case of sulphur dioxide.

The application of heat is the most difficult of the additional treatments that may be needed after applying the reagent to the strips, and suffers from the further difficulty that it is hard to describe accurately since the actual temperature of the paper cannot be measured at all easily. Unless specific measures are taken to control or stop evaporation of the vehicle during treatment with heat (Block, 1950), the dimensions of the heating chamber, the atmospheric humidity, the air velocity, and the nature of the vehicle are all important variables. Even when evaporation is controlled these variables are important and affect, and are affected by, the relative amounts of heat transferred by conduction, convection and radiation. The complexity of this problem is seen in the results of Kochakian and Stidworthy (1952) who found that the intensity of colour obtained with dinitrophenylhydrazine and oxosteroids on paper chromatograms was greater if the reagent was sprayed on to the paper in a warm blast of air than if heat was applied after the paper had been sprayed. A similar effect is seen with the Zimmermann reagent but is difficult to control (J. Swale, personal communication).

There are many empirical solutions to this problem and as with most technical problems a large number of variables should be taken into account, not all of which need necessarily be connected with this particular step in the procedure. If for instance it is found that the background with a particular reagent is high and variable with the minimum amount of heat necessary to secure a complete reaction it is first necessary to examine the various variables mentioned above. If, however, these cannot easily be controlled more closely the problem may often be solved by using different chromatographic systems, by washing the paper more exhaustively before use or by changing the concentration of the reagent or its vehicle. The Zimmermann reaction is again a good example of this sort of difficulty. In general the lower the water content of the solvent system in which the strips have been run, the higher the background optical density is with this reagent. This is just tolerable with unwashed paper using light petroleum —96% methanol as solvent system if the reagent is extremely pure (p. 356), and if heating is both spatially even, and timed for each batch of strips and reagent to within 3 sec (the usual duration is 70–90 sec according to the quality of the reagent and other unknown variables). The background with strips run in this solvent system is reduced very considerably however by dipping them in water and drying at 40–50 °C in a draught of warm air, after the original solvents have evaporated (p. 66). This reduces very considerably the difficulty of controlling the treatment with heat. In recent experiments it has been found that far less heat is needed if the concentration of ethanolic potassium hydroside is raised to $3 \cdot 2$–$3 \cdot 5$ N instead of the usual 3N (Bush and Mahesh, 1959a). The background optical density is smaller and more constant but there have not yet been enough results to assess the molecular proportionality obtained under these conditions.

Radiant heat from electric fires, tungsten ("infrared") lamps, or bench hot-plates have all been used by various workers. An oven is more convenient in that if the heating takes more than a few minutes the actual temperature of the strip is known fairly accurately and the conditions are more easily repeated by other workers. On the other hand it suffers from the disadvantage that in the usual models heat distribution is extremely uneven unless the oven contains a convector, or ports producing a turbulent stream of air through it. The rate of evaporation is then however difficult to control and varies with the detailed dimensions of the oven (not just its "size"), with the thickness, area, number and position of the strips being heated, the air velocity in the oven, the mean temperature of the oven, and the atmospheric humidity, among others. This in turn affects the temperature of the paper due to the latent heat of evaporation of the vehicle and so on. It is difficult to control evaporation in an oven except when the reagent vehicle is largely or completely aqueous (Block, 1950). With radiant heat it is far easier to produce even spatial distribution, constancy with time, the desired degree of convection and air velocity, and to heat strips in vessels controlling evaporation. If necessary an accurate description of the conditions can be obtained with a bolometer but it is usually sufficient to use an ordinary mercury thermometer and record the temperatures reached when only the bulb is exposed to the source in the usual position of the paper strip. Commonly, strips or sheets are heated in the vertical position, with lamps or electric fire elements at distances of about 15–20 cm to obtain more or less even irradiation over the length and width of the paper. This produces a convection effect which leaves much to be desired; the solvent evaporates and ascends over the vertical surface of the paper thus reducing evaporation in the upper part of the strip and with m-dinitrobenzene the reagent also travels in this course because of sublimation. The radiating elements have also to be much longer than the strip to avoid underheating the ends. Experiments are now in progress to improve the original heater of Bush and Willoughby (1957) in the light of this experience by using a horizontal position of the strip and by mounting it close to the surface of a heated metal plate, rather than a longer distance from electric elements, to obtain even rradiation.

(d) *Optical Requirements*

Light-absorbing material in filter paper is probably partly precipitated, partly adsorbed and partly in solid or gel solution. Dry filter paper also scatters and diffracts light on its own account because of its inhomogeneity. In any small element of paper being "scanned" the effective "optical cell" is not only inhomogeneous but also of very short path length, and usually the concentration of the coloured material is much higher than in comparable methods with cuvettes. Thus the peak surface concentration described above for a zone containing 50 μg of steroid on a 2·5 cm wide strip of Whatman No. 2 paper corresponds to a volume concentration of about 1·5 mg/ml or 5mM for a typical 17-oxosteroid; the usual final concentration in a comparable cuvette method is in the range of 1–10 μg/ml or about 3–30 μM. It is to be expected therefore that scanning methods may encounter all the complications of nephelometry and of deviations from Beer's law at high optical densities.

It is usual with the methods for the paper electrophoresis of proteins to detect the proteins by their association with dyes and to scan the strips after impregnation of the dry strip with an involatile organic solvent mixture or oil having nearly the same refractive index as cellulose (Rees and Laurence, 1955). This reduces the scattering considerably but its application to other colorimetric methods is by no means easy since the filter paper contains excess reagent of a different refractive index and of uneven distribution. Both with visible and ultraviolet light the densitometry of unoiled dry strips gives very much larger optical densities of absorbing zones (Grassman and Kannig, 1954; Rees and Laurence, 1955; Price and Ashman, 1955), but the calibration curves are no longer linear with dyed protein zones. This effect has been attributed to multiple internal reflections and scattering (Rees and Laurence, 1955). The same effect is easily demonstrated with the blue tetrazolium reactions for reducing steroids; the calibration curves obtained with wetted or oiled paper have about half the slopes of the curves obtained after the same strips have been allowed to dry (Bush, 1959a).

Experience with the blue tetrazolium and Zimmermann reactions however suggests that deviations from Beer's law with these two reagents are small with dried unoiled strips. It is possible that this is due to the considerable load of excess reagent on the final strip, since Rees and Laurence (1955) observed their deviations from Beer's law with dyed paper electrophoretograms where the dried strips contain relatively little excess reagent. In the two reactions we have studied, on the other hand, the interstices of the paper are loaded with material of high refractive index. The fibres are also probably considerably altered in structure and partially mercerized by the strong alkali in the original reagents; by the sulphur dioxide treatment in the blue tetrazolium method; by the ionic and osmotic effects of the concentrated solution produced during drying; and by mechanical disruption of cellulose fibres by crystal formation.

The need for "oiling" strips before scanning seems reasonable but if, as Rees and Laurence (1955) suggest, scattering is the main source of deviation from Beer's law, simple optical ways of avoiding much of the scattered light are available, as will be seen. In view of the extra complications of oiling it seems best to avoid it if conditions can be found under which Beer's law is obeyed with dried strips over a useful range of surface concentration.

The theory of the scattering of light by asymmetrical particles is both complicated and incomplete so that no quantitative treatment of the behaviour of filter paper is as yet possible. It is useful, however, to give some idea of the factors involved since this may lead to an empirical solution of the problem (see also Crook, Harris and Warren, 1952; Crook, Harris, Hassan and Warren, 1954). It is clear that the nearest approximation to the behaviour of an optical cuvette will be observed with those rays which pass straight through the paper without deflection; some transmitted rays collinear with incident rays may have suffered multiple deflections but on average the majority of such rays will have suffered no deflections, or else reflections at low angles of incidence and hence with little loss of intensity. On the other hand transmitted rays which emerge at large angles with the direction of incident rays are more likely to have taken part of their pathway outside the zone being scanned; they have therefore only passed through part of the element being scanned and are more likely to have suffered losses of intensity due to reflections with large angles of incidence. If the incident light is not monochromatic, as it is not in the EEL Scanner, highly scattered rays will suffer dispersion; different microscopic regions within the element will therefore be scanned effectively at differing wavelengths of light for which the molecular extinction will differ from that at the wavelength of maximum absorption. Under certain conditions these many complicated factors seem to average out in a way giving close agreement with Beer's law, or other regular relationships (Brown and Marsh, 1953), but our results have been much better when the amount of scattered light reaching the photoelectric cell was reduced to a minimum by using a narrow slit over the cell in addition to the slit providing incident light (Bush, 1959a).

The Evans Electroselenium Scanner is a sturdy manual instrument which has served very well for the last five years in our laboratory but needed some modifications for good results. This instrument uses unfiltered light from a tungsten lamp as incident light, emerging from a variable slit about 5 mm from the surface of the paper. The selenium cell has a slot for interposing various glass or gelatine filters in the beam of transmitted light and an exposed surface about 1·5 cm wide (i.e. in the direction of the length of strip). The frame holding the paper is supported in a split channel at its upper edge which allows a lot of rotation of the frame about the long axis of the paper strip. The importance of scattering is easily seen by rocking this frame about the long axis and observing the appreciable change in optical density readings that occur. This can be prevented by adding three or four small roller guides gripping the bottom edge of the frame, mounted on a small plate screwed to the block holding the variable slit. Another feature of this and any scanning instrument is the difficulty of scanning the full width of the strip without getting large errors due to light passing over the edge of the strip. In fact it is impossible to avoid this if the full width is scanned and in practice the only solution is to scan the middle zone excluding the outer 1–2 mm of each edge, with an appropriate mask carried with the paper strip or by careful arrangement of the entry and exit slits.

The preceding account has been based upon the assumption that because of unavoidable slight irregularities of coloured zones on paper chromatograms the only satisfactory way of scanning, if really accurate results are to be obtained, is to measure the optical density of successive small elements of each strip and integrate the areas under the peaks on the scanning record (e.g. Redfield and Barron, 1952). This is analogous to the usual method of determining all the fractions eluted from a column. Other ways have been proposed which use the optical density of the peak of each zone (Block, 1950; McFarren and Mills, 1952; Roland and Gross, 1954) or the optical density of the whole zone or spot when placed under a standard aperture just larger than the largest zone to be measured (Rockland, Blatt and Dunn, 1951; Rockland and Underwood, 1956; Polson et al., 1951). The latter method is also favoured for direct fluorimetry on paper (e.g. Ayres et al., 1957) and for this technique is quite appropriate. For densitometry, however, the method using a standard aperture will only give a linear optical density calibration curve if an integrating photometer is used (Baker, 1955); otherwise a complex curve is given.

Methods of scanning such as those which do not integrate the optical density of the whole of each zone are likely to have a large error since they depend upon absolutely regular zones conforming exactly to the theoretical shape, or at any rate with a constant deviation from it. They further suffer from the disadvantage that even with perfect zones the slope of the calibration curve depends upon the number of theoretical transfers carried out and the R_F value of the substance to be determined. The slopes will therefore vary with the duration and efficiency of the chromatogram, as well as with the position of the zones; in general the higher the R_F value and the longer the run the smaller the slope of the calibration curve. With a family of substances giving similar extinction coefficients this is inconvenient; either a calibration curve has to be obtained for each substance with standards, or correction factors must be obtained for each substance relative to a standard of known R_F value. It is not impossible to satisfy these requirements but the practical difficulties of doing so are considerable, and are greater with the rather less stable solvent systems needed for steroids than with those used in other fields. The only requirements of the integrating method are that Beer's Law should be obeyed and that the reaction in which the coloured compound is formed should be simple and reproducible.

One refinement of the integrating method can be discussed in the light of the last paragraphs. Current methods still require even distribution of material across the width of the strip since it is scanned only along its length; the concentration profile should be rectangular (Fig. 4.7). To avoid

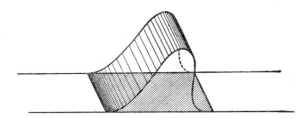

FIG. 4.7. Concentration profile required for unidirectional scanning.

the effects of deviations from this distribution the strip should ideally be scanned by a pencil of light across the strip as well as lengthwise (Wieme, 1958; Bush, 1960a). Wieme (1958) has described a simple instrument for doing this using a curved slit lying across the strip scanned by radial slits in a rotating disc, and shown that improved results are obtained with electrophoretograms of plasma proteins.

To sum up, it appears that the simpler instruments currently used for scanning leave much to be desired. A useful instrument should provide: collimated monochromatic incident light; carefully machined moving parts allowing no play; accurate masks or slits preventing light passing to the photoelectric cell without passing through the paper; a lens and slit limiting the amount of scattered light reaching the cell; and a secondary filter for the cell. A refined instrument should provide in addition some arrangement for scanning across the strip as well as along it. The optical properties of absorbing material enmeshed in filter paper need further investigation and a full theoretical treatment would be of value. For instance, the polarization of light passing through filter paper needs investigation. Much of the scattered light is probably polarized both by multiple reflection and by inhomogeneous distribution of optically active regions in the paper; it is possible that the on-axis scattered light (i.e. that part which is not prevented from reaching the photoelectric cell by slits and collimator) could be eliminated by using plane-polarized incident light and a parallel analyser for the transmitted light.

(e) Geometrical and Planimetric Methods

With an absolutely constant background and well-separated zones, no difficulty is presented. With manual techniques, however, there is always some variation of background along the strip and it is often convenient to "resolve" overlapping zones by a geometrical construction.

Fixing the base line is a necessary preliminary to measuring the area of single or mixed peaks by any method. It is usually easy with standards, but sometimes difficult with natural extracts due to impurities or to traces of material which are too faint to be definitely recognized as discrete zones. It takes some experience to judge the true limits of a zone by eye (e.g. Fig. 4.3), but the best method is to assess these limits by eye and then mark them on the scanning record. The true base line for the zone is then taken to be the line between the minima or inflections within 0·7 cm of the limits judged by eye (Fig. 4.8). This distance is an arbitrary figure which approximates to the largest error made by eye with typical zones on chromatograms of steroids in methanol systems. It is fixed in order to set a limit to subjective judgement of the scanning records. Because of the width of the base of typical peaks the error in measuring the area of a profile is very largely made up of the error of fixing this base line and no pains should be spared in obtaining the most even background and making a careful assessment of the record (Fig. 4.9).

Visual inspection of the strips is also useful in picking out abnormal colours. It is perfectly reasonable for instance to expect a brown or orange coloured zone to affect the "background" of a purple peak with the Zimmermann reagent. Such peaks should be allowed for by treating the specific and non-specific colours as if they were mixed peaks (Fig. 4.9).

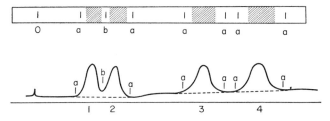

FIG. 4.8. Fixing the base line to a scanning record. Strip and typical scanning record. (a) marks giving visual assessment of limits of zones. (b) marks giving visual assessment of point where optical density is equally due to peaks 1 and 2. (o) origin marked in heavy pencil to obtain correspondence between the strip and scanning record.

It is essential to adopt a rigid convention for resolving peaks so as to avoid subjective bias (Bush and Willoughby, 1957). With absolutely symmetrical peaks the method of triangulation or the drawing of a vertical division at the minimum between the two peaks is a good enough approximation (Redfield and Barron, 1952). Similarly, almost perfect agreement is obtained between areas measured by planimetry and the product of the height and "span" (i.e. width of the peak at half its height (Fig. 4.10). The best method, however, is to plot the peaks out by matching with the profiles of standards and to correct the plots by successive approximation until they agree when carried out in both directions (Fig. 4.11). Thus the tail of the front peak is plotted by drawing the axis of the peak and using one or two constants of proportionality measured from the profile of a single standard of comparable height, taking the front of the peak as "pure". The front of the rear peak is then plotted by subtracting the tail of the first from the scanning record. If a third peak is involved the tail of the second is then calculated and plotted from its (calculated) head by the same method, and then the head of the last peak by subtraction. The profiles of the calculated peaks are then measured to see whether their proportions

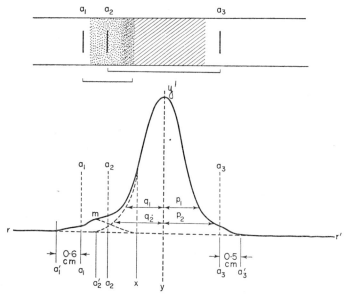

Fig. 4.9. Detail of base-line construction and planimetry. Zimmermann reaction on strip.

Operations:

1. a_1, a_2, a_3 visual estimates of zone boundaries.
2. a_1' and a_3' are taken as true limits of the mixed zone: i.e. inflections within 7 mm of visually estimated limits.
3. a_2 is taken as true limit of the 17-ketone inflection, with impurity in small amount.
4. Using evidence from a standard profile that zones are symmetrical ($p/q = 1 \cdot 0$; small distortions as in head of the main zone are neglected) the minor peak is calculated and plotted as mx.
5. Values along mx are subtracted from the record rr' to give dashed segment of $y'a_2'$.
6. Axis of main peak y' is drawn and symmetry checked by measuring p/q at various positions.
7. Area enclosed by $a_2'y'a_3'$ is measured with planimeter.

agree with the ratios of assymmetry obtained with the single standard peak. The deviations of the mixed peaks are then minimized by replotting the curves in the reverse direction, i.e. starting from the "tail" of the last peak. This technique is limited in its application to zones which are relatively well separated and is likely to give large errors if the profiles are very variable due to faulty chromatographic behaviour.

(f) *Electronic Recording Apparatus*

The integrating scanning method requires that small elements of the strip be passed in front of the photoelectric cell and their optical density be recorded. With manually operated instruments such as the Evans Scanner the galvanometer is fitted with a logarithmic scale giving the readings in optical density units directly. The readings along the strip are plotted on a sheet of graph paper clipped to the body of the instrument so that the eventual record corresponds in length and position of the zones

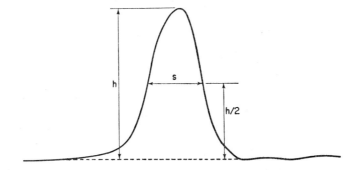

FIG. 4.10. A useful approximate parameter of the area of a peak. The span (width at $h/2$) of a peak is usually easily and accurately measured due to the steep slope of the curve in this region. (Height × span) varies almost linearly with area, the difference is negligible in practice with present colorimetric methods.

with the paper strip that has been scanned. This method is tedious and several workers have devised automatic instruments which feed the strip past the cell mechanically and record the optical density with a pen recorder of some kind. More recently instruments have been devised which provide an integral record of the optical density as well as the direct record (Durrum and Gilford, 1955; Block, Durrum and Zweig, 1958, p. 568). To obtain an output corresponding to optical density (i.e. proportional to the logarithm of the reciprocal of the percentage transmission) an electronic circuit of the type first devised for this purpose by Sweet (1946) is used. An ingenious optical alternative is described by Barrollier, Heilman and Watzke (1958). A good example of a typical instrument used for the paper electrophoresis of proteins is described by Laurence (1954); an integrating instrument by Durrum and Gilford (1955); and the use of such circuits with cross-scanning of the paper by Wieme (1958). The principle of such circuits is that the output of a photomultiplier is fed back on to the control grid of a pentode which in turn controls the potential across the dynodes

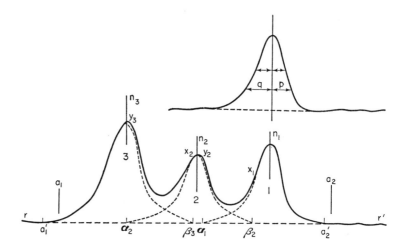

Fig. 4.11. Resolution of overlapping peaks.

Operations:

1. Axis n_1 drawn and $x_1\alpha_1$ plotted by using the ratios p/q from a comparable standard peak shown in the upper record.
2. $y_2\beta_2$ is plotted by subtracting $x_1\alpha_1$ from rr'.
3. $x_2\alpha_2$ is plotted using p/q ratios from the standard.
4. $y_3\beta_3$ is plotted by subtracting $x_2\alpha_2$ from rr'.
5. Axes n_2 and n_3 are drawn and p/q ratios checked neglecting minor distortions if present.
6. If p/q ratios are in error in peaks 2 and 3, the operations are carried out in reverse—i.e. plot $y_3\beta_3$ from p/q ratios, $x_2\alpha_2$ by subtraction, $y_2\beta_2$ by p/q ratios, and $x_1\alpha$. by subtraction.
7. Error is minimized by taking the plots intermediate between those given by the first and second routes of calculation.

a_1, a_1', a_2, a_2' as in Fig. 4.9.

of the photomultiplier. With this negative feedback the anode current of the photomultiplier is kept constant and the feedback current is proportional to the log of the intensity of illumination of the photocathode; it is this current that is measured and fed to the continuous recording instrument (Sweet, 1946).

Various methods of integrating the record electronically are available. The response times of such instruments are rather long but they are still much faster than manual scanners and there is no time spent on the measurement of the areas of the peaks. Integrating recorders giving digital records are now available (Beckman, Spinco Division, U.S.A.; Joyce Loebel Co. Ltd., Newcastle; Sunvic Controls Ltd., London) these give records of the type shown in Fig. 4.12, on which it is only necessary to make vertical projections at the head and tail of each peak on the smooth (direct) record, and count the number of digital units between the projections. These instruments giving integral records have not been in use very long and their errors have not yet been examined in any great detail. The heart of such machines is usually an "integrating motor" which is designed so that its speed of rotation is directly proportional to the input voltage while its rotor has a very low inertia. Only very good motors have a deviation from a linear response of the order of 1–2%, and mechanical attachments to the motor to provide a chart record must be kept to a minimum so as to keep the inertia low. Results so far are encouraging (Fig. 4.13). Much work remains to be done in this field but it has been stimulated by the need for integrating the records obtained with gas–liquid columns, and advances should be rapid.

(g) Fully Automatic Apparatus

The treatment of paper strips with reagents and heat is subject to numerous errors when carried out manually. The author therefore carried out experiments to devise a mechanical method of doing this. Since it is easier to control temperature and humidity over small regions of space, and since a continuous process has its own advantages the machine was based on the passage of the strips of paper at a constant speed through various zones in which reagents and other treatments were applied. The unit for applying the reagent described on p. 227 was used as the first stage and the strip was then passed on belts to a treatment zone in which heat, gaseous reagents and draughts of air could be applied in any desired order. The improvement in the quality of the results and particularly the constancy of the background achieved by this method was immediately apparent even with a crude mock-up. Finally these units were then combined with the recording apparatus, thus removing another source of error, namely the different intervals of time between the full development of colour in any zone of the strips and the time when that zone was scanned. The final apparatus thus passes the strips at constant speed

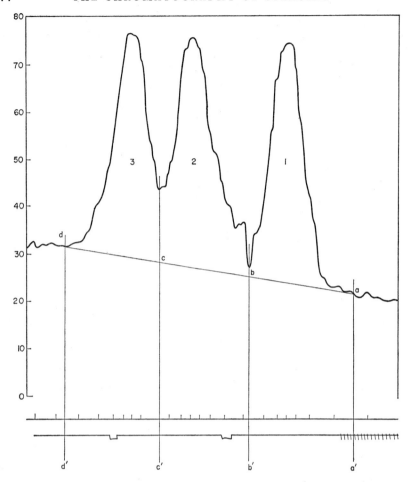

Fig. 4.12. Mechanical scanning with Sweet-type photomultiplier circuit and an integral recorder

Whatman No 2, 5 cm wide strips. Zimmermann reaction: 1. androsterone; 2. aetiocholanolone; 3. dehydroepiandrosterone: approx, 60 μg each. (Group A system, 25·5 °C, 14 hr run.)

The integral record is at the bottom. Lower line of digits reads in units up to a chart reading of about 25% of full-scale: above this in hundreds. The upper line reads always in tens. aa', bb' cc', dd' best projections defining limits of the three peaks. For peak 1 the "baseline" is taken from a calibration curve as the integral/unit length of chart for the mean of the readings a and b, and multiplied by the length a'b'. This is then subtracted from the total integral reading between a' and b' to obtain the "area" of peak 1 in digits.

Sunvic RSP2 recorder on 2·5 mV full-scale range: slight "hunting" due to high speed of scanning (approx. 0·8 cm/sec).

through the treatment units and on at the same speed through the scanning and recording unit which provides a digital integral record as well as a direct record of the peaks.

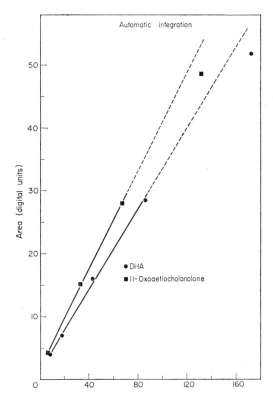

FIG. 4.13. Automatic integration of peak areas. Calibration curve from records similar to those in Fig. 4.12 for two 17-oxosteroids run single-length in a group B system on 5 cm × 40 cm strips of Whatman No. 2 paper (Zimmermann reaction). Non-linearity appears somewhere between 18 and 30 μg steroid/cm width of paper.

In accordance with modern practice the machine is called CASSANDRA (Chromatogram Automatic Soaking Scanning ANd Digital Recording Apparatus). There is a dead time of about 15 min after which each strip and its accompanying record are delivered at about 60 cm/min, or one

40 cm strip every 40 sec. Apart from greatly reduced error the machine is about 15 times as productive as a skilled technician, using the manual methods with planimetry for batches of about 20 strips; with larger batches the increase in productivity is greater due to the decreasing proportion of dead time and the limitations on the working hours available for the manual methods.

7. Logistics of Quantitative Estimations Using Chromatography

It may seem slightly unbalanced to dwell so long on the scanning method in this chapter, and some explanation is needed for the bias of its later parts, since this method has by no means gained general respect in the last five years. Indeed it is somewhat heterodox to suggest that really accurate quantitative methods can be expected unless column chromatography is used coupled with some orthodox method of determination. This general opinion is justified by many published results and the most that could be claimed for the existing quantitative methods based on paper chromatography is that where large numbers of estimations are needed and a moderately large error can be tolerated (say ± 5 to 10%; cf. Miethinen and Moisio, 1953) their convenience and speed is enough to justify their use. The author's position, however, is that although the technical problems of getting really accurate results with paper chromatography are considerable they are not insuperable if their magnitude and sophistication are realized. In skilled hands the error of the scanning methods for reducing steroids and 17-oxosteroids is little if at all greater than that of the comparable column + cuvette methods as far as they can be ascertained from the literature. Since a large part of this error is due to variations in manual skill and in the final planimetry of the records and can be reduced very considerably by automatic apparatus there is every reason to hope that in most cases absorptiometric methods will be adaptable to the direct scanning procedure with little or no loss of accuracy.

At present the automatic apparatus needed for the ideal operation of this procedure is expensive and complicated, but so is a motor car and an infrared spectrophotometer, and with many colour reactions the "dipping" unit (p. 227) is probably sufficient to remove the main source of error in the manual method. The main argument in favour of the scanning method and of the use of paper for quantitative chromatography, however, is still that of productivity and economy of space and time. Steroid biochemistry is *par excellence* a field in which productivity is important; chromatographic methods have made it possible in theory to measure more than a hundred steroid metabolites which are of known or potential importance in endocrine physiology and pathology and in metabolic diseases. With the exception of gas–liquid chromatography the automatic scanning method appears to be the only procedure in which the speed of quantitative determination matches the speed and power of the chromatographic method of separation. Despite the existence of powerful methods covering most of this field for the last five years there is still, for example, only very limited information available on the steroid metabolism of the human being and its variation with age, sex, race, climate, disease and so on, except for a very small range of compounds. It was this sort of consideration which led to our attempts to improve the scanning method despite its first appearance of lack of promise and difficulty.

8. Use of Radioactive Isotopes in Quantitative Chromatography

There are two ways in which isotopically labelled substances may be the objects of quantitative chromatographic methods. In the first the steroid is labelled and used in a metabolic experiment, after which it is necessary to separate and estimate the metabolites and determine their specific or absolute radioactivity, or their enrichment with non-radioactive isotopes such as oxygen or nitrogen. In the second, the steroid is treated with an isotopically labelled reagent and the aim is to measure the absolute amount of radioactivity in the derivative in order to estimate the amount of steroid that was present in the reaction mixture.

The first case has been the subject of numerous papers, and chromatographic separations have merely been a part of a long series of operations designed to obtain pure material of constant specific activity (e.g. Gallagher, Bradlow, Fukushima *et al.*, 1954a; Gallagher, Fukushima, Barry and Dobriner, 1954b; Baggett and Engel, 1957). The standard approach to such "tracer" experiments is to isolate the labelled metabolite by adding unlabelled substances suspected to be present and purifying these substances until repeated measurement of their specific activity is constant in the face of several successive crystallizations and other purification processes. It goes without saying that chromatographic separation is one of the most powerful of such purification processes and in the steroid field is almost essential for such work. It should, however, be no surprise that work with labelled substances often shows up the limitations of the method in general, or of particular procedures. The methodology of tracer work is thoroughly described by Veall and Vetter (1958) among others so that only a few special features concerned with chromatographic methods will be discussed here.

The absolute amount of radioactivity in a steroid is usually determined by one of the planchet techniques after elution from a paper chromatogram or column. Little difficulty is experienced with carbon-14 but with tritium rather special techniques are needed for good results unless very high activities are available. Thus Ayres *et al.* (1958) found that aldosterone and other corticosteroids could only be assayed satisfactorily in small amounts if progesterone were used as a carrier for plating. The gas-flow counter is a moderately cheap and efficient way of counting tritium in solid examples; scintillation counting is favoured by many with larger budgets and the Packard Tricarb Counter has the great advantage of having a particle energy discriminator which enables the simultaneous counting of two isotopes in one sample if the energy difference of their emission is sufficiently great. This is easy with carbon-14 and tritium, thus making double-labelling methods with steroids conveniently rapid. It is sometimes suggested that scintillation counting with the substance dissolved in the phosphor overcomes the difficulty of self-absorption, that is the main problem in the counting of tritium in solid samples, and that this means that impurities in general interfere with the assay less than with the gas-flow counter. Impurities can in fact be as troublesome in scintillation counting as with solid samples but for a different reason, namely alteration of the efficiency of the phosphor, an effect which can either enhance or reduce the apparent counting rate (J. F. Tait, personal communication) or by phosphorescence or light absorption (Veal and Vetter, 1958, p. 69).

The difficulties of counting tritium are outweighed by the very large specific activity that can be obtained. The first steroids labelled with carbon-14 were [21-^{14}C]progesterone (Riegel and Prout, 1948; MacPhillamy and Scholz, 1949), [4-^{14}C]progesterone and [4-^{14}C]testosterone (Fujimoto, 1951; Turner, 1950). Tritiated steroids were synthesized by Fukushima and Gallagher (1952) and Pearlman, Pearlman and Rakoff (1954). Pearlman's method of synthesizing [16-^{3}H]progesterone led to the production of steroids with very high specific activities (Pearlman, 1957a, b). [4-^{14}C]Progesterone was used as a substrate with adrenal cortical tissue to obtain labelled corticosteroids (Plager and Samuels, 1953) and recently tritium-labelled aldosterone has been prepared in this way from [16-^{3}H]progesterone (Ayres *et al.*, 1958).

The first problem in the assay of radioactive steroids after chromatography is largely an exacerbation of the general one discussed in other chapters, namely the predicted distribution of a substance in a chromatographic system and how far this is departed from by the distribution found in practice. The spreading of zones on paper chromatograms beyond the limits suggested by colour reactions described earlier (p. 223) is particularly noticeable with radioactive steroids (Berliner and Salhanick, 1956) and this is generally larger with volatile solvent systems than with impregnated paper (Berliner, Dominguez and Westenskow, 1957) due to the lower capacity of the former (Bush, 1954; Reineke, 1956; Neher, 1958a). Both types of solvent system, however, were found to suffer from the defect that radioactivity was often found in zones other than that occupied by the main carrier identical with the labelled steroid (Berliner and Salhanick, 1956) and this was particularly marked with non-polar steroids. This was put down to displacement of the labelled steroid by association with other lipids. It is not quite clear from this paper, however, whether this was actually proved. In the author's experience such displacements are often due to decomposition of the labelled reference steroid, to overloading the strip, or to failure to achieve complete equilibrium in the chromatographic chamber. Nevertheless such defects are indeed likely to occur in any situation where a large quantity of impurities are present in the sample to be analysed.

Such displacement effects are most likely to be due to the formation of "complexes" with other steroids or lipids; to produce discrete zones they must be relatively stable complexes once formed, since "streaks" would result from reversibly dissociable complexes with even a moderately large dissociation constant. It is impossible to say whether they will be prevented in all cases by using a system of larger capacity but this would be expected; the occurrence of this sort of defect is an indication for the use of column rather than paper methods (Tait and Tait, 1960), at any rate in the early stages of fractionating labelled steroids. Another possible way of overcoming this defect is to add to the extract not only the carrier suspected to be identical with the labelled steroid but additional substances known or likely to compete with the labelled steroid for association with the impurities causing the trouble. The separation of bile acids in the presence of substances capable of forming choleic acids, for instance, would be troublesome since the different choleic acids of one bile acid would have different R_F values, in turn different from the bile acid itself. The problem might be solved either by equilibrating the extract with an excess of a fatty acid giving the stablest choleic acid and running the bile acid entirely in this form; or by adding another bile acid known to form more stable choleic acids than the one to be separated. With other steroids less is known about the formation of complexes of this general type but it would still be profitable to try adding two steroids to the extract, namely the usual amount of the desired carrier and a larger amount of another steroid designed to compete with the former in forming association complexes.

Difficulties of this sort, namely apparent failures of chromatography to live up to its expected standards, are also found with column methods; here again the difficulty seems greatest with rather non-polar steroids. Thus Solomon, Lanman, Lind and Lieberman (1958) found that constant specific activity of labelled 17α-hydroxyprogesterone prepared by biosynthesis was only achieved with great difficulty despite numerous chromatographic separations on alumina columns and paper.

In general then, chromatographic methods used with radioactive steroids are likely to provide disappointments unless the limitations of the method are appreciated (see e.g. Glueckauf, 1955). For reliable results the closest attention to technique is needed and all conditions should be weighted in the direction giving maximum performance. This means that the longest convenient time should be allowed for equilibration when using paper, and slow-running solvent systems used (p. 101) with both columns and paper. Similarly, thicker and wider strips of paper, and more voluminous columns (p. 152) should be used if there is any doubt about the capacity of the system to handle the impurities present in the mixture to be analyzed. All the precautions suggested for quantitative work in this and Chapter 3 apply with even greater stringency to work with radioactive steroids.

The direct scanning of paper strips has been advocated by Berliner and Salhanick (1956) for the assay and identification of radioactive steroids. They emphasize the need for preparing derivatives with different mobilities as a check on identity and, as with absorptiometric scanning, point out that the location of zones and the detection of impurities incompletely separated from the reference carrier is more accurate than by elution. The counting of a succession of derivatives is also far easier with very small quantities than by orthodox carrier methods.

In the identification of labelled steroids the measurement of specific activity in different regions of a zone is a valuable check on identity. This was first done with p-iodobenzenesulphonyl ("pipsyl") derivatives of amino acids using ^{131}I and ^{35}S for double labelling. The zone detected with ninhydrin on a paper chromatogram was cut into four or five areas and the ^{131}I : ^{35}S ratio determined in the several areas (Keston et al., 1950; Udenfriend, 1950). The same method has recently been applied to the "pipsyl" derivatives of esterifiable steroids (Jensen and Bojeson, 1958; Bojeson, 1958). The Taits have obtained the specific activity of aldosterone diacetate in several areas by scanning the paper directly for fluorescence after NaOH treatment and direct counting of the areas. With the high specific activities available it is possible to count tritium-labelled steroids directly on paper in very small amounts (Tait and Tait, 1960). An entirely analogous method has been used on the fractions of a Craig counter-current machine by Baggett and Engel (1957) who then carry out a statistical analysis of the deviations from the mean specific activity of the various fractions. In this way very small differences in the partition coefficients of a labelled steroid and a suspected reference carrier steroid can be detected with confidence.

As with other problems the solvent systems used for separating labelled steroids should be chosen and run for lengths of time so as to give zones in positions that are most sensitive to changes in structure, and which allow accurate elution. The first chromatogram (CHCl$_3$–formamide; 8 hr) shown by Berliner and Salhanick (1956), for instance, gives an ideal position for cortisol, but tetrahydrocortisol is still overlapping the origin line; the second chromatogram of the acetates of the two steroids however (benzene–formamide; 3 hr) gives a useful position for tetrahydrocortisol diacetate but cortisol monoacetate is too close to the origin for characterization. For this sort of work the volatile solvent systems offer considerable advantages because they are more readily modifiable to meet special difficulties of this kind (p. 127).

The second use of isotopically labelled material is more closely related to the subject of this chapter, namely the use of radioactively labelled reagents as a means of estimating unlabelled steroids. The general principle of this method is to couple some labelled group to a steroid in a natural extract, isolate the labelled derivative by chromatography, measure its absolute radioactivity, and thus by means of the known specific activity of the reagent, to obtain the absolute amount of steroid in the original extract. Most of the reactions suitable for this method are rather unspecific and hence the specificity of the method of estimation depends very heavily on the selectivity of the means employed to isolate the derivative.

The first method of this type was the use of labelled p-iodobenzenesulphonyl chloride ("pipsyl chloride") to make the N-"pipsyl" esters of amino acids and purines (p. 250), and the first attempts with steroids used labelled acetic anhydride to make the acetates of hydroxysteroids (p. 323). Later Ruliffson, Lang and Hummel (1953) prepared [^{131}I]acethydrazide-3-iodopyridinium bromide as a reagent for steroids containing reactive ketone groups: the steroid hydrazones were then chromatographed on paper with a butanol–water system. This method seems unlikely to be of much use unless modified since the excess reagent and a labelled impurity in the reagent have R_F values little smaller than the more polar steroid dihydrazones. The most recent methods have taken advantage of the availability of tritium-labelled acetic anhydride of high specific activity (Avivi et al., 1954; Kliman and Peterson, 1957; Whitehead, 1958; Tait and Tait, 1960) but are essentially similar to the earlier versions using ^{14}C-labelled anhydride (Simpson et al., 1954; Berliner, 1957). The "pipsyl" method has also been adapted for use with non-phenolic hydroxysteroids by using the less reactive anhydride of p-iodobenzenesulphonic acid for preparing the esters (Jensen, 1956; Bojeson, 1956; Jensen and Bojesen, 1958). Preliminary accounts of its use have been given (Bojeson, 1958).

The great advantage of the methods using acetates is that the latter are readily run in most of the usual solvent systems; acetylation is in fact one of the best modifying steps available for hydroxylic steroids quite independently of the possibility of using it for labelling with isotopes (p. 283). The polar hydrazones used by Ruliffson, Lang and Hummel (1953) are only separable into groups of mono- and di-hydrazones with their solvent system; the ΔR_{Mr} caused by adding this group is so large that the solvent system required gives much smaller differences in the ΔR_{Mg} values of the typical substituents of steroids. Similarly, first impressions are that the "pipsyl" esters of steroids are not as easy to run on chromatograms as the acetates. Finally the reagent needed, acetic anhydride, is easily available and contains an isotope of long half-life (tritium, 12·3 yr) while the other reagents must be synthesized fairly frequently with fresh isotope (^{131}I, half-life, 8 days) to obtain maximum sensitivity. There remains a need for reagents suitable for steroids lacking esterifiable hydroxyl groups, preferably labelled with tritium because of the high specific activities obtainable, and giving a ΔR_{Mr} value convenient for the handling of the derivatives in the usual solvent systems. There are several obvious possibilities (p. 215) and only time will show which are the most practicable.

The use of double labelling, as mentioned before, has two main advantages. Firstly the selectivity of separations is increased by the fact that quantitative recovery of the derivative after chromatography is unnecessary:

the second isotope acts as an internal control on recovery. This means that much narrower zones can be eluted from paper chromatograms and smaller groups of fractions taken from columns for re-chromatography or counting. Secondly, methods can often be designed where the second label is added in the form of labelled steroid to the material to be extracted; the second isotope now acts as an internal control on recovery throughout all the steps of the procedure, and the error of such methods is confined to the volumetric error of adding the labelled carrier steroid to the material, the error of mixing with the latter, and the error of counting the two isotopes (Ayres *et al.*, 1958; Peterson *et al.*, 1957; 1959). This is subject, however, to the overriding necessity for complete separation of the steroid derivative from the labelled derivatives of other steroids and from other compounds containing the group used for coupling.

9. Complication and Simplicity

Just as any estimation method involves a compromise between the two basic ways of obtaining specificity, so the choice of a method involves a compromise between simplicity on the one hand and accuracy and specificity on the other. This has been put well by Neher (1958b) who points out in connexion with methods needed in clinical chemistry that it is unreasonable to expect specific methods in a field as complicated as the steroids to be easy or simple. Only a small number of situations can be expected to arise in which one steroid is present in a natural source which is free from substances interfering with crude methods of separation and determination. It is, however, reasonable to ask where the line should be drawn when faced with a choice of methods of varying simplicity and supposedly varying efficiency. Some specific suggestions will be made in Chapter 6 but a few general considerations should be made here.

In a few situations it is possible to estimate steroids with one chromatographic step with reasonable specificity. Cortisol in the blood of many species, the six major 17-oxosteroids of urine and the three major reducing metabolites of cortisol in human urine are examples. All these, however, will only be estimated with reasonable accuracy and specificity if the methods of extraction and purification are such as to provide a perfect final chromatogram. Such methods are all likely to be upset badly from time to time by oddities of diet or medical treatment (Neher, 1958b).

In many situations a good method will only be obtained by using several chromatographic steps and in these cases the use of a modifying reaction will usually be simpler than juggling with solvent systems. For many or most of them acetylation will be adequate or ideal, and with the rapid method described (p. 358) it is not inconvenient, taking about three hours, most of it being "waiting time". The extra time involved in a two-chromatogram procedure is not usually a serious disadvantage as long as the fraction taken up by "working time" is reasonable.

The complications of isotopic methods are in some ways not so large as they seem at first sight. Their sensitivity means that very small quantities are needed so that only small volumes of material need be extracted. This in turn means that the weight of material in the extract is suitable for separation on paper without the use of column methods. With the method of Peterson et al. (1957b; 1959) for example, only 10–12 ml of urine are needed for the estimation of aldosterone, and 1–2 ml would suffice for the estimation of cortisol. Only 0·5 ml of human plasma would be needed for the estimation of cortisol. The separations are all carried out on paper. The use of tritium-labelled acetic anhydride is of such general application that this type of method should be contemplated very seriously as potentially one of the simplest methods to adopt; the gas-flow counter or Ekco N664A Scintillation Counter are adequate for most tasks and are less expensive and bulky than the equipment needed for the full range of photometric methods used in steroid estimations.

The use of double labelling, although aesthetically pleasing and undoubtedly capable of precision, seems to the author to give rather small returns for the extra complication. In the first place the full capabilities of the method are only realized when very large specific activities or very long periods of counting are used. Automatic scanners, for instance, take several hours to scan the typical paper chromatogram and it is often inconvenient to bring the statistical error of counting below 5% (Berliner and Salhanick, 1956). Secondly, plating is a time-consuming technique and the error due to this part of tracer methods is of the order of ±3% even in careful hands. Finally it is to be doubted whether the use of the second isotope as carrier and internal control on recovery is really necessary; its only convenience is the simplicity of simultaneous counting in instruments such as the Packard Tricarb Counter. The unique advantage of a labelled carrier is that for very small quantities of steroid it is the only way of adding a detectable amount of carrier which is yet small compared with the amount of material to be estimated by coupling with a labelled reagent (Tait and Tait, 1960); thus it is the only way of providing an internal control for all the steps of a procedure—that is by adding the carrier to the material to be extracted. However, the main losses in such methods usually occur in the later stages of the overall procedure and tracer studies have shown that for at least some steroids $0\cdot01$ μg can be chromatographed on paper without loss in the absence of carrier. In the method of Peterson et al., (1957b, 1959), for example, the carrier is only added after extraction and acetylation of the extracted material; it thus serves only as a control on recovery from the subsequent chromatograms and the modifying step. The same functions would be fulfilled equally well by using a larger quantity of unlabelled aldosterone diacetate as carrier and estimating the carrier by a good photometric method on a part of the final eluate, and assaying the tritium in the remainder of this eluate; or alternatively by carrying out the photometric analysis on the whole sample before counting.

Finally there is the question of complications of apparatus and of materials. A very large factor in deciding what is reasonable is the manpower available; for instance a commercial electrically driven fraction-collector is excellent if local servicing is available, but the time wasted when it goes wrong can be enormous in the absence of a good electrician. In such cases a few days spent making a gravity-driven fraction-collector of the type described by James et al., (1951) would be time well spent. Again it is possible to keep a large number of tanks running for paper chromatography if there is plenty of technical help to keep them filled with purified solvents; with a "one-man show" it is better to run two or three well-tried systems and adapt one's methods accordingly (p. 337). In the author's view column methods suffer from grave disadvantages in terms of the time and space needed for operating them unless the work to be done involves only a small number of estimations per month per worker, or the fractionation is a very simple one. With the methods of determination available today there is rarely any need to use large amounts of material for quantitative estimations, and hence the extracts to be run on chromatograms are rarely beyond the capacity of convenient widths and thicknesses of filter paper (p. 186). Nor is there convincing evidence that the error of column methods for various classes of urinary steroids is much less than that of the best comparable paper methods. It is not unreasonable to hope that this difference, if it exists, will be diminished in the future so that the great productivity of paper methods will be available without any sacrifice of accuracy.

CHAPTER 5

STRUCTURAL ANALYSIS AND IDENTIFICATION OF STEROIDS BY CHROMATOGRAPHY

1. The Basis of the Classical Method

THE CLASSICAL method of establishing the structure and identity of an organic compound consists of a number of procedures which are by now so standardized and familiar that their theoretical and logical basis is often forgotten, and perhaps sometimes never properly learnt. There are several large textbooks of organic chemistry which offer no systematic account of this topic although it is the very basis of the coherence and success of structural chemistry. A brief survey of the classical method must therefore be given so that chromatographic methods can be compared with it on the basis of logic rather than fervour (Bush, 1954a).

The various procedures of the classical method can be divided into two classes. The first is those procedures giving what can be called contributory information, usually of a quantitative nature, and the second is those which aim at giving exclusive information. In the first class come the determination of the empirical and molecular formula, the numerous methods determining the number and type of characteristic groups in the molecule, and the determination of physical properties such as absorption spectra, molecular rotation, and so on. While these methods must give results consistent with the proposed structural formula, and are all-important in discovering the probable structure of an unknown compound, they are rarely if ever exclusive of all other formulae owing to the variety of known or possible organic compounds; they do not "establish" a structure simply because they do not exclude others.

The proof of structure, therefore, must rest on the second class of procedures. This class is distinct from the first in that the aim is to show that the unknown substance is identical with a known compound. The known compound must either be a close derivative of the unknown obtainable by an unambiguous reaction with the latter, or it must be identical with the unknown and obtained from a compound of established identity by an unambiguous reaction. Two properties can be used for comparison of the known and unknown compounds: firstly their melting points separately and mixed; secondly the "finger-print" regions of their infrared spectra. The crux of this method is the comparison of "unknown" and "known" compounds by methods known to detect differences with great sensitivity.

Why is the mixed melting point such an admirable check on identity, and a sharp melting point a good criterion of purity? It is important to recognize that the first and most important answer to these questions is simply that it has been found to be so; in other words the method is justified largely by practice. The reverence accorded to this method is probably due to the terrible mess in which organic chemistry found itself before Chevreul pioneered the use of melting points in 1823. In 1835, Wöhler (Partington, 1937, p. 216) compared organic chemistry to a "primeval jungle", and the whole field was littered with erroneous conclusions largely if not entirely due to misidentification of compounds. In contrast to the properties determined by the first class of procedures the melting point of a pure substance is usually a discontinuous property; its sharpness is extremely sensitive to impurities when these are of molecular weight comparable with the main component, and it is almost invariably depressed by mixture with a different compound. The reason for this is that the lattice energy of most organic crystals is largely due to forces of relatively short range; a very slight disturbance of molecular packing by impurities therefore lowers the lattice energy very considerably and hence the melting point of the mixture. The mixed melting point is taken with the known and unknown substance in approximately 1:1 ratio to obtain the maximum depression, which is usually but not always found near this ratio.

Although the logical difference between the two classes of procedure is great, the undepressed mixed melting point is by no means the "Law of the Medes and Persians" that it is sometimes made out to be. This criterion of identity of two substances is purely pragmatic and based upon an implicitly statistical appraisal of the results obtained with the method. It is of course not applicable to many amorphous substances, substances which decompose before melting, substances lacking solid derivatives and to substances which are not available in more than minute amounts. A few substances give no definite melting point or several melting points, passing through different stable transition states or crystalline forms (e.g. oestrone, m.p. 254°, 256° and 259° C). At least two steroids (cortisone and

11-deoxycortisol and their acetates) are known with indistinguishable melting points and which show no depression of melting point on mixing.

The practical situation in the steroid field however seems by now to be agreed upon, in that undepressed mixed melting points are considered to be reasonably good criteria of identity, but sharpness of melting point is only a fair criterion of purity in most cases. Indeed paper chromatography has become almost a routine procedure in some laboratories for checking the purity of steroids and few chemists can by now have escaped the dismay of seeing a white crystalline material of sharp melting point shown up as two or more components on a paper chromatogram. Much scepticism, however, still remains as to the acceptability of using chromatographic methods to identify steroids *de novo*, particularly if crystalline material has not been obtained or an infrared spectrum has been omitted.

2. The Basis of Chromatographic Identification

The first thing to be noticed is that one feature of the mixed melting point is not only missing from chromatographic identification but is in fact the main reason for the success of partition chromatography; this is the fact that R_F values in partition systems are not affected by moderate amounts of impurities (see p.2). From the formal point of view, therefore, it must regretfully be admitted that any chromatographic analysis falls into the first rather than the second class of classical procedures. In other words the properties measured are continuously variable, largely independent of impurities, and cannot give the exclusive type of information that is given by the mixed melting point.* On the other hand there is a rational way of overcoming this difficulty which can be justified by reconsidering our placing of the infrared spectrum in the second class of properties rather than the first. This classification was made purely because the "finger-print" region, as its name suggests, has rapidly been recognized as quite characteristic of a substance. But it is in fact a continuous function of wavelength and light absorption and subject to quantitative variations of detail. Why has it been so rapidly accepted as a valid criterion of identity? The answer is that it is realized that this range of frequencies is made up of harmonics and compound atomic oscillations whose parameters are determined by almost every conceivable property

---

* With the possible exception of displacement in adsorption chromatography (p. 42).

of every atom and bond in the molecule concerned, so that the probability of finding two molecules giving the same spectrum in this range of frequencies is extremely small, and in practice difficulty in distinguishing between different compounds is encountered only with higher aliphatic homologues. In other words the strength of this criterion is again statistical in origin.

This now gives us the clue to a satisfactory basis for using continuously variable properties of substances as a means of identification; we must measure a large number of them. This has already been the basis of identification of proteins and other substances of large molecular weight and the difficulties inherent in the classical view of chemical identity in this field have been discussed thoroughly by Pirie (1940). By "a large number of properties" is meant a number which is comparable with the number of significantly different properties that can be recognized in a given compound. This entails that all really significant properties should be examined if possible. For instance the identification of a steroid alcohol is improved by showing that both the free steroid and its acetate are chromatographically identical in solvent systems that give low or moderate R_F values, but the finding that other esters such as the propionates and hexanoates are also identical on chromatograms rarely adds to the weight of the evidence. In contrast it should be noted that in most cases the finding that two different esters, rather than one alone, gave no depression of melting point when mixed with their suspected reference esters would definitely enhance the weight of an identification by classical methods.

A further point is that since we are always involved with the first class of properties in chromatographic identification it is essential that they be measured by methods giving the greatest sensitivity to changes in those properties. Thus chromatographic identification of steroids should always be based on runs in which R_F values are low, other things being equal. The range 0·75–1·0 is useless for this purpose except in rare cases. Similarly identification of Δ^4-3-oxosteroids by means of their absorption of ultraviolet light or their reaction with dinitrophenylhydrazine is unsatisfactory unless more specific methods such as the fluorescence with NaOH are also used (see p. 379).

This general principle, namely the statistical nature of evidence of chemical identity, underlay the introduction of a new approach to chromatographic identification by Zaffaroni and his colleagues (Reich, Nelson and Zaffaroni, 1950; Burton *et al.*, 1951a, b; Zaffaroni, 1953) who carried out microchemical reactions and ran the products on chromatograms as well as the original steroid under examination. The principal reactions used were the formation of acetates, the oxidation of the acetates with chromic acid, and the oxidation of α-ketolic steroids with periodate. The author extended this procedure on the basis of the above arguments with the aim of subjecting a large number of properties to scrutiny by a sequence of operations with an unambiguous interpretation (Bush and Sandberg, 1953; Bush, 1954a; Bush and Willoughby, 1957; Bush and Mahesh, 1959b, c). By the use of sodium bismuthate (Norymberski, 1952) it was possible to carry out degradation of 17-hydroxy-C_{21} steroids to 17-oxosteroids and then examine nuclear substituents in turn by esterification and oxidations.

Later in this chapter these methods will be considered in detail but it should be remarked here that the essence of this method is the design of a sequence that subjects each substituent and skeletal feature to separate scrutiny. A possible criticism of the procedure must also be answered at once. It might be objected that although this method depends upon degradations analogous to the classical method for identifying natural steroids, each degradation product is incompletely identified chromatographically, considerable reliance being placed upon circumstantial evidence. In other words it might be said that it is an example of the old fallacy that several weak arguments add up to a strong one, and is therefore subject to the equally old gibe that "an argument is like a chain—it's as strong as its weakest link". This criticism, however, is based upon a mistaken notion of the logical nature of the steps of this procedure. Unlike a classical degradation in which we can say, "An operation A upon the substance P (unknown) gave rise to a substance Q' identical (by classical criteria) with a known substance Q", the chromatographic equivalent yields the statement, "An operation A upon the substance P gave rise to a substance Q' which had the following more or less characteristic properties of a known substance Q". This is not by itself an identification of Q' with Q; it is a simple statement of facts, and hence quite unexceptionable as evidence. Thus a classical degradation yields among other things a number of statements of identity. The logical equivalent of the mixed melting-point test is "Q' is not not-Q'": i.e. another type of null-hypothesis test (Jeffries, 1957). A chromatographic analysis of a degradation yields a number of statements about properties and changes thereof.

One more point about chromatographic identification needs discussing, namely the relevance of circumstantial evidence. In general this is small when the identification of a new steroid, or of a steroid in a new situation,

STRUCTURAL ANALYSIS AND IDENTIFICATION

is in question. When methods of estimation are considered, however, it becomes an important factor which it is perfectly reasonable to take into account. Since a good chromatographic step improves the specificity of an estimation method very considerably, circumstantial evidence has in fact more weight than in the case of a method not using chromatography. Criticisms are sometimes directed at chromatographic methods of estimating steroids which would be appropriate if the question of identification of a new steroid were involved, but which fail to give due weight to circumstantial evidence. No one demands an infrared spectrum to substantiate an estimate of blood sugar concentration by a good standard method. On the other hand the acceptance of standard methods as meaning blood "glucose" concealed for some time the fact that foetal "blood sugar" was largely fructose and not glucose. As with sugars so with steroids: let us not quibble too much about the identification of dehydroepiandrosterone in human urine. On the other hand, we must be on our guard if similar methods are claimed to estimate this substance in the urine of birds or tortoises and we may wish to re-examine our methods when studying babies instead of adults.

The purpose of this chapter is to show how far one can reasonably go in the identification of steroids by chromatographic means. Klyne (1957, p. 191) has given a very reasonable account of the dangers of this approach and recommended that chromatographic identification should be followed up by isolation of visible quantities of material for identification by classical methods. While agreeing with his cautionary views one cannot agree that "visible amounts" are really necessary. Infrared absorption spectra can be obtained of 15–50 μg steroid in many laboratories and recently Dr. W. Mason (Rochester, N.Y., U.S.A.) with a new optical arrangement of the Perkin-Elmer instrument has succeeded in getting excellent spectra with as little as 1 μg of steroid. Spectra in concentrated sulphuric acid (Zaffaroni, 1950, 1953) are almost as characteristic when combined with chromatographic information. Visible amounts are only needed for melting points and rotations: with few exceptions neither of these are necessary for identification when infrared spectra can be obtained. Many problems simply cannot be solved unless identification can be achieved with 100 μg of material or less; others can only be solved by the expensive and tedious extraction of enormous quantities of urine or other material. For instance the classical proof of the identification of aldosterone in normal human blood would require the extraction of about a thousand litres of human plasma (approx. 375 gallons of blood) to obtain about 0·5 mg material—assuming 70% recovery, which would be optimistic for such an operation: and there would still not be enough for a measurement of optical rotation. It is hoped that the examples given below will show that many steroid structures can be determined completely and unambiguously by chromatographic methods when about 100 μg is available, and that further developments can reasonably be expected to improve the scope of this procedure. With smaller amounts of material success would be likely if more use of infrared spectra were made than indicated in the examples given.

3. General Procedure for Structure Determination

The first principle is that the maximum amount of information must be obtained about the substance, and about each of the derivatives which is prepared. On the other hand useless additional information must not be sought or material will be wasted; e.g. if a substance reduces blue tetrazolium strongly in the cold and is oxidized by sodium bismuthate to a substance giving a purple colour with alkaline m-dinitrobenzene it is better to examine the properties of its acetate than to show that it does or does not give the Porter-Silber reaction. The most important practical point is to keep a semi-quantitative check on the yield of each reaction, and in general it is best to carry out such reactions in parallel with a control substance which has the group undergoing reaction, and preferably any other groups which may give minor reactions if the identical reference steroid is unknown or unavailable. For instance some samples of bismuthate oxidize 11β-hydroxyl groups to a minor or even major extent; while such samples should be avoided it is wise to use a substance containing an 11β-hydroxyl group as a control, to keep a check on this undesirable reaction. The reason for keeping a check on yield is that contaminants of the original substance, undetectable by the method used to discover the latter, may give rise to detectable products in the reactions employed. The latter products may then be taken to be derived from the substance that was originally detected. For example the substance X_6 (De Courcy, Bush, Gray and Lunnon, 1953; Gray, Hellman, Lunnon and Weiner, 1955) was tentatively and wrongly given the structure P^4-17α,20α-ol-3, 11-one largely because periodate oxidation apparently gave rise to an incompletely identified 17-oxosteroid. Subsequent isolation with a better separation gave a product which gave no 17-oxosteroid on periodate oxidation (author, unpublished). The assessment of the original oxidation was in error probably because the original sample was contaminated with urinary steroids giving 17-oxosteroids on oxidation with periodate; the error would not have been made if it had been recognized that the apparent yield of the reaction was low, or if the chromatogram of the oxidation product had been examined by both the Zimmermann reaction and the fluorescence reaction with NaOH. The last point is another example of the rule of obtaining the maximum information from every derivative; if there are two detectable groups they should preferably both be followed by using appropriate tests throughout the series of derivatives, whether or not they are changed by the reactions used.

Two other rules are important if mistakes are to be avoided. The first is that some part of each derivative must be run on a single-length chromatogram so that all potential reaction products are inspected. For instance a steroid giving an unexpectedly non-polar acetate might be mixed with one giving a monoacetate; the former might be missed if an overrun chromatogram were used for the esters. The second is that each product should be located by test reactions and eluted with great accuracy (see Chapter 3). In general the substance should be eluted from a chromatogram used for group separation and then re-run for 16–36 hr in a system giving the material an R_F value of 0·1–0·35.

All comparisons of chromatographic mobility must be made with systems giving R_F values in the critical range (0·15–0·75) where possible, and to obtain the best comparisons the unknown should be run on a sheet with suspected reference steroids alongside. If the reactions are done cleanly, so that the extract applied to a chromatogram contains 10–20 μg material, differences in R_F of $\pm 0·01$ will be detectable and significant in the range 0·1–0·6 approximately. The original substance or at least one derivative should be compared with the suspected reference steroid on an overrun chromatogram in a system giving an R_F value of 0·1–0·3 and run for a time sufficient to bring the material at least 10 cm from the origin.

Apart from observing scrupulous cleanliness the only other general rule is that the reactions used for preparing derivatives should be in a stepwise sequence. That is to say that except where it is uneconomical, intermediates should be examined on chromatograms, and a reaction giving D from A should be avoided in favour of a sequence giving A → B → C → D where possible. Thus for instance an unidentified steroid reducing blue tetrazolium and suspected to contain a 17α-hydroxyl group may be oxidized directly with chromic acid in order to obtain a 17-oxosteroid, whose chromatographic behaviour will give a quick indication of the probable number of nuclear substituents. While this is useful in suggesting the best line of further attack, and in bringing to light unexpected complications, it is not of much use on its own for the actual identification since an unknown number of oxidations are involved.

Because of circumstances I have not made use of infrared absorption spectra to the extent which is possible. The following examples should be read with this in mind; an early infrared absorption spectrum would be advisable in any problem where the amount of material seemed likely to be insufficient for a complete series of derivatives to be prepared.

In order to detect major impurities an amount of material is used which would allow the detection of 5% of any impurity giving a product detectable by the colour or other reaction used for the main derivative; 10–15 μg spread over a width of 1·3 cm is sufficient with most reactions and derivatives—the sheet can be cut to split this area in two so that two means of detection can be used. To avoid waste due to unexpected reactions or losses a small amount of material is subjected to each reaction and the products examined on a chromatogram before proceeding with the bulk of the material and isolating the derivative.

The reader unfamiliar with structural chemistry could read Klyne's succinct summary of structure determination as an introduction to the following examples (Klyne, 1957, p. 63).

4. Examples of Structure Determination

(a) Cortisol

Suppose that the extract of a sample of blood from a hitherto unexamined species has given a zone on a chromatogram of ultraviolet-absorbing material having the same R_F as cortisol on a sheet run with a group-C system. If 50 μg were obtainable we could proceed as follows:

The patch is eluted (Chapter 3) and run in a typical system of group B or C for 16 hr, starting from a "lane" 3 cm wide, with 50 μg cortisol as a marker on a similar "lane" 4 cm away on each side. The marker spots are located and a zone centred on the markers is cut out and eluted. The eluate is filtered and dissolved in ethanol at a convenient concentration. A volume containing 10 μg is made up to 3 ml and its spectrum obtained from 210 to 400 mμ. This solution is then evaporated in a small test tube by a jet of filtered air and 3ml of H_2SO_4 added. After stirring with a small glass rod for a few seconds the solution is decanted into a clean cell and the spectrum observed between 210 and 600 mμ after 30 min; or after longer times as long as they are standardized (Zaffaroni, 1953).

A volume containing 2 μg is now taken from the original solution and run for 6 hr in the same group-B system after concentrating to a small spot on the origin (Chapter 3), together with markers of 0·05, 0·5, 1·0 and 2·0 μg cortisol. The sheet is cut into strips, examined under ultraviolet light and treated with the diluted blue tetrazolium reagent (p. 379). After observing the reduction of blue tetrazolium for 5 min the strips are heated and the NaOH fluorescence observed.

Another 2 μg is dissolved in 0·4 ml of acetic anhydride–pyridine (5:1 by vol) and heated at 60–65° C for 15 min (p. 358). After evaporation (p. 358) the residue is dissolved in ethyl acetate and methanol and run as a concentrated spot in a group-B system for 3·5 hr (single length) with four standard amounts (0·1–2·0 μg) of cortisol treated similarly alongside. The strips cut from the sheet are examined under ultraviolet light (2537Å) and by the combined tetrazolium–NaOH fluorescence reaction.

The remainder of the solution is now evaporated and acetylated by the quick method (p. 358). After running in a group-B system for 16 hr the cortisol acetate is eluted after locating with markers and by its own ultraviolet absorption. The eluate is dissolved and divided into two portions in the ratio 2:1. The smaller portion is dissolved in saturated methanolic potassium bicarbonate and the product extracted after 16 hr (p. 359). A volume containing 1·0 μg is run as a concentrated spot for 6 hr in a group-B system and the chromatogram (with the usual standards of cortisol) examined by the combined tetrazolium–NaOH reaction. After confirming that the product has the properties of cortisol in these tests, the remainder of this portion is oxidized with sodium bismuthate and one half of the product run from a "lane" 1·3 cm wide, with two sets of standards of 11β-hydroxyandrostenedione (1,2,5 and 10 μg alongside) for 16 hr in a group-B system. The test strip is cut down the middle and one half (with one set of standards) examined by the NaOH fluorescence reaction; the other half and other standards are treated with alkaline m-dinitrobenzene. After confirming that this product has the properties of 11β-hydroxyandrostenedione in these tests the residual material is acetylated and run as a 2-cm lane for 3·5 hr in a group-B system with 11β-hydroxyandrostenedione as standard. The substance is eluted after confirming that the material absorbs ultraviolet light in the same zone as the standards and has not been altered in this respect by acetylation. It is then oxidized with chromic acid and run for 3·5 hr in a group-B system as above but with adrenosterone as standard. The two halves of the sheet are examined by the NaOH reaction and the Zimmermann reaction as above.

The larger portion of the acetate of the original substance is now oxidized with chromic acid and 5·0 µg is run as a concentrated spot with cortisone acetate standards in system T/75 (single length), using ultraviolet and the combined blue tetrazolium–NaOH reaction for detection. The major fraction of the cortisone acetate is now eluted and subjected to the same procedure used for cortisol acetate except that the amounts are larger to allow for the lower sensitivity of cortisone and adrenosterone in the NaOH fluorescence test. The only difference is that the bismuthate oxidation production is run alongside adrenosterone for 16 hr in a group-B system (acetic acid type) and eluted after location with ultraviolet light. The eluate is then subjected to chromic acid oxidation and the product run for 3·5 hr in a group-B system ("typical"), the strip being examined in two halves by the NaOH reaction and the Zimmermann reaction.

The reactions used in the above procedure are given in Fig. 5.1. The whole procedure is carried out with amounts such that impurities at the 5% level would be detected on every chromatogram used for colour or fluorescence reactions. If a 10% level of impurities was accepted, and the second route to adrenosterone (via cortisone acetate) omitted, 10–15 µg would suffice for the whole series of tests. The practical details of this procedure will be found in Chapter 3 and Appendix II.

It is instructive to neglect for a moment the evidence obtained from spectroscopy in H_2SO_4 (or infrared) and consider the evidence of the chromatograms and degradations in isolation. The original substance (I) contains two detectable groups; one a reducing group, the other an $\alpha\beta$-unsaturated ketone group; the latter must have a bridgehead double bond in a two-ringed alicyclic system. There must be no other double bonds in these two rings and no hydroxyl group at positions 2 or 7. If the substance is a steroid it must be a Δ^4-3-oxo-steroid or a Δ^{14}-16-oxo-D-homosteroid (p. 297). The reducing group reacts with BT at a speed and with an intensity characteristic of α-ketols, α-hydroxyaldehydes, and enediols. It is protected from oxidation with chromic acid by acetylation, but not from oxidation with alkaline BT: and it is removed by oxidation with bismuthate. The last oxidation (S) results in the formation of a substance (V) with an active methylene group. The methylene group gives a colour with alkaline m-dinitrobenzene characteristic of 11β-hydroxy-17-oxosteroids. After oxidation with chromic acid, however, the colour reaction changes to the pinkish colour developing partly in the cold which is characteristic of 11,17-dioxosteroids. The chromatographic mobility and colour reactions of the bismuthate oxidation product are unchanged by acetylation showing that no esterifiable group is present in the nucleus. The ΔR_M on oxidizing the acetate with chromic acid [2] is of the same order as the ΔR_M on oxidizing the nuclear function of V with chromic acid[6]; there can therefore

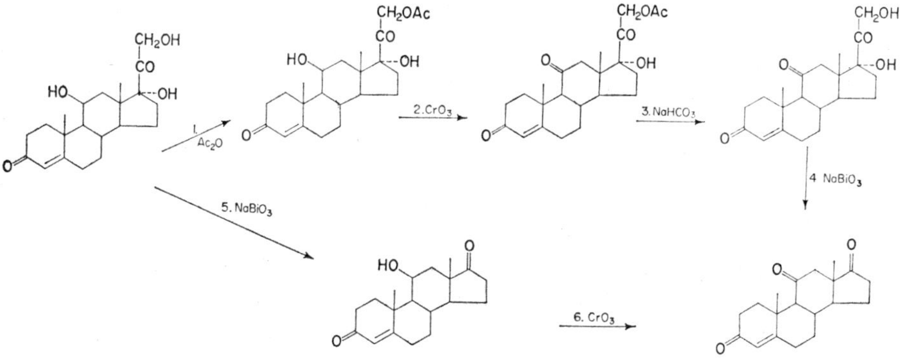

Fig. 5.1. Degradation of cortisol. Reaction 1 obtains the ΔR_{Mr} of the 21-OH group (ketol): reaction 2 the ΔR_{Mr} of oxidising the 11β-OH group: reaction 3 confirms that the reverse of reaction 1 is independent of the group altered by step 2, and that the ester group is easily hydrolysed as in steroid α-ketols. Reactions 4 and 5 provide both the ΔR_{Mr} of the oxidation of a dihydroxyacetone sidechain to a 17-ketone group and the emergence of a new colour-reaction. Reaction 6 confirms that step 2 is independent of the group removed by steps 4 and 5; and that step 1 has not protected more than one group from oxidation with CrO_3; it also produces the characteristic change in the Zimmermann reaction due to the change in the substituent at C-11.

be no extranuclear oxidizable group apart from that which is protected by acetylation and which reduced BT.

In addition to all this the chromatographic mobilities of the substance and its derivatives are identical with those of cortisol and its derivatives obtained in the control sequence of reactions employed. In the case of adrenosterone these mobilities have been compared on overrun chromatograms in two types of solvent system giving characteristic ΔR_{Ms} values of the ketone, hydroxyl, and unsaturated ketone groups.

Two important points should be noted. Firstly it is a good rule to observe the effect of acetylation both on the original substance and on the bismuthate oxidation product, thus obtaining a separate check on esterifiable groups in the nucleus and in the side chain. Secondly it is valuable to observe the effect of hydrolysis of any esters obtained, first with methanolic bicarbonate and later if necessary with sodium carbonate or hydroxide. Only the 21-acetoxyl group is completely removed by bicarbonate in the cold so that typical C_{21} diesters are saponified to monoesters, thus providing another characteristic (Bush and Willoughby, 1957).

The possession of an oxidizable group in the nucleus not protected by acetylation is an almost specific indication of an 11β-hydroxyl group in the steroids; the ΔR_M of oxidation is also correct. Further evidence could be obtained by showing that this group was esterified by acetic anhydride under forcing conditions (Oliveto et al., 1953) or methylsulphonyl chloride (Fried and Sabo, 1957).

The most important aspect of such a sequence of operations is the behaviour of the fully oxidized derivative (i.e. without disrupting the nucleus) and the stepwise route by which it has been obtained. From the information to hand we know in the above case that the final derivative contains at least three groups. One, the $\alpha\beta$-unsaturated ketone group, has been retained (and detected) all along. The second is a group derived by bismuthate oxidation and detected by its activation of a methylene group. The third is derived by oxidation of a group not protected by acetylation. Quite apart from circumstantial evidence the last two groups could only be ketone groups derived from secondary and tertiary hydroxyl groups. Sulphur analogues can be excluded for a number of obvious reasons. If this is so then the mobility of the final derivative must be determined by the sum of the effects of these groups and the hydrocarbon skeleton to which they are attached (p. 111). The NaOH fluorescence tells us that the 3-oxo-Δ^4-octahydronaphthalene structure must be present, and the hindered hydroxyl group requires some rigid structure to provide the necessary hindrance (e.g. even the 11β-hydroxyl group of 19-nor-cortisol is easily acetylated (Zaffaroni, Ringold, Rosenkranz, Sondheimer, Thomas and Djerassi, 1954, 1958)).

It is possible to satisfy these properties by a non-steroidal hydrocarbon skeleton, or at least to suggest structures which could not be excluded as possibilities. However, it is extremely unlikely that such structures would not show differences from the reference steroids (p. 127) of mobility or colour reaction in one or more of the derivatives. Furthermore the influence of the hindered hydroxyl group (and its derived ketone group) on both the ultraviolet absorption of the αβ-unsaturated ketone group, on the sensitivity of the NaOH fluorescence reaction, and on the velocity and type of colour reaction given by the active methylene group with alkaline m-dinitrobenzene (Antonucci, Bernstein, Heller, Lenhard, Littel and Williams, 1953; Wilson, 1954; Bush and Willoughby, 1957) are a complex of properties linking all three groups which are unlikely to be matched by a non-steroidal compound. If this be admitted, then the sensitivity of chromatographic mobility, and the ΔR_M on changing the type of solvent system, is sufficient to justify the identification of the final derivative as adrenosterone. To demonstrate this point the mobilities of some typical steroid polyketones are shown in Table 5.1.

We now come back to the reducing group. It must be linked to the nucleus at the point of attachment of a tertiary hydroxyl group in order to give a nuclear ketone group on oxidation with bismuthate (a, Fig. 5.2) and contains an additional hydroxyl group which is esterifiable. The presence of a third hydroxyl group in the reducing function is quite incompatible with the mobility either of the original substance or of its acetate, thus excluding enediols (b, Fig. 5.2) from consideration. A hydroxyaldehyde (c, Fig. 5.2) would not be protected by acetylation from partial oxidation with chromic acid, leaving only the ketol as satisfying all observed properties (d, Fig. 5.2). Again it is possible to maintain that the addition of a suitable balance of polar and non-polar functions to the side chain would provide a substance with a mobility identical with that of cortisol. It is, however, extremely difficult to suggest such a structure using the ΔR_M values to be expected. It could not contain a hydroxyl group because of the observed ΔR_M on acetylation. Such functions would also require that the 21-hydroxyl group became a secondary hydroxyl with the result that the ΔR_M of acetylation would now be different from that obtained with the primary 21-hydroxyl group of cortisol; the acetate of such a substance would probably be only partially hydrolysed by bicarbonate, and reduce BT less effectively (Meyer and Lindberg, 1955).

The negative evidence provided by the chromatographic information is also extensive. From the spectra and fluorescence with NaOH it is certain that there are no hydroxyl or ketone groups at positions 1, 2, 4, 5, 6, 7 or 12, assuming the steroid nucleus present. There are no double bonds apart from C-4,5 in any of the rings except possibly C-13,14 or C-14,15. However the latter would almost certainly give rise to detectable effects upon chromatographic mobility and changes in the Zimmermann reaction. Carboxyl or other acid groups would both have large effects upon mobilities

Table 5.1. R_M Values of Some Useful Steroid Polyketones

Steroid	Solvent system			
	1. L/85	2. L'T21/85	3. D/A90	4. DX21/A90
αA-3,17-one	−0·05	—	0·52	0·05
βA-3,17-one	−0·08	—	0·43	−0·02
A^4-3,17-one	0·38	−0·25	0·99	0·38
αA-3,11,17-one	0·52	−0·15	—	—
βA-3,11,17-one	0·52	−0·15	—	—
A^4-3,11,17-one	1·10	0·26	—	0·90
A^4-3,6,11,17-one	—	0·54	—	—
A^4-3,6,17-one	0·70	0·05	—	—
P^4-3,20-one	−0·14	−0·50	0·54	0·02
P^4-3,11,20-one	0·87	−0·07	1·45	0·75
P^4-3,6,20-one	0·15	−0·19	—	—
A^4-3,7,17-one	—	—	—	—
P^4-3,7,17-one	—	—	—	—
3MeO-O-7,17-one	−0·40	—	—	—
3MeO-O-17-one	−0·69	—	—	—
αP-3,20-one	−0·43	—	0·11	—
βP-3,20-one	−0·52	—	0·05	—

Solvent systems:
1. Light petroleum–methanol–water (100: 85: 15 by vol.).
2. Light petroleum–toluene–methanol–water (67: 33: 85: 15 by vol.).
3. Decalin–acetic acid–water (100: 90: 10 by vol.).
4. Decalin–xylene–acetic acid–water (67: 33: 90: 10 by vol.). Run at 22–26° C.

and produce serious losses during the alkali washes used in the extraction procedure and in the saponification of the acetates. Aldehyde groups would give carboxyl groups on oxidation with chromic acid, and hence both losses on extraction and large positive ΔR_M values in the methanolic solvent systems and moderate positive values in the acetic acid system. Amino groups would show up by greatly reduced R_F values, by their easy acetylation, and large positive ΔR_{Ms} on changing to the acetic acid system. Lactone groups would be detected by colour reaction with Zimmermann's reagent, if not by ΔR_{Ms} values. Secondary and tertiary amine groups would produce very low R_F values in all the systems used.

STRUCTURAL ANALYSIS AND IDENTIFICATION

Finally it must be emphasized that an infrared absorption spectrum could be obtained upon the original substance and one or more derivatives after chromatographic isolation.

The greatest weakness of this procedure will be noticed in the above example; namely that an empirical and molecular formula are not obtainable by present methods so that one is driven to rely upon the sensitivity of chromatographic mobilities to changes in structure for the final justification of identifying the nucleus as steroidal. While possible in principle

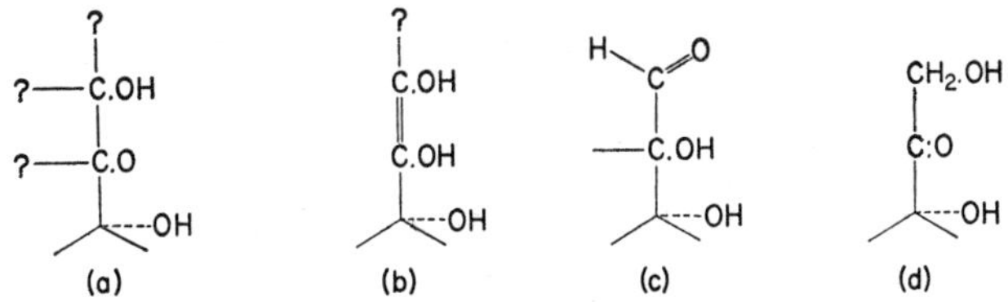

FIG. 5.2. Various reducing sidechains that could "mimic" a steroid α-ketol (dihydroxyacetone sidechain, d). (See text, p. 266.)

it would be laborious in practice to exclude the presence of an ether-linked oxygen atom, for instance, if this was linked to any alkyl radical of the right size; such a "hidden" group would be immediately apparent in the classical method by reason of the molecular formula. Such a group, would, however, also be detected in an infrared absorption spectrum, and a more complete identification of the steroid nucleus is suggested on p. 269.

The following examples will be described in less detail.

(b) $2\alpha\text{-}Me\text{-}P^4\text{-}6\beta,11\beta,17\alpha,20\xi,21\text{-}ol\text{-}3\text{-}one$

Bush and Mahesh (1959c) found this substance in urine after the administration of 2α-methylcortisol to a young male. It was first found lying on the origin of a single-length chromatogram run in system B/50 as a zone absorbing ultraviolet light (2537Å) and giving a brilliant yellow fluorescence with NaOH, The zone was isolated from a large scale chromatogram on Whatman No 3MM paper and a small fraction acetylated. The acetate had an R_F of 0·09 in a group-B system and gave a minor product with R_F 0·24. Oxidation with either bismuthate or periodate gave a substance at 0·4 cm on a chromatogram run for 5·25 hr ($R_M \approx 2\cdot5$) in the same system, absorbing ultraviolet light, giving a yellow fluorescence with NaOH, and a purple colour with alkaline m-dinitrobenzene. A minor

product giving the same reactions was found at 4·7 cm ($R_M \approx 1\cdot0$).*
The two oxidation products were isolated from an overrun chromatogram in the same system and acetylated. The acetylated products both had lower R_M values than the starting materials (R_M 1·5 and 1·0 group-A system) and were apparently single compounds. When these acetates were oxidized with chromic acid the minor product was unchanged but the acetate of the major product was altered to a substance having the same R_F as the minor product's acetate. The minor product and its acetate gave a pink colour with alkaline m-dinitrobenzene; likewise the derivative of the major product now gave a pink reaction with this reagent in the cold while its precursor (the bismuthate oxidation product of the original material) had given only a bluish-purple after heating. These properties of the major product are characteristic of the presence of an 11β-hydroxyl group so it was inferred that the minor product was probably an 11-ketone and the major product the related 11β-alcohol.

There was evidence at this stage, therefore, of a Δ^4-3-ketone group, a glycerol side chain, an 11β-hydroxyl group, and an esterifiable hydroxyl group in the nucleus of the major product. Since no such compound had been observed in control urines, circumstantial evidence suggested that the substance was derived from the administered 2α-methylcortisol.

Chromic acid oxidation of the bismuthate oxidation product yielded traces of less polar substances giving the Zimmermann reaction, and a considerable amount of material remaining on the origin. This could have been due to oxidation of a primary alcohol group to a mixture of aldehyde and carboxylic acids groups, or to ring-opening to form a *seco*-compound. The latter possibility and the known 6β-hydroxylation of C_{21} steroids in man (Burstein, Dorfman and Nadel, 1954) led us to subject the bismuthate oxidation product immediately to spectroscopy in alkali (Meyer, 1955) and to oxidation of the presumed 11-ketone with manganese dioxide (Amendolla, Rosenkranz and Sondheimer, 1954) which is specific for allylic hydroxyl groups. The spectra were clearly those of a 6β-hydroxy-Δ^4-3-ketone. The product of manganese dioxide oxidation was an apparently single substance with an R_F in a group-B system of 0·31 (starting material, R_F 0·1; $\Delta R_{Mr} - 0\cdot6$) and gave a new reaction; when the NaOH fluorescence was carried out the material gave an orange-yellow colour in the cold before the chromatogram was heated to produce fluorescence,

---

* These R_M values are very approximate and are only given for gross comparison with those of the acetates.

a reaction characteristic of Δ^4-3,6-diketones (Savard, 1953; Meyer, 1955). When this oxidation product was isolated from another chromatogram and treated with ethanolic KOH the spectral changes specific for a Δ^4-3,6-diketones were observed (Meyer, 1955). It was further noted that the oxidation only occurred when the chloroform medium was refluxed; 6α-hydroxy-Δ^4-3-ketones are oxidized readily at room temperature, (Amendolla et al., 1954). When this product was subjected to chromic acid oxidation for a short time (p. 363) no change occurred.

We now had evidence that the final oxidation product had the probable structure 2α-Me-A^4-3,6,11,17-one but not enough material with which to synthesize this compound for comparison. In order to obtain a direct comparison with an available reference steroid the original bismuthate oxidation product was reduced with zinc dust and acetic acid, a method specific for the reductive elimination of allylic hydroxyl groups (Fieser, 1953; see Appendix II). The product had the chromatographic properties of 2α-methyl-11β-hydroxyandrostenedione*. On oxidation with chromic acid a substance with the chromatographic properties of 2α-methyladrenosterone was obtained in good yield (Fig. 5.3).

Once again stepwise degradation revealed in turn the nuclear substituents, and the final derivative could be compared directly on a chromatogram with an available reference steroid. As with the case of cortisol both circumstantial and chemical evidence made the identification of the final product as 2α-methyladrenosterone extremely probable. However, it should be noted that the inference of the presence of the 2α-methyl group makes the case weaker than that of cortisol simply because of the lack of certain necessary reference substances. Thus for instance it is not certain that a 2β-methyl isomer would behave differently. Similarly reference steroids were not available for direct comparison with either the original metabolite or any of its intermediate degradation products. However, with the available material we were able to prepare a number of 2α-methyl-steroids in small amounts which were enough to show that the 2α-methyl group gave similar ΔR_{Mg} values in four 2α-methyl-Δ^4-3-ketones and did not affect the ΔR_M values of the 11-ketone, 11β-hydroxyl, or 17-ketone groups, or of the dihydroxyacetone side chain.

No attempt was made to assign an orientation to the 20-hydroxyl group because too little material was available for synthesizing the necessary 20-epimers, and the main point of the investigation was to determine what

---

* A small amount of the 17ξ-alcohol was also obtained (see original paper).

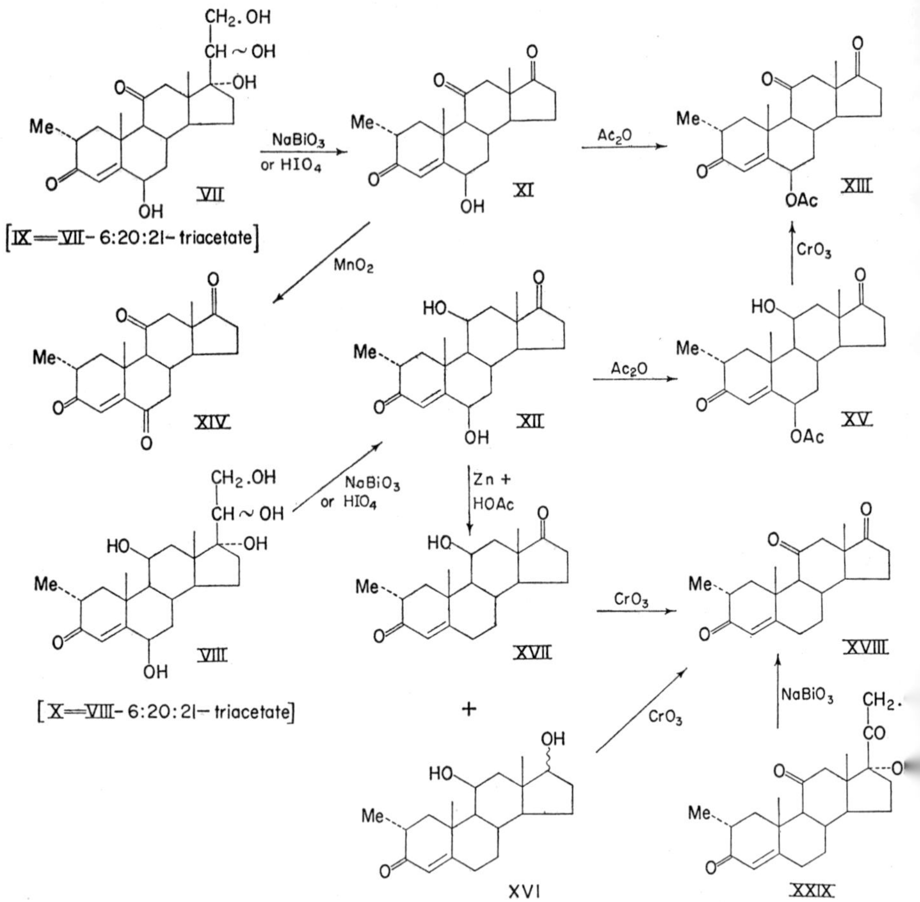

Fig. 5.3. Degradation of a 6β-hydroxysteroid metabolite of 2α-methyl-cortisol (Bush and Mahesh, 1959c. See text, p. 271). Reproduced by kind permission of the Editors of the *Biochemical Journal*.

was happening to the 11β-hydroxyl group. The evidence for β-orientation of the 6-hydroxyl group is threefold, namely:

(a) The 6β- and 6α-hydroxy-Δ⁴-3-ketones are distinguishable in Meyer's procedure of spectroscopy in ethanolic alkali;
(b) The oxidation with manganese dioxide did not proceed at room

temperature while typical 6α-hydroxy-Δ^4-3-ketones are oxidized completely at room temperature in 0·5 hr;

(c) The ΔR_M of the bismuthate oxidation product of the metabolite compared with 2α-methyladrenosterone was close to that of a typical 6β-hydroxyl group and too small for the 6α-hydroxyl group (Savard, 1954). The ΔR_{Mr} on acetylation of this product was too small for an equatorial hydroxyl group, again suggesting the axial 6β-orientation.

This example has been given to illustrate two main points. First, that a steroid suspected to contain a group close to another important functional group can only be studied effectively by sequential degradation when a reasonable number of similarly substituted steroids have been examined to determine the ΔR_{Mg} values for a number of other substituents. Secondly, that the final degradation product should be directly comparable in its chromatographic behaviour with an identical derivative prepared by an unambiguous reaction, in this case 2α-methyladrenosterone prepared by bismuthate oxidation of 2α-methylcortisone.

(c) 9α-F-αP-3α,11β,17α,21-ol-20-one

On preliminary fractionation of a urine extract from a subject who had taken 9α-fluorocortisone by mouth, four main zones reducing blue tetrazolium were found (Bush and Mahesh, 1958a). One band had the same mobility as tetrahydrocortisol but was suspected to be a mixture of tetrahydrocortisol and 9α-fluorocortisol. On acetylation two compounds were found on a single length run in a group-B system, one of them in the position of tetrahydrocortisol diacetate and the other, the major component, with a lower mobility. On oxidation of the original mixture with bismuthate (but not with periodate) a zone with the mobility of 11β-hydroxyaetiocholanolone and giving a purple colour with alkaline *m*-dinitrobenzene was obtained. When this material was acetylated two 17-oxosteroids were obtained, the minor product having the mobility of 11β-hydroxyaetiocholanolone 3-acetate, and the other a lower mobility. When the latter was oxidized with chromic acid a new product was obtained giving a pink colour in the cold with alkaline *m*-dinitrobenzene and having an R_F value greater than that of 11-oxoaetiocholanolone 3-acetate. The former was oxidized similarly to a product with the R_F and pink colour reaction of 11-oxoaetiocholanolone 3-acetate itself.

The ΔR_{Mr} on acetylation of the original product and of its oxidation product was that characteristic of a 3α-hydroxy-5α(H) steroid (p. 283).

The ΔR_{Mr} on oxidation of the hydroxyl group which was not protected by acetylation was that characteristic of an 11β-hydroxyl group with an α-fluoro-substituent. (The 9α-chloro-, 9α-bromo-, and 9α-iodo-steroids all show different behaviour in this respect.)

The mobilities of the original compound and its bismuthate oxidation product were those expected for 9α-F-αP-3α,11β,17α,21-ol-20-one and 9α-F-αA-3α,11β,ol-17-one so that the expected final degradation product obtained by complete oxidation with chromic acid should be 9α-F-αA-3, 11,17-one. A small amount of this substance was therefore prepared by reducing 9α-fluorocortisol with hydrogen and platinum oxide and oxidizing the product with bismuthate and chromic acid in turn. When the bismuthate oxidation product of the urinary fraction was isolated chromatographically (via the acetate to remove the small amount of 11β-hydroxyaetiocholanolone in the mixture) and oxidized with chromic acid a product was obtained with the same mobility and colour reactions as the reference steroid. As is characteristic of 9α-fluoro-11-ketones the R_F value was *greater* than that of the related unsubstituted 11-ketone (αA-3,11,17-one).

Finally, a specific indication of the presence of the 9α-fluorine atom was obtained by treating the final trione (and the reference trione prepared from 9α-fluorocortisol) with collidine at 100° C for 15 min (Fried and Sabo, 1957). The products were identical in R_F, colour reaction, and in the appearance of absorption of ultraviolet light (2537 Å) due to the production of the $\alpha\beta$-unsaturated $\Delta^{8(9)}$-11-ketone group by the dehydrohalogenation. The R_F value of this last derivative was characteristic in that owing to the introduction of unsaturation it was *smaller* than that of the related saturated trione αA-3,11,17-one. These reactions are summarized in Fig. 5.4 and the outstanding chromatographic changes shown diagrammatically in Fig. 5.5.

This is another example of a substance in which characteristic changes in mobility of various derivatives and colour reactions were used to characterize it, and only two reference substances were available for direct comparison with derivatives of the unknown.

(d) 18-Hydroxyoestrone

An instructive example of the use of chromatographic methods together with chemical conversions is the identification of 18-hydroxyoestrone in pregnancy urine by Loke, Marrian and Watson (1959). An apparently homogeneous spot on a paper chromatogram gave two after reduction with borohydride. On the other hand the component of principal interest was unchanged in R_F value by the reduction although it was clearly a ketone because of previous Girard separation. Here we see a typical puzzle which might crop up in any kind of structural investigation, and which needs just the same kind of "chemical intelligence" in either classical or chromatographic methods. The ketone group could hardly be capable of condensa-

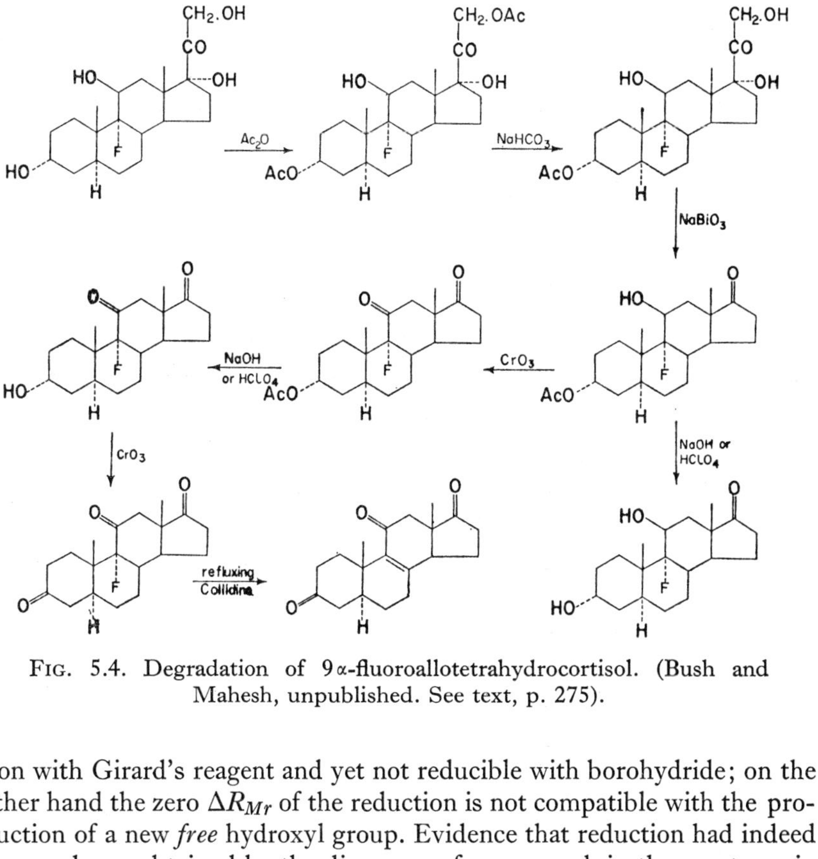

FIG. 5.4. Degradation of 9α-fluoroallotetrahydrocortisol. (Bush and Mahesh, unpublished. See text, p. 275).

tion with Girard's reagent and yet not reducible with borohydride; on the other hand the zero ΔR_{Mr} of the reduction is not compatible with the production of a new *free* hydroxyl group. Evidence that reduction had indeed occurred was obtained by the discovery of a new peak in the spectrum in sulphuric acid (approx. 460 mμ) and a change in the spectrum given with the Kober reagent.

Circumstantial evidence and the results at this stage suggested that the most reasonable structures were x-hydroxyoestrone or x-oxo-oestradiol, or epimers thereof. The most reasonable explanation of the zero ΔR_{Mr} on reduction was that the new hydroxyl group interacted with the first sterically or by hydrogen bonding (cf. p. 94); it could not, however, be α to

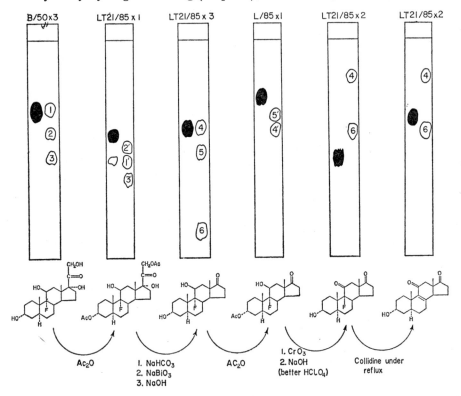

Fig. 5.5. Chromatographic steps in the degradation of 9α-fluoroallotetrahydrocortisol. (Bush and Mahesh, unpublished. See text, p. 275). 1, Tetrahydrocortisol: 2, Allotetrahydrocortisol: 3, Tetrahydrocortisone: 1′, 2′, 3′, Diacetates of 1, 2, 3: 4, 11β-Hydroxyaetiocholanolone: 5, 11β-Hydroxyandrosterone: 6, 11-Oxoaetiocholanolone: 4′, 5′, 3-Acetates of 4, 5. ●, Urinary substance and derivatives.

the original hydroxyl group since the original ketol did not reduce blue tetrazolium to any significant degree. If one of the two groups was at C-17 hydrogen bonding could occur with the other hydroxyl group placed at C-18 (for a 17β-hydroxyl) and with a 14α- or 15α-hydroxyl group (for a 17α-hydroxyl). The infrared adsorption spectrum of the original ketol showed a flat peak at 1725 cm^{-1} attributed to a 17-carbonyl group and supported by the presence of the 1408 cm^{-1} band; and it also lacked a band at 1375–7 cm^{-1} which has been attributed to the 13-methyl group.

It was then shown that four hours' treatment with N-NaOH at room temperature gave a nearly theoretical yield of formaldehyde and a compound which now had the same R_F as oestrone and was considered to be 18-nor-oestrone. Finally, synthetic 18-nor-oestrone was compared with the final product and shown to be identical with it by infrared spectroscopy and mixed melting points (Fig. 5.6).

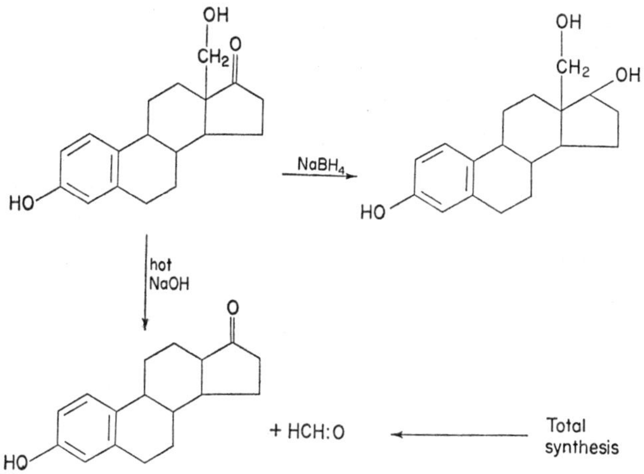

FIG. 5.6. Degradation of 18-hydroxyoestrone. (After Loke, Marrian and Watson, 1959: see text p. 276).

This example has been given at length because it shows up both the power of a combination of chromatographic, microchemical and spectrophotometric methods, and also two traps which might have misled anyone who tried to follow this sort of procedure without using chemical insight, and without carrying the investigation to its logical end. First the failure to obtain a significant ΔR_{Mr} on reduction with borohydride (p. 277); second the identical R_F values of 18-nor-oestrone and oestrone. The latter finding is probably due to two effects giving opposed ΔR_{Mg} values; first the loss of a methyl group should give a definite positive ΔR_{Mr} as in the 19-nor-steroids (p. 95). However, the 18-methyl group will have a positive inductive effect easily transmitted to the 17-ketone group and thus increasing its polarity; on removing this methyl group there will be a resultant decrease in the polarity of the ketone group which should give a negative ΔR_{Mr}. The two effects apparently cancel one another out. On the other hand this should be detectable chromatographically by reducing the ketone group to the 17-hydroxyl group, for in this case the polarity of the hydroxyl group will be weakened by the positive inductive effect of an 18-methyl group. The removal of the methyl group should therefore give rise to an increase in the polarity of the hydroxyl group and an additional positive ΔR_{Mr} superimposed on that due to the removal of the methyl group itself. 18-nor-oestradiol should therefore be separable from oestradiol and have a lower R_F value than the latter. The expected ΔR_M should be approximately 0·25–0·35.

5. Systematic Investigation of the Unknown Steroids

The previous examples given in this chapter have been of steroids in which a fair amount of circumstantial evidence could be used to help in the identification, and in which substituents at relatively common positions were involved. It may be asked how chromatographic methods would fare with steroids containing unusual substituents, and substituents in unusual positions, or which failed to give an easily recognized derivative at the end of several degradations. Can one formulate a general procedure capable of elucidating the structure of steroids and other substances which is based entirely upon chromatographic and spectroscopic methods, and which could succeed with "invisible" amounts of material? In the author's view this will soon be possible without too much difficulty. For the moment one can only suggest the lines along which such a procedure might evolve and describe those cases in which experiment has already shown that possibilities are also actualities.

The basis of such a procedure is to find a series of operations with an unambiguous relation to chemical structure which does not involve the use of more than a fraction of a milligram of material, and which thus excludes the determination of an empirical and molecular formula, the taking of melting points, and the determination of specific molecular rotations. It is the aim of the rest of this chapter to show that there is already enough information to hand to devise a general approach of this sort and that most classical operations in structural chemistry have reliable chromatographic equivalents. This is not in any way to disparage the classical method of structure determination since most of the operations to be used are based on the findings of the classical method. Rather it seems to the author that there are many problems, particularly in biological chemistry, which need to be solved if possible without going to the labour and expense of isolating "visible" amounts of material. In view of the possibilities of the chromatographic method it is unjustifiable to adhere dogmatically to the requirements of the classical method. The irrelevance of these requirements today is perhaps best indicated by reminding ourselves of their uselessness in protein chemistry, which has progressed largely on the basis of quite different criteria. The crystallization of some of the enzymes by Northrop was indeed a landmark in protein chemistry but not to be compared with the elucidation of the structure of insulin by Sanger and his colleagues which was achieved almost entirely by means of chromatographic methods (Sanger, 1955).

As with the classical method the aim is to arrive at a proposed structural formula and then compare the unknown material with a reference substance synthesized by unambiguous reactions. It was shown in the first two chapters that the chromatographic properties of steroids can be related to those of other classes of organic compounds on the basis of the theoretical approach of Martin, Bate-Smith, Pierotti and their colleagues (p. 7). Particular substituents and even the hydrocarbon skeleton of organic compounds can be recognized by characteristic changes in R_M values on changing from one solvent system to another (ΔR_{MS} values). The only thing peculiar to the steroids is the pronounced influence of steric factors on their R_M values; it was seen, however, that even this was an exaggeration of trends seen with simpler compounds such as flavonoids (Bate-Smith and Westall, 1950) and polycyclic hydrocarbons (Pierotti et al., 1959.) Just as a secondary hydroxyl group in a steroid has a lot of chemical properties typical of all secondary hydroxyl groups, so are its chromatographic "properties" to a large extent general properties and independent of the type of molecule in which it is substituted. Nor is the problem necessarily more difficult with large and complicated molecules since it has been seen that more and more striking and useful steric effects become apparent as the complexity of the molecule increases.

The best way of recognizing reactive substituents chromatographically is to measure the ΔR_M caused by specific reactions altering such substituents in known ways (ΔR_{Mr} values). Unreactive parts of the molecule have to be recognized by the ΔR_{Mg} value calculated for them by calculation from the R_M value of the substance, the ΔR_{Mg} values of the reactive substituents, and by the ΔR_{Ms} values obtained on appropriate changes of the solvent system. There is still a lot to be done in this field in that most unreactive parts of organic molecules are non-polar and much more is known about ways of altering the interactions of polar groups than ways of altering the magnitude of non-polar interactions. The studies of Pierotti *et al.*, however, suggest that numerous possibilities are available and that pairs of solvent systems giving large ΔR_{Ms} values for different types of non-polar structures will be found in the near future (see also Fig. 2). At present the effects obtainable with non-polar groups in liquid–liquid systems are rather small, although already sufficient for the distinction of a large number of polynuclear hydrocarbons (p. 304).

Another powerful application of chromatographic properties is analogous to what might be called the method of chemical tautology. In the classical method this consists of altering a substance A to another substance B and then carrying out the reverse reaction by a stereospecific reaction to obtain A'. A' and A are then shown to be identical or not identical, and in the latter case the direction and magnitude of the change of properties such as light absorption or molecular rotation are observed. This is used most often with asymmetric substituents to decide what configuration is present. The same method can be applied chromatographically because of the easy differentiation of equatorial and axial substituents (p. 284). The logic of the argument is, however, slightly different. In the classical method we can state that A' either is or is not the same substance as A. With the chromatographic method we must be careful to avoid saying "A' is the same as A" when A and A' have the same R_M value (p. 260), although we can be correct in saying "A' is not the same as A" when A' does not have the same R_M value as A. Instead we must say in the former case that the overall ΔR_{Mr} of the forward and reverse operations is zero. In the presence of additional information, however, such as the magnitude of the ΔR_{Mr} of the forward reaction, the nature and specificity of the reactions used, and so on, it will usually be possible to make the reasonable inference from this factual statement that the group altered by the forward reaction has been restored to its original configuration by the reverse reaction. In practice we shall not often transgress the allowable limits of inference if we act as if the identity of A and A' were in fact established: in the present state of our knowledge, however, it is good to remember that we can, if necessary, phrase our inferences in a way that does not exceed the strict limits of our information, and admit that this is less complete than with the classical equivalent of these operations.

At the end of a series of degradations it should be possible to write a structural formula using the least possible amount of circumstantial evidence. The final stages consist of the comparison of the properties of the synthetic compound with this formula with those of the "unknown" material. The best procedure is to carry the authentic compound through as many of the steps used for the "unknown" as possible and using as many different types of solvent systems as are needed to give characteristic ΔR_{Ms} values for different parts of the compound. It is a good rule to use in addition some changes of solvents systems giving large ΔR_{Ms} values for groups *not* supposed to be present in the "unknown". Thus the failure to obtain a large positive or negative ΔR_{Ms} on changing from an equivalent acidic to a basic solvent system is unshakeable evidence that the substance does not contain free acid or basic groups.

In the following account the majority of reactions have been applied by the author or by others to actual chromatographic work; that is to say it is already known that these reactions give reasonable yields on the microgram scale and only minor trouble from side reactions. In some cases reactions are included which have not yet been used in chromatographic work to my knowledge but which are known from spectroscopic evidence to give reasonable yields with microgram quantities, or which are so clear cut on the macroscopic scale that it is reasonable to assume they will be equally useful on the micro-scale. In a later section some more questionable possibilities will be raised which will almost certainly need careful investigation to establish the conditions needed for success on a micro-scale. The use of specific enzymic reactions should always be considered (Eppstein, Meister, Murray and Peterson, 1956; Hurlock and Talalay, 1956).

6. General Properties of Simple Functional Groups

(a) Changes by Chemical Modification—ΔR_{Mr} Values

(i) Hydroxyl and related groups—Hydroxyl, carbonyl, and carboxyl groups are best considered together since their interconversions form the basis of identification. The most useful reactions for hydroxyl groups are acetylation under mild and forcing conditions, oxidation with chromic acid under mild and vigorous conditions (Burton *et al.*, 1951a, b; Zaffaroni, 1953) and dehydration (p. 368): for carbonyl groups ketalation (Reineke, 1956), and reduction with and without ketalation (Bush and Mahesh, 1959b, c); for aldehyde groups reduction and oxidation: and for carboxyl groups esterification (Haslewood, 1954, 1958). These reactions and their ΔR_{Mr} values are summarized in Figs. 5.7, 5.8. It is seen that primary, secondary, tertiary, equatorial, and axial hydroxyl groups are readily

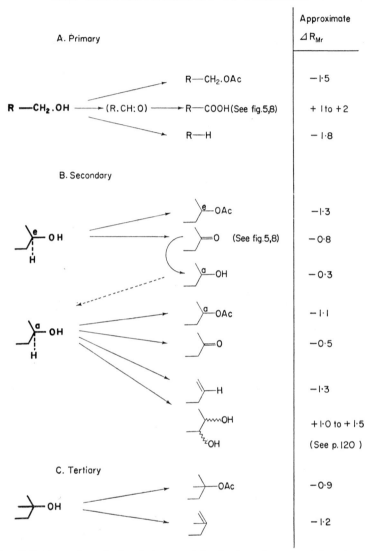

FIG. 5.7. General methods of characterizing hydroxyl groups. The ΔR_{Mr} figures are for relatively unhindered groups in "typical" solvent systems and should only be taken as rough guides for any particular position. Differences between axial and equatorial groups tend to be smallest at positions 3 and 17 and largest at positions 6 and 11 (cf. Barton and Cookson, 1956). ΔR_{Ms} values are also of general value particularly with changes to t-butanol systems (p. 307).

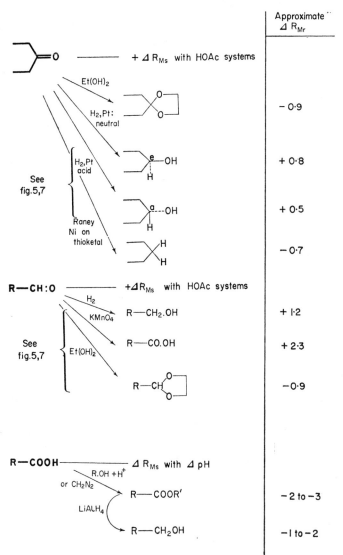

FIG. 5.8. Characterization of groups related to hydroxyl groups. The same considerations apply as in Fig. 5.7, but ΔR_{Ms} effects are included in this figure. In strongly acid systems the (unionized) carboxyl group is rather less polar than a hydroxyl group and exact ΔR_{Mr} values for this group are meaningless without specification of the pH of the solvent system. The values given are for a neutral unbuffered system.

distinguished. In a later section specific features of groups in different positions and conformations will be dealt with as a separate topic.

(*ii*) *Phenolic hydroxyl groups*—The best conversion is probably the methylation procedure of Brown (1955) using borate buffer medium (p. 367). This gives a very large negative ΔR_{Mr}.

(*iii*) *Unconjugated double bonds*—Numerous possibilities are available but the simplest and best tried is the formation of an epoxide with a peracid (Seebeck and Reichstein, 1943; Rubin et al., 1953) causing a small positive ΔR_{Mr}. This reaction may give trouble with side reactions with certain steroids, although it gives a fairly clear-cut product with unsaturated 17-oxosteroids (Rubin et al., 1953) so that it would be an advantage to confirm the formation of an epoxide by converting it in turn to a diol or ketol (Fig. 5.9), in the latter case introducing a group not only giving a large additional positive ΔR_{Mr} but also reducing tetrazolium salts (Fotherby, Colas, Atherden and Marrian, 1957). Spectroscopic methods are also of value if conjugation can be established.

(*iv*) *Amino groups*—Primary amino groups are easily detectable by formation of dinitrophenyl derivatives with fluorodinitrobenzene (Sanger, 1945) with a large negative ΔR_{Mr} value in acidic or neutral solvent systems; primary and secondary amines give acetates with large negative ΔR_{Mr} values in acid or neutral solvent systems. Tertiary amino groups will be distinguished by the ΔR_{Ms} of a strong base (see below) coupled with the absence of any effect on acetylation.

Primary and secondary amines should be easily distinguished by a chromatographic application of the macroscopic test of Hinsberg. Both form benzenesulphonamides with benzenesulphonyl chloride which would give a large negative ΔR_{Mr} in acidic solvent systems: on the other hand the derivative of a primary amine is an acid and forms salts while that of a secondary amine does not. The result would be a large positive ΔR_{Ms} for the derivative of the primary amine on changing from an acidic to a basic solvent system, and a small or zero ΔR_{Ms} for the derivative of a secondary amine.

(*v*) *Halogen groups*—Dehydrohalogenation with pyridine or collidine (e.g. Fried and Sabo, 1957) is the best reaction to use for suspected monohalides and is known to give reasonable yields on the microgram scale in some cases (Bush and Mahesh, 1958b). The olefin can then be investigated further (see Fig. 5.9, p. 287). In most cases it is likely that the halogen will be part of a complex function (q.v.). With simple halides, however, the reaction should give a small to moderate positive ΔR_{Mr} in the order $I > Br > Cl > F$.

STRUCTURAL ANALYSIS AND IDENTIFICATION

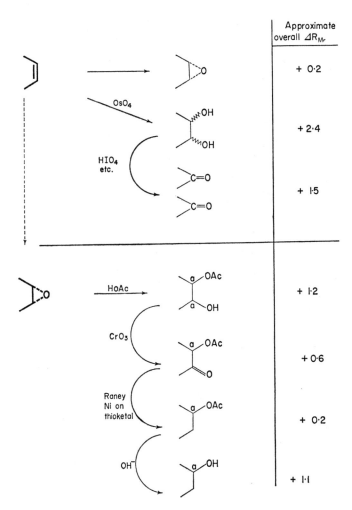

FIG. 5.9. Characterization of double bonds. In each case the ΔR_{Mr} value is with respect to the *original* material in the left-hand column, i.e. olefine or epoxide. The values will vary characteristically with the position and orientation of the final substituents. The general procedure is of course open to many variations in specific cases and many other reactions of epoxides could be exploited similarly (see e.g. Klyne, 1957, p. 76).

(vi) *Alkoxyl radicals*—Ether-linked oxygen is the biggest nightmare in the identification of steroids by chromatographic methods unless an infrared spectrum is available. Simple alkoxyl groups could probably be split off and recovered with hydriodic acid; the alkyl halide would be easily identified by gas–liquid chromatography (James and Martin, 1954), but the larger part of the steroid would be unlikely to survive the reaction in recognizable form. One must rely on spectroscopic evidence and ΔR_{Ms} values (q.v.) for the detection of such groups.

(vii) *Hydrocarbon radicals*—With very few exceptions the hydrocarbon skeletons of steroids are not very amenable to modifications unless unsaturated or substituted points of attack are present. The exceptions are not at present likely to be useful for chromatographic identification because they are of relatively low yield and give numerous different products. The recognition of hydrocarbon skeletons therefore depends mainly upon spectroscopic evidence and the ΔR_{Ms} values of solvent changes. An important advance in this field, however, is the identification of hydrocarbon fragments derived from the side chain from pyrolysis of lophenol (Djerassi, Mills and Vilotti, 1958), and of $\Delta^{8(14)}$-ergostenol, $\Delta^{8(14)}$-stigmastenol and cholesterol (Cox, High and Jones, 1958) using gas–liquid chromatography.

(b) *Solvent Changes—ΔR_{Ms} Values*

(i) *Hydroxyl and related groups*—Water, methanol, ethanol, isopropanol, acetic acid, ethylene glycol, formamide, and propylene glycol show small or negligible differences in their interactions with steroid hydroxyl groups unless the latter are very strongly hindered or in complex functions (q.v. p. 290). Changes of solvent system using these substances as stationary phase will therefore not give large ΔR_{Ms} values for hydroxyl groups in general. On the other hand, it is clear that t-butanol systems give exceptionally low ΔR_{Ms} values for all the hydroxyl groups of steroids that have been studied, and a reason for this was suggested in Chapter 2. In general, therefore, the change from one of the commoner systems to one based on t-butanol will be the best way of obtaining large ΔR_{Ms} values for hydroxyl groups. Less general but similar effects are seen with systems based on dioxan and acetic acid but these show special effects with certain complex functions or particular positions, and may be confusing, although often useful (p. 98).

Carbonyl groups interact strongly with all the above solvents and it was seen in Chapter 2 that the ΔR_{Mg} values were very similar with the exception of the very useful increases found with concentrated acetic acid used as a stationary phase. All these solvents possess protons capable of hydrogen bonding and the evidence suggests that the principal factor determining the ΔR_{Mg} values of steroid ketone groups is the charge on the carbonyl oxygen, which will be the main factor determining the energy of hydrogen bonding with amino or hydroxyl groups (p. 276). Large negative ΔR_{Ms} values for ketones would therefore be expected on changing from solvents with available protons to those lacking them. The t-butanol systems all contain too much water or methanol to show such an effect. Although not yet tested, systems based on low concentrations of water and high concentrations of tertiary amines as stationary phase might give the desired effect of reducing the ΔR_{Mg} of ketone groups while retaining high ΔR_{Mg} values for hydroxyl groups.

(*ii*) *Phenol groups*—These will be detectable by reason of small ΔR_{Ms} values for a change from neutral to acidic solvent systems, coupled with large positive ΔR_{Ms} values on changing to a strongly alkaline system (p. 32).

(*iii*) *Carboxyl and basic nitrogen groups*—These will probably both give rise to streaking in neutral systems as long as ionization is not suppressed by a low dielectric constant of both phases. Carboxyl groups cause a large positive ΔR_{Ms} on changing to basic solvent systems, of which those based on ammonium hydroxide are the most convenient. A large negative ΔR_{Ms} will occur on changing to an acidic system of which several varieties are useful. Systems based on formic or acetic acids suppress the ionization of most carboxyl groups sufficiently to give symmetrical zones with little or no tailing, but R_F differences correlated with the pK_1 of carboxyl groups in the solutes concerned have been observed with simple organic acids (Cassidy, 1957, p. 385). On the other hand very little suppression of ionization of stronger acid groups such as ester sulphate occurs (p. 32). The latter only gives a large negative ΔR_{Ms} on changing to solvent systems based on a mineral acid of which hydrochloric is most convenient. Similar behaviour would be expected for sulphonic and certain activated carboxylic acid groups (e.g. taurine, α-fluoro-carboxyl groups).

Basic groups will show the ΔR_{Ms} values of opposite sign to those given by acid groups. This has not been studied closely with steroids, but results with other families of compounds are sufficiently clear cut to justify this inference (Martin and Synge, 1941). In anhydrous systems the distinction between primary and secondary amines which have protons available for hydrogen bonding, and tertiary amines which do not, might well be apparent in the same way as seen in gas-liquid systems (James, 1952).

(*iv*) *Double bonds*—Little systematic work has been done on unconjugated double bonds in steroids and some of them show no detectable ΔR_{Mg}

values in the common solvent systems (Reineke, 1956). The work of Pierotti et al. (1959), however, suggests that detectable effects should be obtainable by changing from saturated aliphatic to unsaturated or aromatic solvents as mobile phase. This matter is at present under investigation. Gas-liquid systems have already been found giving regular ΔR_{Ms} values for unsaturated hydrocarbons (James and Martin, 1954).

Conjugated double bonds are an easier matter. The ΔR_{Mg} values of unsaturated ketone groups in steroids are similar in the common systems used. Large positive ΔR_{Ms} values are however obtained with $\alpha\beta$-unsaturated 3-ketones on changing from these systems to equivalent systems based on acetic acid or nitromethane (p. 307). Little, however, is known about the behaviour of other types of conjugated double bonds and the effects seen with the last two types of solvent system must be considered as special effects at present related to the complex function as a whole rather than to the double bond itself. The large ΔR_{Mg} values for 3-, 17- and 20-ketone groups suggest that terminal groups are particularly prone to interact strongly with acetic acid.

(v) *Other groups*—Lactone and alkoxyl groups have been insufficiently studied to make any sound generalizations. It is to be expected, however, that the interaction mainly responsible for the positive ΔR_{Mg} value of these groups is hydrogen bonding restricted to solvents with an available proton. As with ketones and ethers one would expect a moderate to large negative ΔR_{Ms} on changing to a system based on a stationary phase lacking phase protons available for hydrogen bonding.

In general the commoner groups in steroids can all be characterized by ΔR_{Ms} values although the evidence for ketone, lactone, alkoxyl and olefine groups is at present largely negative in the general case. This situation is probably capable of improvement but in view of the importance of hydroxyl groups, and of the difference in the ΔR_{Mg} values of the latter four types of group, the available solvent systems already enable us to obtain a lot of information by the measurement of ΔR_{Ms} values. It is clear, however, that the use of chemical modifications and the measurement of the resulting ΔR_{Mr} values is in general a more powerful method.

7. Special Properties of Simple and Complex Functional Groups

(a) ΔR_{Mr} *on Modification*

There are so many possibilities to be explored here that only a few examples can be discussed. It should, however, be clear that any reaction giving reasonable yields is a possible candidate for chromatographic demonstration. Numerous examples of obvious importance will occur to the reader but those given below are largely confined to those which have been thoroughly studied in chromatographic work.

The series of reactions using oxidation with bismuthate, either alone or in conjunction with reductions with borohydride, introduced by Norymberski and his colleagues (p. 292), have been the basis of many chromatographic identifications (Bush and Sandberg, 1953; Bush and Willoughby, 1957; Bush and Mahesh, 1959b, c). These reactions enable various types of side chain commonly found in adrenal steroids and their metabolites to be degraded to the easily detectable 17-ketone group. Table 5.2 summarizes the chromatographic changes that are obtained with different types of side chain.

Steric hindrance of hydroxyl groups and differences in type give rise to numerous characteristic differences in the ΔR_{Mr} values of the general reactions for hydroxyl groups. Acetic anhydride at room temperature with pyridine as catalyst fails to acetylate 11β-, 15β- and 12ξ-hydroxyl groups, and tertiary hydroxyl groups. Acetic anhydride and pyridine at 60° C esterifies 3ξ- and 21-hydroxyl groups completely in 10 min but esterification of the 18-hydroxyl group of aldosterone (hemiacetal) only occurs to the extent of about 5%: complete esterification is achieved in 90–100 min. Saponifications also show useful differences in rate. Hydrolysis of 3α,21-diacetates with potassium bicarbonate at room temperature yields about 5% of the fully hydrolyzed 3α,21-diol and 95% of the 3α-monoacetate (Bush and Willoughby, 1957). Other reactions showing useful differences in rate with particular positions and orientations of the substituent under attack are dehydrations and oxidations.

Certain *cis*-diols are readily characterized by forming the acetonides (Reichstein and von Euw, 1940; Fried, Borman, Kessler, Grabowich and Sabo, 1958). This is probably also subject to useful differences in rate with different conditions. Thus triamcinolone (9α-F-P^4-11β,16α,17α,21-ol-3, 20-one) gives a 16,17-acetonide very readily with perchloric acid as catalyst (Fried *et al.*, 1958). This reaction goes readily on the microgram scale and gives a ΔR_{Mr} which is too large to measure accurately in any of the usual solvent systems; the reaction is complete in 15 min at 45° C (p. 367). On the other hand 20ξ,21-diols do not react under these conditions (author, unpublished).

Aldosterone gives characteristic ΔR_{Mr} values with several reactions (Fig. 5.10). Partial and complete acetylation (above) and partial and

Table 5.2. Oxidations of 17β-Side Chains

	NaBiO$_3$		HIO$_4$	
	Product	ΔR_{Mr}	Product	ΔR_{Mr}
R.CO.CH$_3$	unchanged	0	unchanged	0
R(OH).CO.CH$_3$	unchanged	0	unchanged	0
R.CO.CH$_2$OH	R.COOH	+++*	R.COOH	+++
R.CH(OH).CH$_2$OH	R.CHO	−2·0 to −2·4	R.CHO	−2·0 to −2·4
R(OH).CO.CH$_2$OH	R:O	−1·3 to −1·5	R(OH).COOH	+++
R(OH).CH(OH).CH$_3$	R:O	{20α −1·1 to −1·3 20β −0·9 to −1·1}	R:O	{20α −1·1 to −1·3 20β −0·9 to −1·1}
R(OH).CH(OH).CH$_2$OH	R:O	−2·2 to −2·5	R:O	−2·2 to −2·5

* Dependent upon pH (see ΔR_{Ms} effects, p. 23) ΔR_{Mr} values, approximate, for "typical" systems, e.g. LT21/85.

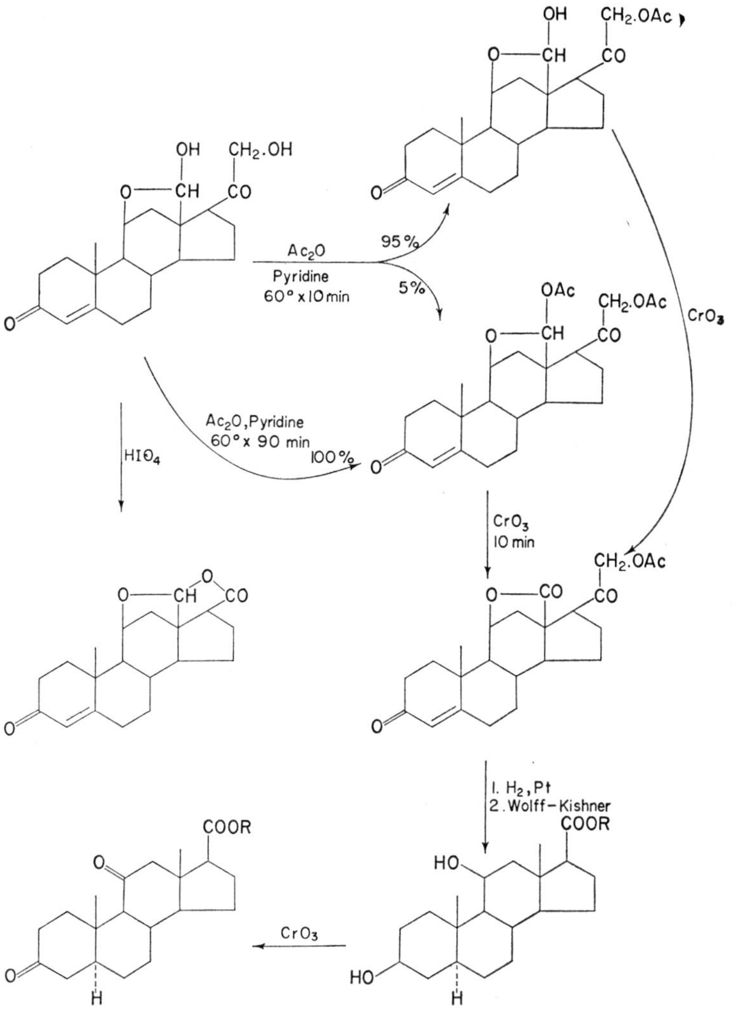

FIG. 5.10. Characterization of aldosterone. All the reactions shown can be used on the microgram scale and give rise to derivatives easily characterized on paper chromatograms. (Based on Simpson et al., 1954; Peterson et al., 1957b; Bush and M. Reich, unpublished.)

complete hydrolysis of the diacetate (Simpson *et al.*, 1954) are easily carried out on the microgram and smaller scale. Similarly the diacetate gives a highly characteristic partial hydrolysis and oxidation to a lactone on treatment with chromic acid in acetic acid (Peterson *et al.*, 1957b, 1959). This reaction has reasonable yields with as little as 0·01 μg (see p. 255).

The 5-ene-3β-ol structure common to many important steroids is readily characterized by the sequence shown in Fig. 5.11 which gives reasonable although incomplete yields (unpublished experiments by Miss M. Gale).

(b) ΔR_{Ms} on Change of Solvent System

The hemiacetal group of aldosterone shows numerous characteristic ΔR_{Ms} values on changing solvent systems. Since these are of some practical importance they are summarized in Table 5.3.

The saturated 3-ketones epimeric at C-5 are a specially interesting case. Several pairs of this type are easily separated on alumina (p. 101) but only two types of partition system are known which are capable of separating them. The first is the methylcyclohexane–butane-1,3-diol system (Axelrod, 1953) and the second the systems based on 90–95% acetic acid (Bush, 1959b). These epimers are all but inseparable in all other systems that have been tried (p. 116). In both systems the 5α-compound has the higher R_M value. The resolution is better with Axelrod's system (Mobility ratio approximately $5\beta/5\alpha = 1\cdot5$) than with the acetic acid systems*.

The acetic acid systems not only give a resolution of the epimeric 3-ketones but also produce an atypically large ΔR_{Mg} value for all the saturated and unsaturated 3-ketones that have been tried, and for 17- and 20-ketones. On the other hand the ΔR_{Mg} value for ketone groups in other positions is only slightly larger than obtained with aqueous methanol systems (p. 100). The 3-, 17- and 20-ketones therefore can be characterized by a moderately large positive ΔR_{Ms} on changing from a methanol to an equivalent acetic acid system.

The 20,21-diol group appears to have a much larger ΔR_{Mg} value in aqueous methanol systems than in propylene glycol systems (Neher, 1958a, p. 89) thus affording a useful ΔR_{Ms} for this grouping. Similarly the 17α-hydroxyl group added to a 20,21-ketol has a larger ΔR_{Mg} in propylene glycol systems than in aqueous methanol systems, which is in turn larger than that in t-butanol systems (p. 115). This is useful in the separation of corticosterone and 17-hydroxy-11-deoxycorticosterone, the largest ΔR_{Ms}

---

* See however p. 101.

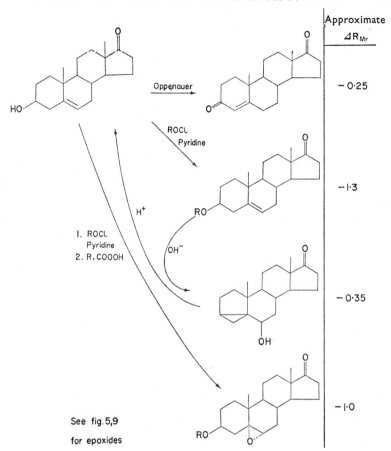

Fig. 5.11. Characterization of Δ^5-3β-ols. Dehydroepiandrosterone is used as an example. ΔR_{Mr} values are for "typical" systems and with respect to DHA itself for each derivative. Scheme based on Hershberg et al. (1941), Teich et al. (1953), Wallis et al. (1937), Shoppee and Summers (1952), and unpublished experiments on the microgram scale by Miss Muriel Gale and the author. R = p-toluenesulphonyl.

for this pair being obtained by changing from propylene glycol to certain t-butanol systems.

Certain *cis*-diols including steroids form complexes with borate and show moderate negative ΔR_{Ms} values on changing to a solvent system incorporating boric acid (Wachtmeister, 1951; Isherwood, 1954; Brooks et al., 1957).

Table 5.3. Behaviour of Aldosterone in Various Solvent Systems

Solvent system (22–26° C) (α)	Cortisol I		Cortisone II		Aldosterone III		ΔR_{Ms} for $\Delta(1 \to x)$	ΔR_M (II–III)
	R_F	R_M	R_F	R_M	R_F	R_M		
(1) Benzene–MeOH–H$_2$O (2:1:1 by vol.)	0.27	0.44	0.48	0.03	0.37	0.21	—	−0.18
(2) Toluene–EtOAc–MeOH–H$_2$O (95:5:50:50 by vol.)	0.23	0.52	0.40	0.17	0.23	0.52		−0.35
(3) Isoöctane–t-BuOH–H$_2$O (10:5:9 by vol.)	0.50	0.00	0.59	−0.16	0.36	0.25	+0.04	−0.41
(4) Petroleum–t-BuOH–H$_2$O (10:6:8 by vol.)	0.48	0.04	0.54	−0.07	0.40	0.17	−0.04	−0.24
(5) Chloroform–formamide								
(6) Toluene–propylene glycol	0.15	0.75	0.25	0.47	0.26	0.45	+0.24	+0.02
(7) Toluene–acetic acid–H$_2$O (10:9:1 by vol.)	0.40	0.17	0.42	0.15	0.47	0.05	−0.16	+0.10
(8) Toluene–acetic acid–H$_2$O (100:75:25 by vol.)	0.165	0.70	0.21	0.59	0.245	0.49	+0.28	+0.10
(9) Benzene–acetic acid–H$_2$O (10:7:3 by vol.)	0.19	0.64	0.23	0.53	0.28	0.40	+0.19	+0.13

The mobility of aldosterone relative to that of cortisone is expressed in the last column as the ΔR_M between the two steroids.

Solvents: (1) Simpson and Tait (1953); Neher (1958a); Author (unpublished).
(2) Simpson et al. (1954): Neher (1958a).
(3) Eberlein and Bongiovanni (1955) 37° C → falling.
(4) Author, 22° C. Unpublished. (High temperature appears to improve the separation (cf. Neher, 1958a))
(5) Mattox and Lewbart (1959); Brooks (1960).
(6) Grundy, Simpson and Tait (1951); approximate only.
(7, 8, 9) Author (unpublished).

8. Spectroscopic Properties, Colour, and Fluorescence Reactions

These are summarized in Appendix III and only a few points will be dealt with here. Spectroscopic work is admirably summarized and reviewed by Dorfman (1953) for the ultraviolet; Diaz, Zaffaroni, Rosenkranz and Djerassi (1952); Arrogave and Axelrod, (1954); Bernstein and Lenhard (1953, 1954) for the ultraviolet in sulphuric acid; Jones and Herling (1954, 1955, 1956) for the infrared; and by Meyer (1955) for the ultraviolet in alkali.

In conjunction with the observation of chromatographic properties the use of colour reactions and fluorescence and spectroscopy is obviously of great value. However, it should always be remembered that in the absence of circumstantial or other evidence, these reactions, however specific for one particular group of steroids as distinct from other groups of steroids, cannot be considered specific with respect to other classes of organic compound. Thus it is sometimes forgotten that the Zimmermann reaction is not a "reaction for ketosteroids" but a reaction for certain active methylene groups; that α-ketol groups are not the only ones reducing tetrazolium salts; and that the sodium hydroxide fluorescence reaction is not specific for Δ^4-3-oxosteroids but probably for the Δ^4-3-oxo-octahydronaphthalene group (Bush, 1954a) or groups converted to this structure by alkali (Neher, 1958a, p. 57). If, however, a substance reduces blue tetrazolium rapidly at room temperature; gives a monoacetate with a ΔR_{Mf} typical of primary α-ketols, which is hydrolyzed completely by cold potassium bicarbonate; and is oxidized with bismuthate to a substance giving a typical purple colour with the Zimmermann reagent and a ΔR_{Mf} typical of the oxidation of a dihydroxyacetone side chain to a 17-ketone, then it is not an unreasonable inference that the active methylene group in the last derivative is in fact activated by a ketone group, probably in a five-membered ring (Callow, Callow and Emmens, 1938).

Some important new features of the specificity of the sodium hydroxide fluorescence reaction have come to light recently. Neher has found that the reaction is not given by 7α-hydroxy-Δ^4-3-oxosteroids, probably because of their easy dehydration to the 4,6-dienones during the reaction, and is very weak with 2α-hydroxy-Δ^4-3-oxosteroids. Similarly the reaction is given by saturated 3,6-diketones probably because of enolization to the 6β-hydroxy-Δ^4-3-ketones already known to give the reaction (Balant and Ehrenstein, 1952; Neher, 1958a, p. 57).

The author also found that the reaction with 2α-hydroxytestosterone was extremely weak and attributed this to dehydration to the 1,4-dienone or oxidation to the 2,3-diketone. It is not the result of blocking the 2-position, since 2α-methyl steroids all give the reaction, albeit at a greatly reduced sensitivity (Bush and Mahesh, 1959c) (approximately 1/5 that of the unsubstituted analogues). The earlier finding that 3β-benzoyloxy-cholest-5-en-7-one failed to give the reaction was taken, with reservations, to exclude the possibility of 5-en-7-ones reacting as a general rule (Bush, 1954a). In view of the finding with 7α-hydroxy-Δ^4-3-ketones and 2α-hydroxy-Δ^4-3-ketones it seems wiser to withdraw this supposition until further work has been done, since dehydration to the 3,5-diene is an obvious possibility under such drastic conditions, or alternatively conversion to the 3,5-cyclo-6β-hydroxy compound with subsequent dehydration.

The mechanism of the reaction is still a mystery and its sensitivity is subject to numerous variations which are hard to understand. Reineke (personal communication) noticed early on that the sensitivity was very much greater with very polar steroids such as 6β-hydroxycortisol, and the author found that it was very much less with non-polar steroids such as progesterone and cholestenone. (A few have had great difficulty in getting any fluorescence with progesterone). Lately it was found that the sensitivity with 9α-fluorocortisol and 9α-fluorocortisone is about twice that with cortisol and cortisone (Bush and Mahesh, unpublished). Dr J. F. Tait (personal communication) has found that the sensitivity with aldosterone diacetate is very much greater than with the free compound. With four pairs of 11-ketones and related 11β-hydroxysteroids the former gave one half the sensitivity of the latter under the author's conditions (p. 381). The sensitivities with cortisol and corticosterone are identical. These ratios hold both for the minimum detectable amounts under standardized conditions with spots of equal area, and for the relative intensities with larger amounts as judged by direct fluorimetry of the paper strip or visual matching with standards.

The 11-oxygen function also produces highly characteristic changes in the Zimmermann reaction with 17-oxosteroids. Wilson (1954) found that 11,17-dioxosteroids gave greater extinction coefficients than 11-deoxy-17-oxosteroids, which were in turn greater than those of 11β-hydroxy-17-oxosteroids. The same thing is found with the reaction as done on paper chromatograms. In addition it is observed that the reaction with the 11,17-diketones is noticeable in the cold and gives a pink rather than a purple colour. The reaction with 11β-hydroxy-17-ketones is slower and not even noticeable until after 40–50 sec heating; the final colour is slightly bluer than 11-deoxy-17-ketones but this is not always easy to see. This reagent is also useful for detecting the lactone group of the cardiac aglycones (Schindler and Reichstein, 1951) but a more permanent colour is given with alkaline *m*-dinitrobenzoic acid (Bush and Taylor, 1952). The blue colour given with 3-ketones is also subject to characteristic variations. This colour is given by non-polar steroids such as testosterone and progesterone, and by steroids such as corticosterone and cortisone; it is distinctly weaker with the more polar compounds however and rapidly becomes a dirty grey. In two series it has been observed regularly that the 5α-3-ketone gives a bluish-violet colour which remains so for an hour or more while the 5β-epimer gives a similar colour at first which changes to a dirty grey within minutes.

Fluorescence with phosphoric acid (Neher and Wettstein, 1951) and with trichloroacetic acid (De Courcy, 1956) is extremely useful for many steroids on paper chromatograms. The only trouble with such reactions is that they are not easily related to specific groups; one is never quite sure that different colours may not be produced by the interference of impurities or by non-steroidal substances. Other ranges of colours are obtained with concentrated or fuming sulphuric acid (Axelrod, 1953, 1954). Another useful reagent of low specificity is phosphomolybdic acid (Kritchevsky and Kirk, 1952a). The conditions described on p. 378, using n-propanol as the vehicle (R. P. Martin, 1957), give a much greater sensitivity than the usual strong heating after applying an ethanolic solution of the reagent. The n-propanol evaporates slowly and prolongs the reaction (cf. p. 379). On the other hand the usual conditions of strong heating are useful in that 5-en-3-ols give a much more rapid and intense colour than any other class of hydroxysteroids; this is easily seen if the chromatogram is heated within view rather than in an oven. The Folin–Ciocalteu reagent is useful for phenols but is of very low specificity (Mitchell, 1952).

On the whole, colour and fluorescence reactions should be used as methods of detection and estimation (Chapter 4) and only used as evidence of structure in conjunction with chromatographic properties such as ΔR_{Ms} and ΔR_{Mr} values.

Ultraviolet spectroscopy in sulphuric acid and in ethanolic potassium hydroxide (p. 272) are of considerable value since most workers have an ultraviolet spectrophotometer while few have their own infrared apparatus. Certain pairs of substances with epimeric hydroxyl groups can be distinguished easily, but others such as allotetrahydrocortisol cannot (Zaffaroni, 1953; Bush and Willoughby, 1957). It is probable that these spectra are mainly due to dehydrations of hydroxyl groups to olefinic groups, and to pre-existing olefinic groups (Zaffaroni, 1953); epimeric hydroxyl groups differing considerably in their ease, or principal direction, of dehydration will therefore be distinguishable if the unsaturated centres give strong bands. Hydroxyl groups lacking such differences or giving rise to unsaturated centres with weak absorption will be indistinguishable.

9. Future Developments

Some reactions are at present of unknown value in chromatographic identification. These reactions are either difficult to control on the microgram scale because they depend upon the use of accurate molar ratios of steroid and reagent, or because they give low yields and many side reactions. There is every possibility, however, that conditions will be obtainable in which the former type of reaction can be controlled by careful timing, and in which the latter give better yields. Some of these reactions have been the basis of many important degradations and it can reasonably be inferred that they would be easily characterized by ΔR_{Mr} values (Fig. 5.12). Since these reactions would complete the parallelism between classical and chromatographic methods they are worth considering as possibilities for the future.

Some complex steroids would be easier to characterize if certain groups could be completely eliminated. The best candidates for use on a microgram scale would seem to be Wolff–Kishner reduction of ketone groups; dehydration of hydroxyl groups, and the elimination of ketone groups by reduction of the thioketals with Raney nickel (e.g. Haslewood, 1958).

Many degradations involve the transfer of substituents from one position to another. There seems little reason why these should not work on the microgram scale (Fig. 5.13).

Complete reduction to a hydrocarbon is an unequivocal method of proving the nature of the hydrocarbon skeleton of a steroid (e.g. Steiger and Reichstein, 1938). Since this is a reaction requiring forcing conditions it should be applicable on the microgram scale; now that hydrocarbons have been shown to be easily separable in gas–liquid and some liquid–liquid systems it becomes a feasible step in chromatographic work. It is seen from the work of Wieland and Kracht (1957) (Table 5.4) for instance that the resolution of different basic hydrocarbon skeletons with a simple liquid–liquid system on paper is extremely good, and in particular phenanthrene and chrysene are easily separated from other polycyclic structures.

It is not, however, certain that epimeric saturated hydrocarbons such as pregnane and allopregnane would be resolved by such systems although it is likely that they would be resolved by gas–liquid systems.

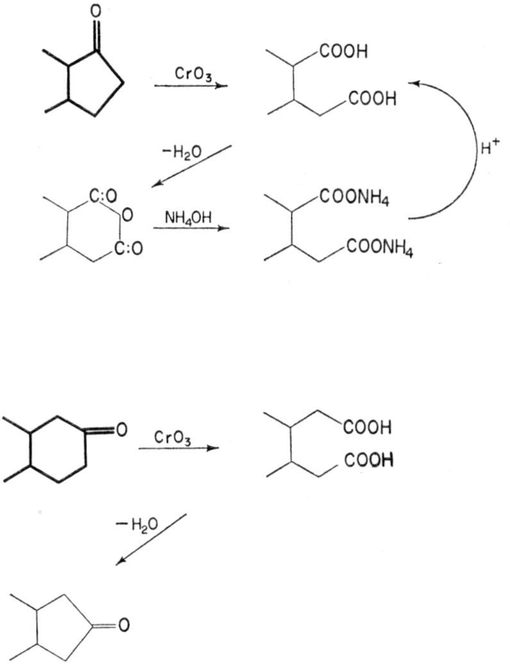

FIG. 5.12a. Classical degradations probably amenable to chromatographic analysis. (a) *Blanc rule*: 1:2-, 1:3-, 1:4-, 1:5-dicarboxylic acids dehydrate to anhydrides on distillation and will regenerate the original acid.

1:6- or higher dicarboxylic acids dehydrate and decarboxylate to give a cyclic ketone. The cyclohexyl ketone (II) will give a ketone as final product with overall ΔR_{Mr} equal to that of losing one methylene group. The dicarboxylic acids will have obvious chromatographic properties (p. 285).

The rule only holds when at least one carboxyl group is not attached directly to a ring.

Other possibilities giving large ΔR_{Mr} values are given in Figs. 5.12, 5.14.

Another field in which there is plenty of scope for development is the measurement of reaction rates by chromatography (Zaffaroni, 1953).

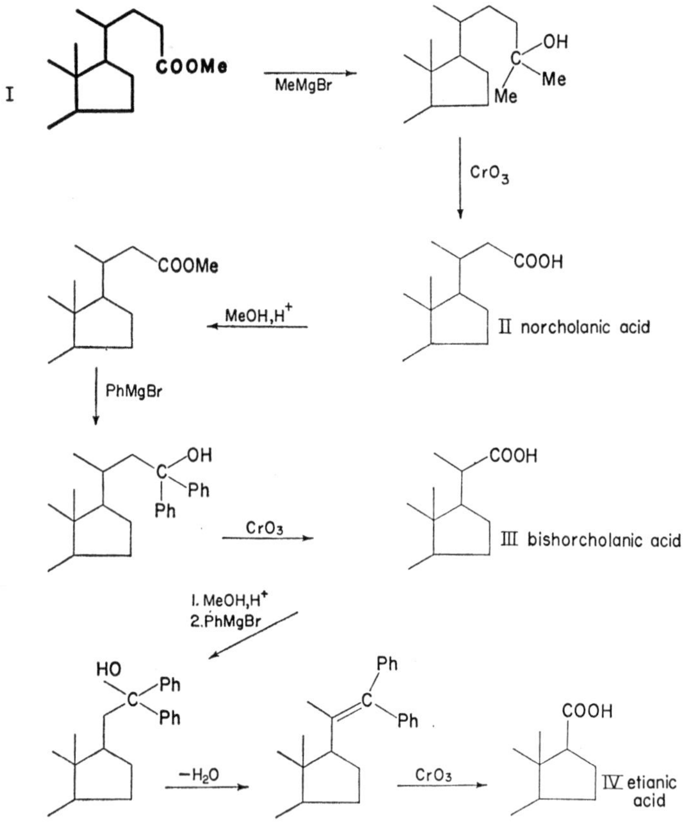

$I \rightarrow II$: ΔR_{Mr} = +0·2 (HOAc system)
$II \rightarrow III$: ΔR_{Mr} = +0·2 (HOAc system)
$III \rightarrow IV$: ΔR_{Mr} = +0·4 (HOAc system)

FIG. 5.12b. The Barbier-Wieland degradation. Although it has not been worked out for the microgram scale, this classical degradation would be readily followed by chromatographic methods (see text, p. 300).

Rates of esterification and saponification, of dehydration, and of oxidation, to name only a few, are all easily measured on the microgram scale, and with suitable radioactive labelling techniques on the submicrogram scale. Such reaction rates are often of great use in confirming or elucidating the conformations of particular groups. In Fig. 5.15 is shown the result of

(a) 7-Ketone → 6-Ketone (Haslewood, 1958)

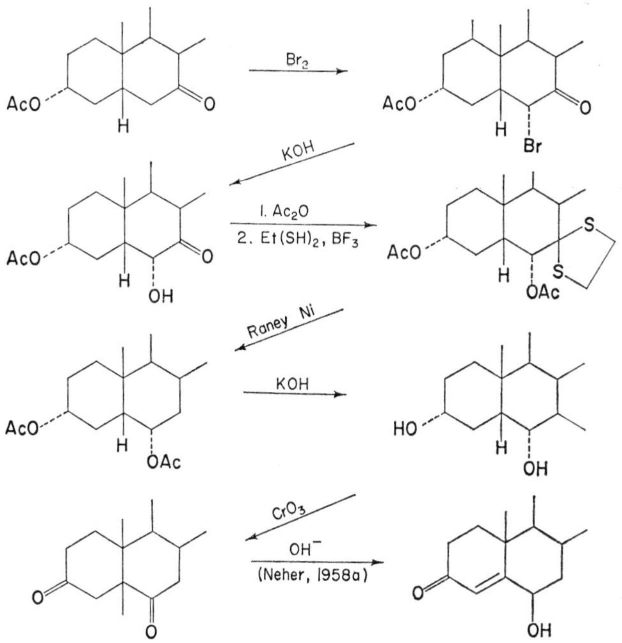

(b) 5α-hydroxyl → 6α-hydroxyl: (Herzig and Ehrenstein 1951; Amendolla *et al.*, 1954,)

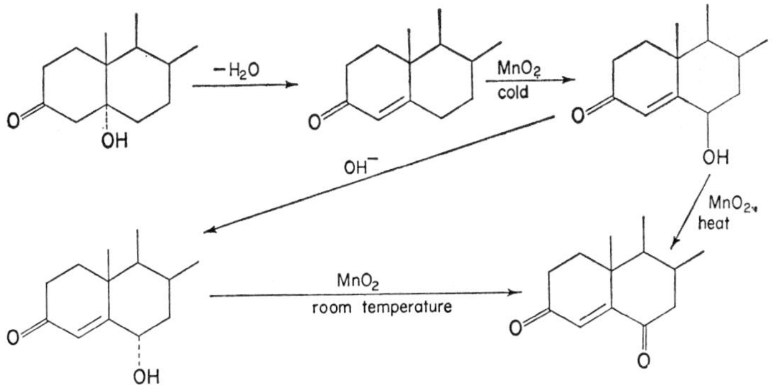

FIG. 5.13. Characterization by transfer of substituents to new positions. In both cases a slightly "awkward" substituent is transferred to the 6-position. Such 6-oxygen functions are both more consistent in their chromatographic behaviour (p. 87) and have extremely valuable spectroscopic properties (Meyer, 1955). It should be easy to see that each reaction shown has a distinctive ΔR_{Mr} value: in practice it would suffice to examine a few key derivatives because of the highly characteristic spectral and fluorescent properties of the 6-ketones.

Table 5.4. PAPER CHROMATOGRAPHY OF
POLYNUCLEAR HYDROCARBONS

Acetylated paper (Schleicher and Schull, 2043B).
Solvent system MeOH–ether–H$_2$O (4: 4: 1 by vol).

	R_B	
Carbazole	5·35	
Anthracene	4·5	
Phenanthrene	4·12	
Fluoranthrene	4·0	Wieland
Perylene	3·3	and
3,4-Benzpyrene	1·0	Kracht
Chrysene	1·45	(1957)
Triphenylene	4·2	
1,2-Benzanthracene	2·9	
Pyrene	3·78	

$$R_B = \frac{\text{mobility of 3,4-benzanthracene}}{\text{mobility of substance}}$$

(i.e. the reciprocal of the usual expression).

comparing the rates of acetylation of androsterone (3α-hydroxyl, axial) and aetiocholanolone (3α-hydroxyl, equatorial); the results were obtained by scanning (p. 244) but in many cases conditions can be obtained in which the different rates are so large as to be obvious on visual examination of the chromatograms. Esterification of carboxyl groups would also be amenable to measurement in this way (Barton and Schmeidler, 1948; 1949).

10. GENERAL PROCEDURE USING ΔR_{Ms} AND ΔR_{Mg} VALUES

At the moment, large and characteristic ΔR_{Ms} values are only obtainable for certain types of substituent so that in general the quickest results will be obtained by concentrating on ΔR_{Mr} values. However, ΔR_{Ms} values are also of great use because of the fact that large values can be obtained for some of the commonest groups in steroids. Again although large ΔR_{Ms} values relatively independent of steric factors are only obtainable for hydroxyl groups as a class, such groups are so common in steroids that it is very useful to be able to obtain an estimate of the total contribution of hydroxyl groups to the R_M value of an unknown substance that is independent of whether or not they are esterifiable. The special groups for which large ΔR_{Ms} values are available are summarized in Table 5.5. It is

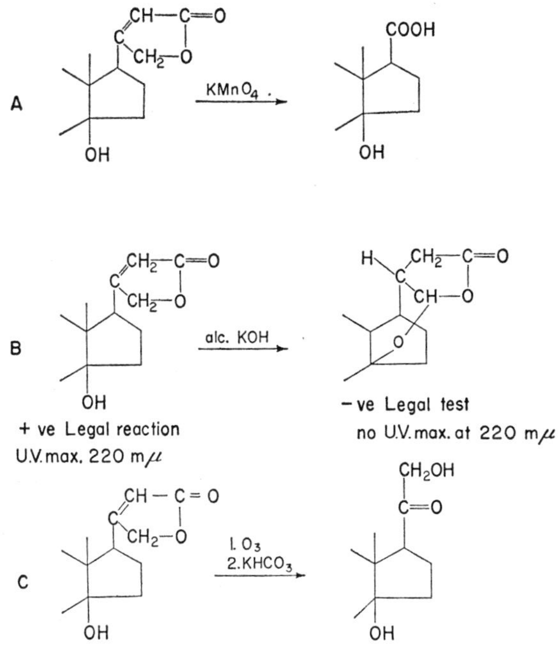

FIG. 5.14. Further reactions expected to give characteristic ΔR_{Mr} values. A is a useful general degradation giving 14-iso-etianic acids. B is useful in that a moderate negative ΔR_{Mr} (ca. $-1\cdot4$) is expected and that 17-iso-cardenolides do not give this reaction. C gives the readily characterized ketol sidechain.

seen that both chloroform–formamide and t-butanol systems give low ΔR_{Mg} values for 20ξ-hydroxyl groups compared with methanol systems as judged by the ΔR_{Mr} value for the conversion 20-ketone → 20β-hydroxyl (the ΔR_{Mr} for 20α → 20β-hydroxyl is well established, p. 112): the formamide system, however, gives insignificant ΔR_{Ms} values for the other hydroxyl groups on changing from a methanol system (p. 87). Similarly the t-butanol systems and the acetic acid systems give exceptionally large ΔR_{Ms} values for the saturated and unsaturated 3-, 17- and 20-ketones relative to any related saturated 3-hydroxysteroids (the 5β(H)-3α-OH configuration has been arbitrarily chosen as the reference for comparison here, but the same relative differences are seen with any configuration; see e.g. p. 109). Again, however, the acetic acid systems show only small

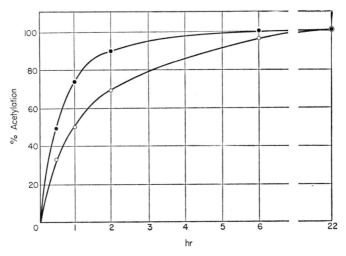

FIG. 5.15. Characteristic rates of reaction. Acetylation of androsterone (axial hydroxyl group; ○) and aetiocholanolone (equatorial hydroxyl group, ●) at 22° C. 200 μg of each steroid were dissolved in 0·9 ml acetic anhydride and 0·05 ml pyridine was added (author, unpublished experiment).

or negligible change in the relative ΔR_{Mg} values of ketone and hydroxyl groups at most other positions.

These groups are very common in steroids of interest. Suppose for example that a steroid for which there was no spectroscopic or other information gave a ΔR_{Ms} of +0·4 on changing from a methanol to an equivalent acetic acid system; this is unlikely to be due to any other group than a Δ⁴-3-ketone group if the material is a steroid. If then on changing from a methanol to a t-butanol system the ΔR_{Ms} value was close to that obtained with a reference non-hydroxylic steroid Δ⁴-3-ketone it would be strong evidence that the unknown contained no hydroxyl groups, and probably contained a Δ⁴-3-ketone group. If on the other hand the

Table 5.5. Solvent Systems Known to Give Consistently Large ΔR_{Ms} Effects

Group or structure	Solvent systems	ΔR_{Ms} relative to equivalent aqueous MeOH system
—OH (equatorial)	t-BuOH type	−1·0 to −1·4
—OH (axial)	t-BuOH type	−0·8 to −1·2
	Pyridine type	−0·6 to −0·8
>=O (3,17,20)	HOAc type	+0·4 to +0·6
	Nitromethane type	
>=O (3,?general)	HOAc type	+0·6 to +0·9
	Nitromethane type	
17α,20ξ,21-triol	Propylene glycol type	−0·4 to −0·6
	Formamide type	−0·2 to −0·5

Smaller but still useful ΔR_{Ms} effects can be obtained with changes to dioxan or dioxan–methanol systems (Peterson *et al.*, 1958; author, unpublished), and furfural or aqueous acetone systems (author, unpublished). These and the formamide systems (Layne and Marrian, 1958) should be considered whenever the problem arises of separating steroids with substituents giving complicated or distinctive steric interactions, e.g. *cis*-diols, lactones, 16- and 7-ols.

unknown gave a large negative ΔR_{Ms} on the change methanol → t-butanol system but no significant ΔR_{Ms} on the change methanol → formamide or propylene glycol system it would indicate strongly that it contained hydroxyl groups but that there was no diol group at C-20. The carboxyl, phenol and basic groups are easily detected by ΔR_{Ms} values (p. 286).

The measurement of ΔR_{Ms} values with accuracy is often difficult in practice because of the instability of the volatile systems, variable technique of impregnating paper, and gradual "running down" of the tanks, which it is only possible to avoid by replenishing them sufficiently frequently to be expensive whether or not the solvents are recovered. The best way in practice is to run the unknown with a reference substance of similar R_M value and to measure the difference in the ΔR_{Ms} values of the unknown and the reference substance; the actual ΔR_{Ms} values are then easily calculated from figures for the reference standards that have been obtained under ideal conditions with fresh tanks and maximum periods of equilibration (Bate-Smith and Westall, 1950). Once again it must be emphasized that relative R_F values or R_t values are very inadequate indeed for this sort of work, and a lot of accuracy is lost even in the intermediate range of R_F values (p. 75).

The general procedure then will start by an estimate of the probable polar groups in the unknown in terms of its R_M value in one of the typical solvent systems and some ΔR_{Ms} values on changing to atypical solvents. This will take the form of a list of the possible common combinations based on varying hydrocarbon skeletons, and spectroscopic or circumstantial evidence may give evidence at an early stage which limits the number of reasonable possibilities to one or two general types such as

"Either $C_{21}O_4$; ketol side chain + 1 axial + 1 equatorial hydroxyl group:
or $C_{21}O_5$; ketol side chain + 2 axial hydroxyl groups + 1 ketone group:
or $C_{21}O_4$; ketol side chain + 1 equatorial hydroxyl group + 1 unsaturated group"
and so on.

An acetate is then made by the room temperature method and the ΔR_{Mr} measured in one of the "typical" systems. The ester is then oxidized with chromic acid to detect unesterifiable secondary hydroxyl groups. If there is evidence of a two-carbon side chain an attempt is made to oxidize the original substance to a 17-ketone (or etianic acid); the latter is then also subjected to esterification and oxidation with chromic acid. The oxidized 17-ketone ester is then saponified and oxidized again. The ΔR_{Mr} values of these simple operations coupled with evidence from spectra, fluorescence, and colour reactions will in most cases give a pretty good indication of the nature of the substance, particularly if the final oxidized product has an R_M value identical with that of a known reference compound. It will almost certainly decide between the general possibilities drawn up at the start of the investigation.

Certain difficulties may arise, however, in the placing of one or other of the nuclear substituents. These will require further degradations of which reduction of ketone groups to hydroxyl groups by stereospecific means is the first to be explored. The 16-ketone group for instance will probably be indistinguishable from the 15-ketone group if present in the original unknown substance unless the reference substances are available and are known to show spectroscopic differences or differences in colour reactions. Reduction with platinum and hydrogen in acetic acid and with sodium borohydride should yield predominantly axial and equatorial hydroxyl groups respectively. In the case of the 16-ketone, however, it is to be expected that because of the intermediate conformations of both 16α- and 16β-hydroxyl groups (Klyne, 1957, p. 34) the ΔR_{Mr} will be similar for both reactions and the two epimers will be separated with difficulty.* Similarly the ΔR_{Mr} on acetylation of both epimers will be very similar. With the 15-ketone, however, the epimeric hydroxyl groups will have distinctly axial and equatorial conformations and, in accordance with expectations, it is already known that epimeric 15-hydroxysteroids are easily separated (p. 88): the ΔR_{Mr} for the reductions will therefore differ by about 0·2, and the ΔR_{Mr} of acetylation of the two epimeric hydroxy-derivatives will also differ in the expected direction (ΔR_{Mr} axial < equatorial, p. 284).

There are other troubles which may beset the chromatographic method which would be absent in the classical method because of the availability of empirical formulae. Thus, for instance, if the original substance was an ester, chromic acid oxidation would in some cases hydrolyse the ester (e.g. aldosterone diacetate, Peterson *et al.*, 1957b) and oxidize the resulting free hydroxyl group to a ketone, aldehyde, lactone, or carboxyl group. The last case would be easily spotted because of the very large positive ΔR_{Mr} value which would conflict with all the possible formulae mooted at the beginning of the degradations. In the former three cases, however, it might not be spotted until later in the procedure and give rise to much puzzlement. Thus oxidation would give rise to a much lower negative ΔR_{Mr} than expected if other free hydroxyl groups were present, or to a moderate positive ΔR_{Mr} if no other hydroxyl groups were present: with lactone formation an unexpectedly large negative ΔR_{Mr} would result. The case of reduction of 18-hydroxyoestrone has already been given (p. 276). In all cases the final fully oxidized derivative would have an R_M value too large or small for any of the originally proposed formulae; even if it were not identical with a known reference substance the inference could then be made that either one of the polar groups was derived from, or had given rise to, a group combined with a non-polar radical of which the obvious possibilities are ester, acetal, ketal, lactone, lactol or lactal or hydrogen-bonded group. These could then be distinguished by the ΔR_{Mr} values of appropriate means of hydrolysis, followed by the ΔR_{Mr} values of subsequent esterification, condensation, or oxidation reactions. Hydrolysis of lactones would give rise to an obvious very large positive ΔR_{Mr} value for the

---

* This has now been described for one pair (Neher *et al.*, 1959).

released carboxyl group, which would also show the characteristic ΔR_{Ms} values on changing the pH of the solvent system.

It is clear that the same sort of abnormal courses and rates of reaction that occur with the classical method may upset the chromatographic procedure. Some of these may be more difficult to spot in the latter method because of the absence of the guidance of empirical formulae, however, most of them are easily detected by ΔR_{Ms} values and it is only the occasional example by which the chromatographic worker may be in danger of being badly fooled. It is best to leave the other possibilities of chromatographic identification until further work has been done, Meanwhile, however, it is hoped that this section has shown that most of the characteristic changes used in the classical degradation of an organic substance can be identified fairly unambiguously by chromatographic methods. A lot more work needs to be done to extend and confirm the generalizations about solvent systems used in this chapter, but it is already clear that steroids behave in accordance with general rules and it is to be hoped that such work will also be extended to other related organic compounds.

CHAPTER 6

SOME TYPICAL ANALYTICAL PROBLEMS OF STEROID BIOCHEMISTRY

1. Introduction

IN THIS chapter some specific problems needing chromatography for their solution will be discussed. In some cases published methods will be suggested as suitable solutions or criticized as probably inadequate; in others unpublished or suggested solutions will be described. One of the main purposes of this chapter is to show that in this field as in many others "there is more than one way to skin a cat". Finally I shall try and outline a general scheme for the chromatographic separation and estimation of the major steroid metabolites of urine based entirely upon paper chromatography, since this problem is one which is still of great importance in medicine and of interest to many people in this field. Preference will be given to methods which embody the principle of two chromatograms with a modifying reaction between them, but an effort will be made to show when and how this general procedure can be circumvented. I shall also neglect adsorption methods since there seem now to be few cases in which they are really necessary; secondly they are less easily reproducible by those who have not had a considerable experience of the purchase and standardization of adsorbents; thirdly they involve columns rather than paper (p. 220). Lastly there are several groups of steroids for which they are definitely inferior to partition methods, and most workers find it better in the long run to concentrate upon one general type of method for as many purposes as possible; since partition methods must be used for these steroids it is sensible to try and apply similar methods to the other groups of steroids. Many instructive alternatives to the methods discussed below will be found in Neher's valuable reviews (Neher, 1958a, b, 1959).

2. Cortisol in Human Blood

The methods of Peterson *et al.* (1957a) and Silber and Busch (1956) show that with careful solvent fractionation and appropriate blank solutions as controls chromatography is not really necessary in most cases. However, interference from drugs (Neher, 1958b) and other exogenous substances is always a danger with such methods. The methods of Bondy *et al.* (1957) and of Lewis (1956, 1957) use a single chromatogram in a volatile solvent system followed by fluorimetry with potassium t-butoxide in the first case, and with sulphuric acid in the second. Because it is of very general application, fluorimetry with potassium t-butoxide after elution, or its equivalent with NaOH on the paper (Ayres *et al.*, 1957a, b) seems to be the best choice. To cope with sporadic interference from drugs and foodstuffs or (in addition) to introduce a tritium-labelled acetyl group, acetylation has been found useful, as a modifying step, in the author's laboratory. This technique is also easily extended to other body fluids (Baird and Bush, 1960) and to other classes of steroids such as corticosterone and aldosterone. These procedures are summarized in Fig. 6.1. There is very little reason to doubt that such methods give a highly specific estimation of cortisol in human blood; with the modifying step similar specificity should be obtained for cortisone and corticosterone.

Small Celite columns were used by Morris and his colleagues (Morris and Williams, 1953; Morris and Williams, 1955) to separate corticosteroids in blood, and polarography of the Girard hydrazones or other methods were used for their determination in the eluates. This method gave values similar to other methods for cortisol (Bondy *et al.*, 1957) but results which were too high for corticosterone and 11-dehydrocorticosterone (Bush and Sandberg, 1953; Peterson, 1957; Neher, 1958b). Despite the elegance of many features of the technique it is complicated and cannot be recommended unless much more specific means of determination are applied to the eluates. Equivalent paper chromatographic methods (e.g. Bondy *et al.*, 1957; Lewis, 1956) are both more accurate and far simpler to carry out.

3. Oestrogens in Human Urine

To estimate the very small quantities of these substances in non-pregnant women and in men needs a very specific method of which the only examples at present are those of Brown (1955) and Bauld (1955). The newly discovered D-ketolic oestrogen metabolites (Layne and Marrian, 1958; Marrian, Loke, Watson and Panattone, 1957) need not be considered at the moment since their excretion rates are small even in pregnancy. It has been shown by Short (1959) that the Kober reaction in its modified and greatly improved form (Bauld, 1955; Brown, 1955) can

	Lewis (1956) and Bondy et al. (1957)	Ayres et al. (1957)	Peterson et al. (1957a)	Bush and Mahesh (1959a), Baird and Bush (1960)
First chromatogram	Defatting, only (see p. 143)	Typical system, group C, Celite	Typical system, group C, paper	Typical system, group C, paper
Modifying step	None	Acetylation	None	Acetylation
Second chromatogram	Typical system, group C, paper	Typical system, group C, Celite	None	Typical system, group B, paper
Third chromatogram	None	Typical system, group B, paper	None	None
Method of determination	Elute: fluorimetry in H_2SO_4 or K-t-butoxide	Photoelectric fluorimetry on paper: NaOH fluorescence	Porter–Silber reaction	Visual, NaOH fluorescence on paper
Internal control	None (Lewis) ^{14}C-labelled steroid added to plasma (Bondy and Upton)	Labelled steroid added to second column	Labelled steroid added to plasma	None

FIG. 6.1. Typical methods for estimating cortisol in blood plasma (concentration range about 2–20 μg/100 ml). The amount used for estimation is usually in the range 0·1–1·0 μg for all these methods. The method of Peterson et al. (1957b) is a general one and applicable here, but for convenience is considered under aldosterone in Chapter IV, p. 205. The methods given here are much simpler and quite adequate for most purposes.

be applied readily to eluates from paper chromatograms. It seems reasonable therefore to suggest a paper partition method for the three common human oestrogens based on this reaction and upon the use of methylation (Brown, 1955) as a modifying reaction. It is probably best to carry out the simple partition step of Brown to provide a fraction of the urinary extract containing oestrone and oestradiol, and another containing oestriol. The former pair can be separated easily with a group-B ("typical") system; the latter moves about 12 cm in 6 hr on 3MM paper at 25 °C. The three compounds can therefore be eluted separately from reasonably characteristic positions and the eluates evaporated and methylated by Brown's method. The methyl ether of oestrone could then be separated in a number of systems of group A (acetic acid type) of which the best would seem to be decalin–acetic acid–water (100: 90: 10 by vol.), because of the change in relative R_M values of ketone and hydroxyl groups, and the sensitivity of the cyclic solvents to interactions with unsaturated rings. Oestradiol methyl ether could be run in similar group-A systems. Oestriol methyl ether would be best run in a slightly more polar system of the group B; otherwise a slightly longer run in an A system would be suitable. The zones could then be eluted and determined by the Kober method, preferably after partitioning the eluates between ether and water (p. 186).

This procedure is really nothing more than a partition equivalent of Brown's method. It is however likely to gain specificity from the suggested placing of the modification step between two chromatograms (Fig. 6.2).

There remain many problems for future work to solve. For instance it is not known how many of the numerous oestrogen metabolites found in the urine of other species, notably the cow and the mare, can be separated by partition systems. A large number of these differ only in the number and position of double bonds, most of them conjugated with the aromatic ring; these are most likely to be separated by using cyclic non-polar solvents (decalin and tetralin, p. 101) and by changing from methanol systems to one or other of the "atypical" systems (p. 96). The rather special types of interaction of certain polar groups (such as those of the isomeric oestrogen D-ketols) with butane-1,3-diol (Axelrod, 1953) and with formamide (Layne and Marrian, 1958) are worth remembering when setting out to estimate the more difficult members of this group. Similarly the formation of an acetonide or borate complex with a *cis*-16,17-diol would be a useful modifying step (Hoffman and Darby, 1944; Hoffman and Lott, 1947 (see p. 291)).

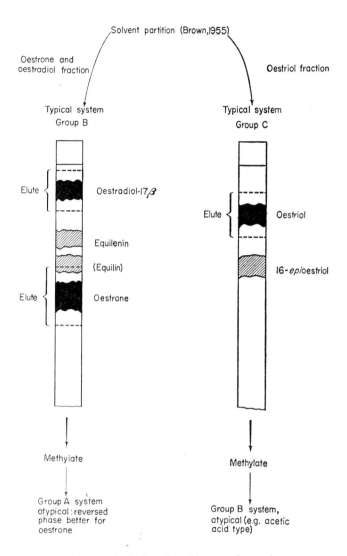

Fig. 6.2. Feasible method for the three major urinary oestrogens. This is really a paper-partition analogue of Brown's (1955) method: based on unpublished experiments of the author.

4. Progesterone in Blood or Urine

Short's method (1957) suffers from the lack of any systems giving low R_F values for progesterone and the absence of any modifying step. Sommerville and Deshpande (1958) described a method using alumina chromatography and ultraviolet absorptiometry of the isonicotinic acid hydrazone which was sensitive to $1 \cdot 0$ µg of progesterone. Zander et al. (1954) and Edgar (1953a, b) had previously described methods similar to Short's using ultraviolet absorptiometry of the progesterone itself at 240 mµ, after chromatography on paper. The following method is based on some recent unpublished work and should give a much more specific separation.

The plasma or urine extract is run on a 5-cm wide strip of Whatman No. 3MM paper in the system light petroleum–methanol–water (100:95:5 by vol.), or decalin–methanol–water (100:85:15 by vol.) at 18–25 °C until the dye F63 (p. 181) is just at the tip of the 45-cm strip. The progesterone zone (R_F 0·48 and 0·57 in the two systems at 22–25 °C) is eluted and treated with the quick acetylation method (p. 358), this modifying step being aimed at contaminants rather than progesterone itself. The acetylated eluate is then run on Whatman No. 2 paper in the system decalin–, methycyclohexane– or light petroleum–acetic acid–water (10:9:1 by vol.) for 16–36 hr at 18–25 °C. These systems give R_F values of approximately 0·2 for progesterone thus allowing considerable overrunning with improved resolution. They also have the advantage of giving considerable changes in the relative ΔR_{Mg} values of ketone and hydroxyl groups relative to the methanol systems (p. 99). If necessary the ammoniacal modification of the methanol system could be used to provide the pH change recommended on p. 214 (Fig. 6.3).

Determination of progesterone could then be carried out on the appropriate eluate by one of the above published methods. These leave something to be desired in terms of sensitivity and it would be valuable to develop an isotopic method for this substance and other non-alcoholic ketones such as ketalation with [^{35}S]ethanedithiol or [^{3}H]ethylene glycol (p. 363). The ketals would be extremely non-polar and best separated on a reversed-phase system such as R. P. Martin's (1957). In this case the best method would be to start with the overrun chromatogram in decalin or petroleum–acetic acid–water; substitute ketalation for acetylation; and finish with a reversed-phase chromatogram. A slightly more complex scheme is suggested in Fig. 6.4 and dealt with in more detail later (p. 326).

5. Aldosterone in Urine

The best method of avoiding the use of acetylation seems to be that of Mattox and Lewbart (1958, 1959). This is summarized in Fig. 6.5. The difference in R_M values relative to those of the main potential contaminants

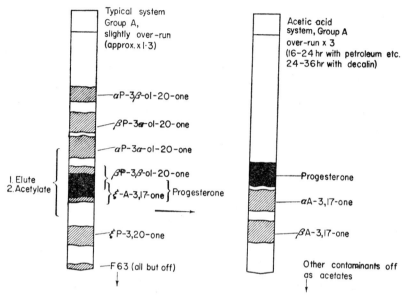

Fig. 6.3. Separation of progesterone from a complex mixture. The scheme shown would usually suffice for estimating progesterone because specific methods of determination are available. Complete separation from androstanedione, e.g. for isotopic work, could be achieved by re-running in a different or reversed-phase system or by one of the schemes shown in Fig. 6.4. Note the degree of over-running possible with the acetic acid system (Group A) in which progesterone has an $R_F = 0.2$. F63 is sudan-type dye (see Table 3.3).

of aldosterone are larger than in the equivalent method of Neher and Wettstein (1956a). The speed of the first chromatogram fully justifies the use of an impregnated-paper system; also it is better to use such a system in the first stage because of its greater capacity. The two main types of method using modifying steps are those of Ayres et al. (1957) and of Peterson et al. (1957b) previously referred to. In the author's view these methods err on the side of complexity in a way that is almost certainly

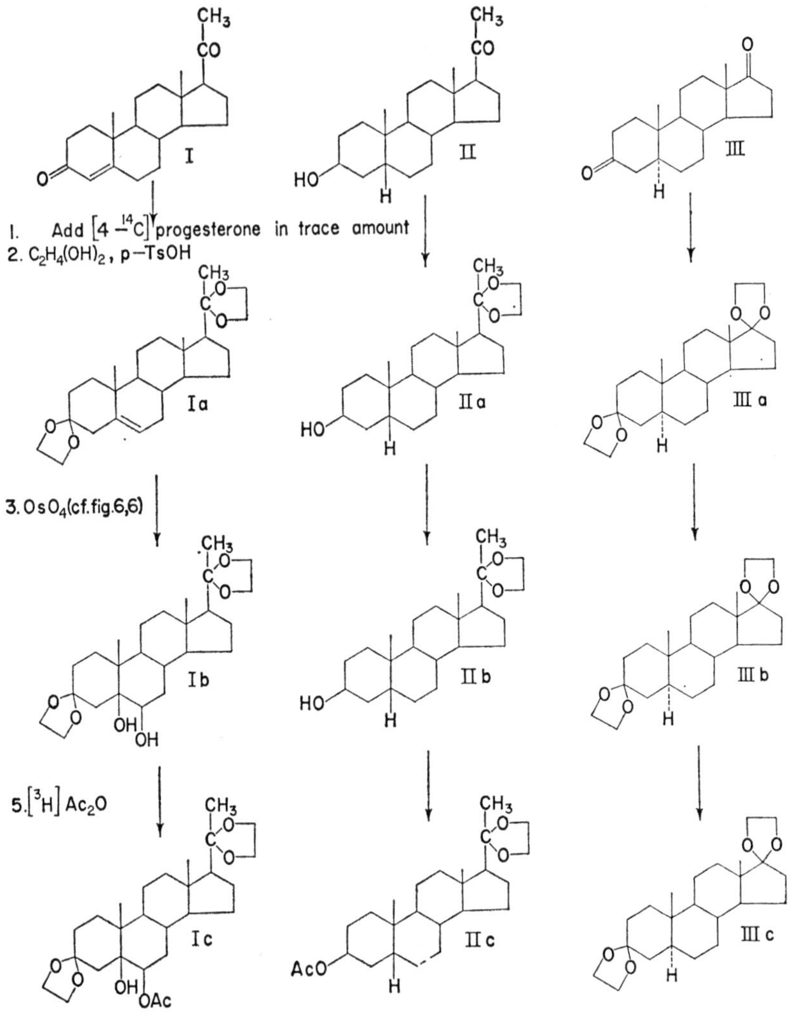

FIG. 6.4. Introducing new substituents to achieve specificity of estimation. Progesterone (I) and the two main contaminants (II and III) in the eluate of the first chromatogram of Fig. 6.3 are treated by a more complicated sequence (Bernstein et al., 1955) using labelled [4-^{14}C] progesterone as internal control. This scheme has many variants but is based on the general idea that simple substances with very few polar groups will usually gain highly specific characteristics if they are converted to *more* complicated derivatives. This in turn depends on the fact that current solvent systems are more sensitive to differences in the type and conformation of polar groups than to differences in the hydrocarbon skeletons of organic molecules (p. 127). The derivatives Ib, IIb and IIIb or Ic, IIc, and IIIc will be very easily separated in almost any solvent system in group A. (See also **Fig. 6.6.** and p. 326.)

unnecessary (p. 253); the first in using columns to obtain a separation that is no better than can be gained with thick paper and a smaller quantity of extract; the latter in using a radioactively labelled internal control. A workable method should be possible with a simpler technique based upon unpublished work with Miss Irene Hughes and Miss Magda Reich. The simplest method would use the fluorimetric determination of Ayres *et al.*;

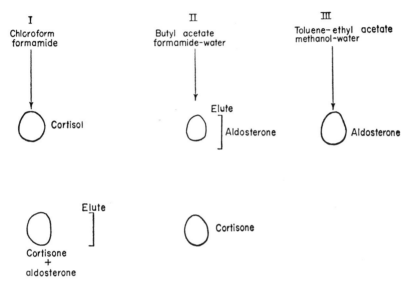

FIG. 6.5. Separation of aldosterone from its common contaminants (see text, p. 317).

i.e. development and measurement of the fluorescence produced on a paper chromatogram by the NaOH reaction without elution. The analogous method using potassium t-butoxide is probably slightly more sensitive (Abelson and Bondy, 1955) but also more expensive and difficult. The best method would be to use tritiated acetic anhydride as combined modifying and labelling steps as in the method of Peterson *et al.* (1957b). The separation techniques vary according to which of these methods of determinations is to be used.

(a) Direct Fluorimetry on Paper

The optimum quantity required for this method of determination is about 0·25 μg with aldosterone diacetate and 0·04 μg can be estimated to within $\pm$ 10% (Tait and Tait, 1959). A normal 24-hr collection of urine contains about 8 µg aldosterone in a form released by continuous extraction (see p. 198) at pH 1·0 (Ayres et al., 1957) so that using this method one-twentieth of a 24-hr specimen would be ample for the sensitivity of the fluorimetric method. This will be a volume of about 80 ml and the "continuous extraction" can easily be carried out by shaking the acidified urine with methylene dichloride in a 500-ml conical flask on a mechanical shaker instead of a Cohen extractor (Mattox and Lewbart, 1959). The washed extract may be cleaned up by the use of a small silica gel column and the ethyl acetate–methanol eluate is clean enough to run on a 5–10-cm wide strip of Whatman No. 3MM paper. With the formamide system the smaller width would be sufficient. Two systems seem suitable for the first chromatogram; the first has the advantage of speed and is the single length run in chloroform–formamide used by Mattox and Lewbart (1958) which takes 2·5 hr with no need for an equilibration period, and which leaves aldosterone and cortisone together at an R_F of about 0·6. The second possibility for those used to volatile systems is the system B/50, equilibrated 3–16 hr and run for 6 hr; with 3MM paper this leaves cortisone at about 35 cm from the origin of a 40-cm strip, cortisol at 17 cm, and aldosterone at 25 cm, being an approximately threefold overrun chromatogram. A faster separation can be obtained with one of the C systems but is less overrun and therefore gives less resolution.

The aldosterone zone is eluted and acetylated by the quick method (p. 358) and then run in a group B system for 3–6 hr after 5–16 hr equilibration, or in a A system (Brooks, 1959) for 16–24 hr. The final chromatogram is then treated with NaOH (p. 379) and the fluorescence measured by the method of Ayers et al. (1957b).

Alternative systems for the two chromatographic steps are given in Table 5.3 (p. 296) which compares the R_F and R_M values of aldosterone, cortisone, and cortisol. The main contaminants of the aldosterone zone on the first chromatogram ("X6" from citrus fruits (De Courcy et al., 1953; Gray et al., 1955) and a blue fluorescent material) are both removed by the acetylation step. The minor contaminants found by Nowaczinski et al. (1957) also seem to be removed by acetylation. The greatest change in the relative R_M values of aldosterone and cortisol and cortisone is achieved by starting with the system benzene–acetic acid–water (10 : 7 : 1 by vol.) (p. 296) and then using Lewbart's second system of butyl acetate–water–formamide. An almost equivalent change is obtained by using the t-butanol system of Eberlein and Bongiovanni (1955) (E$_2$B) for the second run. These are much larger than the analogous changes obtained by previous methods using toluene–propylene glycol followed by the system toluene–ethyl acetate–methanol–water (95:5:50:50 by vol.).

(b) *Fluorimetry in Potassium t-butoxide*

Since this method involves elution from the final chromatogram and hence needs a larger separation of aldosterone from contaminants, method (a) may not be sufficient. There are two ways of overcoming this difficulty. Either method (a) is used and radioactive [16-^{3}H] aldosterone (Ayres *et al.*, 1957b, 1958) or aldosterone [*carboxy*-^{14}C] diacetate is added at appropriate stages as internal control, so that the final elution can be partial and include only the centre of the final zone: or it may be easier to add a third chromatographic step before acetylation such as Lewbart's second system or the E_2B system.

(c) *Assay of Tritium-Labelled Acetoxyl Groups*

About one-hundredth of the urine sample suffices (Peterson *et al.*, 1957b), which is usually about 16–20 ml, and 50 µg each of cortisol and cortisone are added before it is extracted. It is extracted as in method (a). The added steroids are to serve as markers on the first chromatogram, to compete with the much smaller amount of aldosterone for reactive points on the glass of the distillation flask or tube in which the extract is evaporated, and to compete similarly for impurities and strongly adsorbing points in the filter paper of the first chromatogram. It is not altogether easy to avoid disproportionate losses of steroids of this type when the amount being handled is less than 0·5 µg; these losses occur mainly in the evaporation and storage of the extract and seem to be less with glassware that has not been cleaned with chromic acid nor heated to dryness (p. 355); see also Lewbart and Mattox (1959). It is possible to run 0·01 µg of steroid without carrier on paper without losses when the steroid is pure; in the presence of impurities it is safer to add a similar steroid to compete for traces of destructive reagents, complexing substances, or adsorptive material.

The first chromatogram is the same as in method (a). After this the aldosterone zone is eluted using the cortisol and cortisone areas as internal markers and detecting them by brief examination under ultraviolet light (2537 Å). The eluate is evaporated and acetylated with [^{3}H] acetic anhydride (100–150 mc/mmole) by the quick method in a 10 ml graduated centrifuge tube (care with stoppering; use fume cupboard!). After cooling the tube 5 μg unlabelled aldosterone diacetate is added to the tube in 0·2 ml ethanol, and 0·8 ml water added. The solution is extracted in the tube with methylene chloride (5–8 ml) and the aqueous layer sucked off and discarded (Peterson et al., 1957b). The methylene chloride is washed once with 1 ml water and evaporated with a jet of filtered air with the tube standing in a beaker of water at about 40 °C. The residue is dissolved in methanol–methylene chloride and put on a 5-cm wide strip of Whatman No 2 paper over a lane 2·5 cm wide. This strip is then run in the second system of method (a) and the aldosterone zone detected by the ultraviolet absorption of the added carrier. The visible zone is cut out and eluted without any added safety margin (cf. p. 223) and one-tenth of the eluate removed and re-run in the same system for the same time. The major part of the eluate is conserved in the refrigerator. The chromatogram of the minor part of the eluate is then treated with the NaOH fluorescence reaction and estimated fluorimetrically by the method of Ayres et al. (1957b). At the same time it is examined visually for the presence of any other fluorescent zones. This estimation serves to measure the recovery of the diacetate during the chromatography and partial elution of the final zone; it also affords a visual check on the purity of the material with respect to other Δ^4-3-oxosteroids and "X6" (citrus fruit fluorogen). The major part of the eluted aldosterone diacetate is then counted using the plating techniques of Ayres et al. (1958) or in a liquid scintillation counter. Knowing the specific activity of the acetic anhydride and the recovery from the partial elution the amount of aldosterone originally present in the original extract can be calculated.

The method could be simplified by counting the paper directly and correlating this with the scanning record of the NaOH fluorescence reaction (Tait and Tait, 1960). With adequate apparatus this would enable the determination of specific activity in different parts of the aldosterone zone thus greatly increasing the probability of correct identification (p. 42).

This suggested method makes no use of an internal control on recovery, except for the final partial elution for which it is essential. The extraction and first chromatogram are thus carried out "carrier-free" with all the concomitant dangers of disproportionate losses. In the author's experience errors from this cause should be small if the glassware is carefully selected, and they are reduced by the presence of added competing substances and ethylenediaminetetra-acetic acid (Lewbart and Mattox, 1959). Aldosterone [^{14}C]diacetate could be used as an internal control for the last chromatographic step as in the method of Peterson et al. (1957b) if the necessary apparatus were available for counting the two isotopes involved. The above method has been suggested for those who want the maximum simplicity compatible with a reliable method. The main point is to retain the possibility of using the highly specific acetylation step by omitting an internal control in the first chromatogram; to avoid the use of an isotopic compound as the second internal control; and to avoid the use of the highly characteristic but tricky oxidation with chromic acid used by Mattox et al. (1956) and Peterson et al. (1957b). This reaction has a variable yield, and since the lactone acetate gives rather a low intensity of fluorescence with NaOH it makes the use of a second isotope essential as an internal control. Another not inconsiderable advantage of avoiding this reaction as the modifying step is that it reduces the necessary specific activity of the tritiated acetic anhydride to one half of that used in the original method since oxidation to the lactone involves the loss of one of the labelled acetate groups that were introduced. It should be noticed that Peterson's use of aldosterone [^{14}C]diacetate as an internal control is largely due to the lack of aldosterone labelled in the ring with ^{14}C.* As with its similar use by Ayres et al. (1957a) it only suffices to control recovery over the later parts of the procedure, in itself strong evidence that with careful technique disproportionate losses of this steroid can be avoided during the extraction and subsequent handling of the order of 0·2 μg or less.

Much previous work on the paper chromatography of aldosterone has conveyed an excessively pessimistic view of the situation. Much emphasis has been laid on its property of running with cortisone in toluene–propylene glycol and with cortisol in the toluene–ethyl acetate–methanol–water system, and on its being contaminated on chromatograms of urinary extracts with numerous compounds known and unidentified (Neher, 1958b; Neher and Wettstein, 1956a; Moolenaar, 1957; Nowaczinski et al. 1957). Simpson and Tait (1953), however, showed early on that aldosterone ran with an R_F value exactly the mean of those of cortisone and cortisol in system B/50. As expected from their results it was found by the author that excellent separation of aldosterone from these two steroids and from several (but not all) urinary contaminants was obtained by overrunning the B/50 system until cortisone was 35–40 cm from the origin. This is quickly done without losing resolution by using the thick Whatman No 3MM paper; this paper also has the advantage of

---

* *Added in proof:* Peterson's final version uses [^{3}H]aldosterone and [^{14}C]acetic anhydride and provides full control of recovery: Kliman & Peterson (1960) *J. biol. Chem.* **235**, 1639.

high capacity. These experiments formed the basis of the methods suggested above, which are otherwise largely similar to those of Ayres et al. (1957), Peterson et al. (1957b), Brooks (1960) and Mattox and Lewbart (1958, 1959); from these papers an attempt has been made to combine those features making for speed and simplicity in a way which retains their sensitivity and specificity. All the chromatographic steps with the suggested volatile systems have been checked with both urine extracts and pure steroids but no recoveries or quantitative measurements attempted.

6. Testosterone in Blood or Urine

No method specifically designed to estimate testosterone has been published* although chromatography has been used for its demonstration in blood (West, Hollander, Kritchevsky and Dobriner, 1952; Hollander and Hollander, 1958). It is instructive to consider the problems involved in designing a specific method in the light of the general principles of Chapter 4. The difficulty here is that the two polar groups of testosterone are at two of the least hindered positions of the steroid nucleus so that we shall expect that the ΔR_{Mg} values of similar unhindered polar groups in a hypothetical non-steroid contaminant on the first chromatogram will be similar to the ΔR_{Mg} values of the polar groups in testosterone itelf. Similarly any contaminating monohydroxyketone will probably give similar ΔR_{Mr} values with any simple modifying step we may employ. Thus any simple monohydroxylic contaminant will give a closely similar ΔR_{Mr} on acetylation, and any simple monoketonic contaminant will give a similar ΔR_{Mr} on ketalation. The possible procedures (Fig. 6.6) must be considered in the light of this awkward fact.

Derivatives I and II in Fig. 6.6 could be estimated by one of various spectrophotometric methods for the Δ^4-3-ketone group, or by the Zimmermann reaction. Derivative III could be estimated by the Zimmermann reaction, and derivative IV by labelling with [^{3}H] acetic anhydride. With the first derivative the danger of contamination by a non-steroid monohydroxy-compound is considerable so that two steps would be taken to minimize this danger. Firstly the final chromatogram would be on an acetic acid system thus giving a large positive ΔR_{Ms} to the Δ^4-3-ketone group; secondly one would use the highly specific fluorescence in potassium t-butoxide (p. 381) or with NaOH for determination. The latter,

---

* Added in proof: See however Finkelstein, Forchielli & Dorfman (1961) J. clin. Endocrin. **21**, 98.

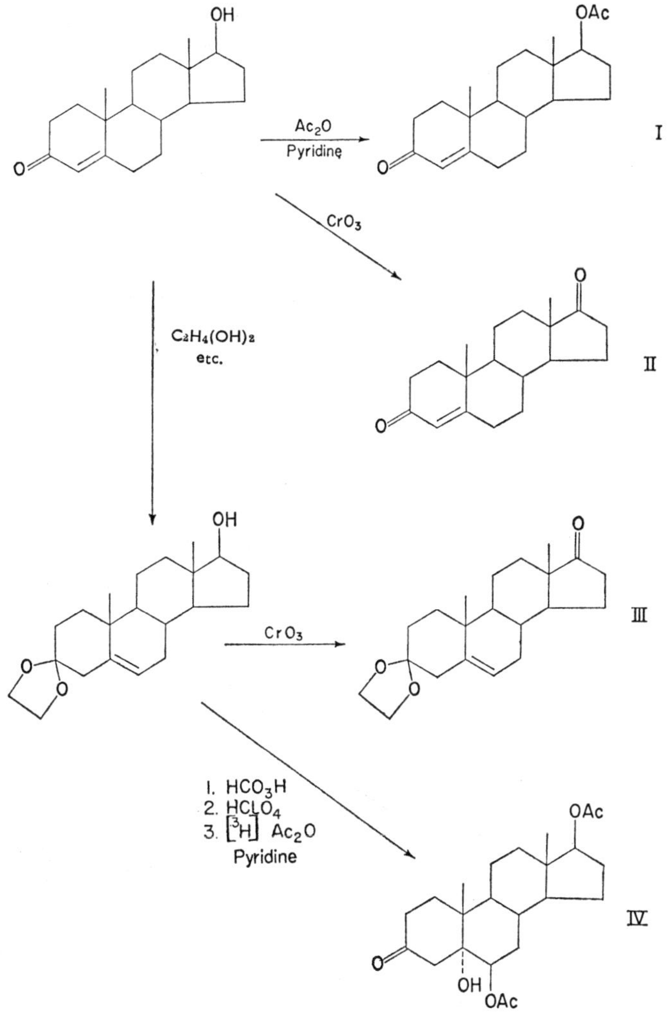

Fig. 6.6. Modifying steps for estimating testosterone. This is similar to the suggested scheme for progesterone (Fig. 6.4) but obtains the 5α,6β-diol in derivative IV (Amendolla et al., 1954). Derivatives I, II and III are useful but may give inadequate specificity in some situations. Other reactions available for "complicating" modifying steps are the direct introduction of 2α-OH via Pb(OAc)$_4$ (Sondheimer); of 6β-OAc by dehydration of IV (Sondheimer et al., 1954) or by oxidation with MnO$_2$; and of 7α-OH via bromination of the ketal (Bernstein et al., 1959).

done by direct fluorimetry on the paper, would provide the best chance of detecting contaminants by inspection of the scanning curve for inflections or asymmetry of the peak. The second derivative could also be handled this way or by the Zimmermann reaction. The latter, however, is dangerously non-specific in this case since the absorption curve would be complex in the presence of the Δ^4-3-ketone group and would not be easy to correct by Allen's method (Callow et al., 1938). The Zimmermann reaction would be better applied to derivative III where a good correction could be obtained.

Neither of these methods of determination is very sensitive with this type of compound, however, and such methods would not work to much below 0·5 μg. Further sensitivity would be obtained by labelling with [^{3}H] acetic anhydride but, as we have seen, this would run the danger of failing to eliminate monohydroxy-contaminants if it were the only modifying step employed. By using the ethylene ketal (Sondheimer et al., 1954; Bernstein and Lenhard, 1955), however, we can introduce the very characteristic 5α,6β-diol group into the molecule giving a large positive ΔR_{Mr}. Any simple "testosterone mimics" are extremely unlikely to give a similar ΔR_{Mr}, even if they possess a double bond since the latter is unlikely to have the same degree of hindrance as the 5,6-position in the steroid. Since this reaction, although reliable, is unlikely to be completely quantitative we should add unlabelled testosterone-3-ketal as a carrier to the reaction mixture, or [4-^{14}C]testosterone at an earlier stage if we wished to use a labelled internal control, and correct our tritium assay by a standard spectrophotometric estimation of the final product or by assay of ^{14}C.

This reaction sequence may seem complicated but has several advantages in this case. Not the least is that a fairly polar product is obtained that can be run on the more convenient "straight" partition systems using a considerably overrun chromatogram. More important is the fact that, being presented with a compound whose only polar groups are in positions where conformational differences have the smallest ΔR_{Mr} values, we have succeeded in introducing two groups in positions where it is already known that conformational differences produce large ΔR_{Mr} effects (Chapter 2). The possibility of "mimics" having groups introduced with exactly the same conformational properties by this sequence is sufficiently remote to give us confidence in using the otherwise rather non-specific labelling method for the final determination.

This example can be considered a special case of the general principle of using modifying reactions, and could be extended to all compounds in which the polar groups are few and in unhindered positions. It is an unavoidable corollary to the R_M rule that simple modifying reactions applied to such groups will be less successful in eliminating contaminants than with groups in more hindered positions. The remedy is to use a special type of modifying reaction in these cases when possible; namely the introduction of new groups in positions which do show marked conformational effects on ΔR_M values.

In the present case it is interesting to note that with the method using derivative IV the order of solvent systems should be the reverse of that used with the other derivatives. Thus the first chromatogram should be with an acetic acid system to use the atypically large ΔR_{Mg} of the Δ^4-3-ketone group, and the second with a "typical" system to obtain large and characteristic ΔR_{Mg} values for the diol group. There would be a strong case here for preferring formamide or propylene glycol to methanol systems because of the greater chance of specific steric effects on the interactions of 1,2-diol and 1,2-ketal groups with the former systems (e.g. Layne and Marrian, 1958).

7. Difficult Separations of Non-Polar Steroids

The volatile systems based on acetic acid (p. 109) give exceptionally large ΔR_{Mg} values for 3- and 17-ketone groups and allow the separation of the epimeric 5ξ(H)-3-ketones. Any steroid as polar as, or more polar than, androstanedione can therefore be handled easily by one or other of the volatile systems. Pregnanedione (5β(H)) has an R_F of 0·45 in methylcyclohexane–90% acetic acid and cannot be subjected to chromatograms overrun by more than a factor of about 1·7. This type of compound, the non-polar sterols, and the esters and ketals of compounds like androsterone can only be separated conveniently with reversed-phase systems or in some cases with phenyl-cellosolve (Neher and Wettstein, 1952), nitromethane (p. 60), or carbitol systems (Reineke, 1956). The phenyl-cellosolve and published reversed-phase systems all suffer from defects of one sort or another with the result that except for simple separations of a few widely different substances they cannot be held to compete seriously with adsorption methods. The use of phenyl-cellosolve is tricky and methyl-cellosolve was found better by Loomeijer and Luinge (1958) who made a very careful study of the technical details that affect the results. These and the experiences of Rubin et al. (1953) are rather a discouragement to the use of this type of system. Furthermore neither this system nor the adaption of Mills's system (Mills and Werner, 1955) by R. P. Martin (1957) (reversed-phase, with odourless kerosene as stationary phase) give sufficiently large ΔR_{Mg} values to the differences in the hydrocarbon skeleton of non-polar steroids which are so important in this class of compound. This type of system gives ΔR_{Mg} values for polar groups which are similar (though of course of reversed sign) to those found with "straight" systems based on similar solvents: it is easy therefore to separate, say, cholest-4-ene-3-one from cholesterol, and the latter from 7-oxocholesterol. However, most of the structural variations that are interesting in this class of compound are in the epimeric centres and degrees of unsaturation of the hydrocarbon skeleton; many of these give minimal ΔR_M values with solvent systems of this type.

It may well be that liquid–liquid systems will never be able to equal adsorption systems for this class of compound (Martin, 1950) but results obtained with gas–liquid systems encourage one to believe that liquid–liquid systems for non-polar steroids are capable of great improvement. It seems well worth-while considering much more exotic non-polar solvents for the stationary phase of such reversed-phase systems, in particular highly asymmetric molecules such as naphthalene, phenanthrene, anthracene, or 6,8-benzoquinone (Desty et al., 1959) dissolved in decalin or tetralin at the highest concentrations compatible with reasonable viscosity.

At present, therefore, partition chromatography cannot be recommended for the separation of very non-polar steroids unless the mixtures to be handled are relatively simple. The adsorption methods used for carotenoids and sterols are still the best solution for problems of this type (Strain, 1942). On the other hand there is room for further development of liquid–liquid systems for this class of compound and the field should be both profitable and interesting. Most of these substances are potentially separable by gas–liquid systems but so many of them are labile that it is not likely to be the most hopeful line of attack in general.*

8. Conjugates or Urinary Steroids

The steroid hormones and their metabolites are excreted largely in the form of conjugates with glucuronic acid, sulphuric acid, and possibly other unknown substances. One of the major problems of estimating these substances has been their quantitative hydrolysis without producing artifacts (Lieberman and Teich, 1953; many papers in *Recent Progress in*

---

* *Added in proof:* Considerable success has now been achieved by Vanden Heuvel, Sweeley & Horning (1960) *J. Amer. chem. Soc.* **82**, 3481.

Hormone Research, **9**, 1954). This has naturally led several groups of workers to attempt the difficult problem of evolving methods for the estimation of the conjugates themselves. Since these are not all known, the work is still in the exploratory stage but considerable progress has been made by Schneider and his colleagues (Lewbart and Schneider, 1955; Schneider and Lewbart, 1956, 1959) who have isolated several of the most important conjugates of human urine in crystalline form or as crystalline derivatives. Lewbart and Schneider (1955) described several solvent systems for steroid conjugates applicable to paper, columns, or counter-current apparatus. Some valuable new systems are described by Schneider and Lewbart (1958). Alternatives were also described by the author (Bush, 1957) and several methods of separating the glucuronosides as a group from the sulphates have been described (Cavina, 1955; Barlow, 1957; Crépy, Jayle and Meslin, 1957a; Crépy, Malassis, Meslin and Jayle, 1957). High-voltage electrophoresis on paper has also been used for the latter group separation (Cavina 1955b; Schroeder and Voigt, 1955). The very non-polar nature of the ammonium salts of the sulphates, however, makes this group separation very easy, and paper or column chromatography in ammoniacal systems (Bush, 1957) seems more convenient than the other methods (Fig. 6.7). The R_F values of some typical conjugates in such systems are given in Table 6.2. For non-polar sulphates (e.g. of dehydroepiandrosterone) this property is so distinctive that chromatography is not really necessary. Thus DHA sulphate, aetiocholanolone sulphate and androsterone sulphate can be completely recovered from urine by half-saturation with ammonium sulphate, making to 0·5 N with ammonium hydroxide, and extracting with ethyl acetate. With suitable washes the extract contains none of the glucuronosides of androsterone, aetiocholanolone and other more polar steroids (Bush and Gale, 1960).

Roy (1956) also devised a simple way of separating steroid sulphates using the solubility of the methylene blue complexes in chloroform (Vlitos, 1953). This method works very well in the author's experience. However, it is interesting that in the acetic acid–t-butanol systems the complex (with DHA sulphate) dissociates so that the chromatogram is indistinguishable from that of the steroid conjugate itself. Schneider (personal communication) has found that the free acid form of DHA sulphate is extremely unstable in the dry state. It is important to keep this type of conjugate in salt form whenever extracts have to be evaporated or stored.

The ammoniacal systems provide a useful method of sub-fractionation of the common urinary conjugates, particularly since some common dyes and indicators can serve as excellent markers for separating the most important groups from one another. With typical urinary extracts this can be done on thick paper (Whatman, Nos 3MM, 7 and 17) or Celite columns (Fig. 6.7).

For the separation of individual conjugates the acid systems are usually the best. Many of these conjugates can be separated very well with the butyl acetate–acetic acid–water systems of Schneider and Lewbart (1955).

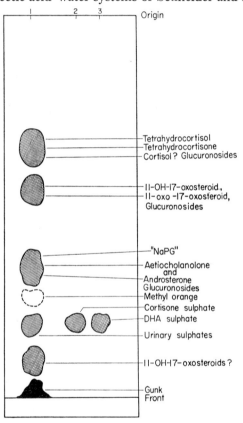

FIG. 6.7. Group separation of some steroid conjugates. Note the high R_F values and lack of separation of the sulphates: also the convenient marking of the urinary groups with methyl orange.

Promising separations were obtained with chloroform–acetic acid on silicic acid columns by Edwards and Kellie (1956). The author is more familiar with a family of systems based on acetic acid, t-butanol and water, together with a suitable non-polar solvent as the main component of the mobile phase. The use of t-butanol was adopted because it is the only convenient alcohol which is not esterified in systems containing large amounts of organic acids: it therefore provides systems of reasonable stability which are not subject to the changes of composition caused by the esterification of primary or secondary alcohols (Hanes and Isherwood, 1949). This set of systems was devised before the results of Eberlein and Bongiovanni (1955) with t-butanol systems for neutral steroids were published, which was unfortunate since the poor separations of 11-hydroxysteroid conjugates from their 11-ketonic analogues were first taken to be due to unavoidable difficulties. It was found empirically, however, that these separations were improved by using smaller quantities of t-butanol in the system, and this became readily understandable when Eberlein and Bongiovanni's paper was published.

The main 17-oxosteroid glucuronosides are readily separated by these systems (Figs. 6.8, 6.9) and are only slightly more difficult to separate than the analogous free steroids. Separation of the sulphates is more difficult and there is room for improved systems for these conjugates (Fig. 6.10).

Schneider and Lewbart (1959), however, include among their newer solvent systems two which represent an important advance in the separation of the steroid sulphates. These are essentially based on an ether (isopropyl or n-butyl) and ammonia–water–t-butanol. The reasons for their undoubted superiority over previous systems for these difficult compounds need a careful discussion since one cannot agree with some of the reasons given by these authors; in particular their suggestion that the alkalinity of such systems means that their mobile phases can be less "viscid" than with acid systems is difficult to understand. The earlier suggestion that acid systems were preferred for separating individual conjugates (Bush, 1957) was correct for the family of systems described; it was not meant to imply that alkalinity or acidity had any special virtues in themselves, a suggestion which would be erroneous in view of the general rules for chromatography of acids and bases (p. 32). One can certainly agree with them, however, that alkaline systems, if of adequate resolving power, are indeed preferable for the steroid sulphates because of the danger of their hydrolysis or solvolysis in acid systems (Burstein and Lieberman, 1958b).

The reason for the superiority of these new systems can in fact be shown to be due to the main *non-polar component* of the system. This struck the author immediately because he had previously observed the poor separation of non-polar steroid sulphates in systems of identical composition to those of Schneider and Lewbart (1959, systems 1 and 2) except that

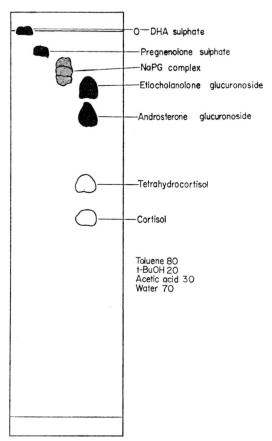

Fig. 6.8. Chromatogram of some conjugated steroids. Note the low R_F values and separation of the sulphates compared with the BuOH/NH$_4$OH system in Fig. 6.7.

ethylene dichloride was used in place of isopropyl ether (Bush, 1957). The latter solvent had been avoided with alkaline systems because of its peculiarly strong tendency to form peroxides. Bush and Gale (1960) were able to confirm the excellent separations of simple steroid sulphates ($C_{19}O_2$ family) with isopropyl ether systems but were also able to show that a similar superiority, in terms of the ΔR_{Mr} value (aetiocholanolone

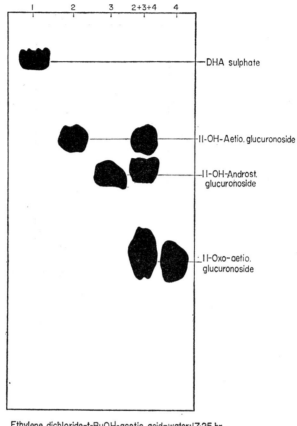

Fig. 6.9. Separation of $C_{19}O_3$ glucuronosides. Careful equilibration (see p. 170) is necessary with this system.

sulphate – androsterone sulphate), was shown by isopropyl ether–acetic acid-t-butanol–water systems as was shown by the analogous ammoniacal systems (Bush, 1960b). It appears then that the improved separation of steroid conjugates with isopropyl ether systems is another example of the fortunate and rather exceptional existence of subtle steric factors in the interaction of the main component of the mobile phase with the rather rigid structure of the steroids (p. 128). It is quite clear that this effect is seen with

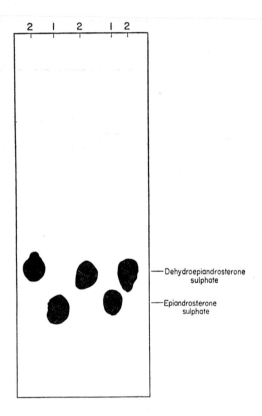

Fig. 6.10. Separation of sulphates. The system is not ideal and considerable over-running is needed (see p. 331).

both sulphates and glucuronosides, and in both acid and alkaline systems. It remains to be seen whether this effect is enhanced by possibly specific interactions between the t-butanol and the ether or not.

In view of this analysis, the author and Miss M. Gale have now checked that the addition of topanol A (2,4-dimethyl-6-t-butylphenol, 0·01%;

see Appendix I) to such solvent systems does not affect the chromatographic results, and successfully suppresses the formation of peroxides. Following this it was shown that the substitution of isopropyl ether for ethylene dichloride in the earlier acetic acid–t-butanol–water systems (Bush, 1957) gives slightly better separations of $5\alpha/5\beta$-epimers of steroid glucuronosides as well as of the sulphates. A slightly larger proportion of t-butanol is needed in each case to provide a system equivalent to a given ethylene dichloride system so that the range of mixture extends from 67:33:70:30 to 90:10:70:30 by vol. of isopropyl ether–t-butanol–acetic acid–water. It is also easier in practice to use such systems, in which the less polar phase is lighter than the more polar phase (p. 170).

Schneider (personal communication) has found that typical concentrates of urinary conjugates are difficult to separate into pure fractions chromatographically and finds that counter-current apparatus is very useful, particularly in the early stages of fractionation. This difficulty is particularly marked with some compounds which, although free of other steroid conjugates after many chromatographic steps, still retain unidentified impurities. In some cases this may be due to slight decomposition of the conjugate during the procedures. This brings up the main difficulty of this field, namely the preparation of extracts that are pure enough to be put on chromatograms.

A general method for extracting the glucuronosides of organic metabolites was first devised by Kamil, Smith and Williams (1953) and applied to a wide variety of conjugates. Edwards, Kellie and Wade (1953) found that this method achieved good recovery of the majority of steroid conjugates as judged by estimating the free steroids after hydrolysis of the extracts by suitable methods. They also found that the extracts obtained by this method were reasonably clean and free from salts, in particular from inhibitors of sulphatase (Dodgson and Spencer, 1953). Other workers have not had equal success (personal communications) and the method involves rather large volumes of solvent. In the author's experience neither this nor any other method provides a really convenient method of extraction when the routine chromatographic estimation of steroid conjugates is under consideration. The extracts provided by butanol extractions (Venning, 1937, 1938; Reddy, Jenkins and Thorn, 1952) of urine are the only ones that are clean enough for the immediate application of partition chromatography, but these are inconvenient for routine estimations because of the low volatility of the solvent.

In spite of previous failures to obtain quantitative recoveries of steroid conjugates with ion-exchange resins (Anderson and Warren, 1951) an attempt to overcome this difficulty by the use of a new anion-exchange resin ("Decolorite"; Permutit Ltd., Chiswick, London) was made. After numerous experiments it was found that all the known common urinary conjugates could be recovered by a relatively simple process if the resin was exhaustively and somewhat vigorously washed before use (Bush and Gale, 1957; Bush and Mahesh, 1959b, c). The free steroids are also recovered quantitatively and the resin has a large capacity. It is clear therefore that non-ionic adsorption is the main factor in this method of extraction and this was not unexpected since Anderson and Warren (1951) had previously shown that strong cation-exchange resins removed steroid conjugates completely from urine, even though the conjugates could not be recovered quantitatively from the resin.

The extract provided by this method contains rather more impurities than that obtained by the method of Edwards et al. (1953) but, being in a moderate volume of methanolic ammonia solution, is easily and quickly concentrated. The concentrate is then easily extracted with small, and more conveniently evaporated, volumes of the more specific solvents available for extracting this rather difficult group of compounds.

It is already possible to estimate the less polar steroid glucuronosides of urine as a routine method, since extracts of this group of compounds are easily prepared and can be purified by convenient washes to a state where they can be run immediately on paper chromatograms. The more polar steroid conjugates, however, require an inconvenient number of purification steps before the extracts are clean enough for chromatography. For some time to come, hydrolysis of the conjugates of at least some of the major steroids of human urine will be necessary for routine methods of estimating them. The methods for the less polar conjugates will therefore be left to future publications and we shall now consider the fractionation and estimation of the major urinary steroid metabolites of man.

9. Fractionation and Estimation of Urinary Steroids

For most clinical studies it is sufficient to measure the excretion of androsterone, aetiocholanolone, dehydroepiandrosterone, 11-oxo- and 11β-hydroxy-aetiocholanolone, 11β-hydroxyandrosterone, pregnanetriol, cortisol, cortisone, tetrahydrocortisone, allotetrahydrocortisol, and tetrahydrocortisol. It is sometimes interesting or necessary to measure 11-oxo- and 11β-hydroxy-pregnanetriol (Zondek and Finkelstein, 1954), cortol and cortolone—and their 20β epimers—(Fukushima, Leeds, Bradlow, Kritchevsky, Stokem and Gallagher 1955) corticosterone and its three principal tetrahydro-metabolites (Engel, Carter and Springer, 1954; Dohan, Touchstone and Richardson, 1955; Richardson, Touchstone and Dohan, 1955), and possibly the 17-hydroxysteroids of the C_{19} series. We shall treat the second group of compounds as special cases and deal first with a method suitable for the routine estimation of the first group. The suggested method is based on those described by Bush and Willoughby (1957) and Bush and Mahesh (1959a), and resembles to some extent the method of Brooks (1958). Other schemes have been used successfully by Neher (1958a) and Nadel, Young, Hilgar and Burstein (1958). It has the advantage over some other methods (e.g. Rubin et al., 1953) that it avoids the need for digitonin or Girard separations; it avoids the displacement effects of impregnated-paper methods with non-polar steroids; it avoids the production of 9(11)-anhydro-derivatives of 11-hydroxysteroids by acid hydrolysis; and it provides an extract which is clean enough for the immediate application of paper chromatography. This makes the method one of the simplest comprehensive methods available for clinical and metabolic studies. As will be seen later it can easily be simplified or made more sophisticated to suit special needs.

One-twentieth of a 24-hr specimen of urine collected under toluene is taken. This is either made directly to pH 4·8–5·0 and 0·5 M of acetate; or the conjugated and free steroids are extracted by the resin method (Bush and Mahesh, 1959a) and dissolved in 20 ml 0·5 M acetic acid–sodium acetate buffer and made to pH 4·8–5·0 with 0·5 M acetate or acetic acid. Succus entericus of the Roman snail (*Helix pomatia*) containing 80–100 $\times$ 10^3 units/ml of β-glucuronidase is then added to the urine or buffer solution of conjugates in an amount of 1 part by vol. to 50 parts of buffer, obtaining a final concentration of the enzyme of $1\cdot6$–$2\cdot0 \times 10^3$ units/ml. This quantity automatically provides sufficient sulphatase to split steroid sulphates completely in the absence of inhibitors (Henry and Thevenet, 1952; Dodgson and Spencer, 1953). The urine is heated on a boiling water-bath for 20 min before adding buffer and enzyme if the resin method is not employed. (The rationale of the extraction method is described fully in the papers quoted above.)

The urine or buffer solution is then extracted twice with three volumes of ether–ethyl acetate (2:1 by vol.) and the combined extracts washed three times with 0·05 vol. of N-NaOH and twice with the same volume of water. The extract is then evaporated to dryness in the usual way after adding 2–3 drops of glacial acetic acid. A slightly cleaner extract is obtained if it is dried with sodium sulphate before evaporation but this is not essential.

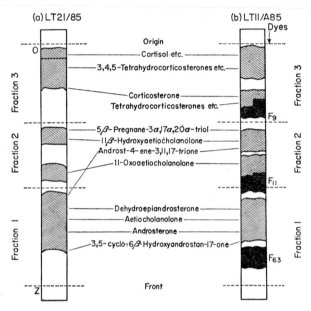

Fig. 6.11. Pre-fractionation of urinary steroids. (a) Method of Bush and Willoughby (1957) which should be compared with that of Brooks (1958). (Reproduced by kind permission of the *Biochemical Journal*.) (b) Similar method (author, unpublished) which has been introduced recently and gives much cleaner fractions 2 and 3. It also incorporates the principle of introducing a change in pH of the solvent systems and is conveniently combined with the ammoniacal variants of the aqueous methanol systems (see p. 213). It also has the advantage (for pre-fractionation) of being usable by the orthodox or impregnation technique (p. 167).

The extract is then deposited over the full width of a 5 cm wide strip of Whatman No 3MM paper and "run up" in the usual way. This chromatogram is equilibrated for 4 hr or overnight and then run, in a group-B system, for a single length which takes about 1·8 hr at 25·5 °C with a 40–45 cm length of paper. The strip is then cut to provide three fractions for elution (Fig. 6.7) using either marker dyes (p. 181) or standards run on control strips to obtain the positions of the cuts. If there is any doubt the cut between fractions 1 and 2 is placed nearer to the centre of the latter fraction rather than in the position indicated by dyes or standards. In this way any error will tend to leave some 11-oxoaetiocholanolone in fraction 1 which will be detected and estimated on the final chromatogram of this fraction. An error the other way, however, would place some of the DHA in fraction 2; since the final chromatogram of this fraction is considerably overrun, DHA would be lost in the run-off and not estimated. The allowable error in the cut between fractions 2 and 3 is small only when it is desired to make sure that all the tetrahydro-metabolites of corticosterone and dehydrocorticosterone are obtained in fraction 3. In the routine method these are not measured and the allowable margin of error is moderately large.

(a) Fraction 1

The eluate is dissolved in 1·0 ml ethanol and 0·2 ml ($\equiv \frac{1}{100} \times$ 24 hr sample) put on each of two 2·5 cm[1] strips of Whatman No 2 paper; preferably washed (p. 160). This strip together with standards is equilibrated for at least 5 hr at 25·5 °C and run in the system L/85 for 12–14 hr. This is awkward for those living far away from the laboratory; running times varying from 16 to 30 hr can be obtained by substituting mixtures of decalin and light petroleum (b.p. 100–120 °C) for the light petroleum of the original mixture (e.g. LD11/85 is excellent), the proportions of decalin varying from 20% to 100% by volume. One strip is used for the Zimmermann reaction, either by elution (Brooks, 1958) or by direct scanning (Bush and Mahesh, 1959a). The other is treated by the phosphomolybdic acid colour reaction (p. 378) to pick up pregnanediol and any unusual compounds. Three-fifths of this fraction are then available for other chromatographic tests if they are desired.

(b) Fraction 2

This eluate is dissolved in 1·0 ml ethanol and 0·8 ml (four-fifths) taken off and evaporated to about 0·1 ml with a jet of filtered air. The extract is transferred with two washings of methanol and ethyl acetate to another 2·5 cm strip of Whatman No 2 paper and the chromatogram equilibrated and run for the same times as used for fraction 1 but using the solvent system LT21/85 or an equivalent. As with fraction 2 any convenient running time can be obtained by using mixtures of decalin and/or xylene in place of light petroleum and/or toluene; the complete substitution of

---

[1] See however p. 244.

these two solvents gives solvent system DX21/85. This chromatogram is used for the Zimmermann reaction as for fraction 1, and the principal compounds found will be 11-oxo- and 11β-hydroxy-aetiocholanolone, and 11β-hydroxyandrosterone. The remaining one-fifth of the fraction is subjected to bismuthate oxidation (p. 365) and the products run as for fraction 1. Pregnanetriol (βP-3α,17α,20α-ol) will be estimated as aetiocholanolone on this chromatogram (cf. Herrmann and Silverman, 1957). If allopregnanetriol (αP-3α,17α,20α-ol) is present in the original extract it will be estimated as androsterone. The 20-epimers of these two compounds will not be distinguished by the overall method (see p. 112). One-fifth of fraction 2 will be sufficient to detect pregnanetriol if its excretion rate is more than 0·1–0·2 mg/24 hr. This sensitivity is ample for picking up this compound in the adrenogenital syndrome (Bongiovanni, 1953); it is not in our experience found in normal human urine although by their method (Cox, 1952) Cox and Marrian (1953) found the excretion to be about 1–2 mg/24 hr in normal human urine. This discrepancy remains to be explained.

If a series of urine samples has been collected it is useful to use this one-fifth of fraction 2 for pregnanetriol estimation in half the samples, and in the other half to put it directly on a 2·5 cm strip and run it in the same system as the main part of fraction 2 but for slightly longer. On treating with phosphomolybdic acid, androst-5-ene-3β,17β-diol,17α-hydroxy-pregnenolone, and other hydroxylic steroids may be detected.

(c) Fraction 3

One-fifth of this is run for 6–8 hr in system T/75 and treated with NaOH to elicit the yellow fluorescence of Δ^4-3-ketones such as cortisol cortisone, 20-dihydrocortisone and 20-dihydrocortisol (De Courcy et al., 1953; Kandrac, 1958; Bush and Mahesh, 1959c). This is usually adequate but in some cases it is better to acetylate this sub-fraction and run the acetates in the system used for fraction 2 (Baird and Bush, 1960). Another one-fifth is overrun for 16–20 hr in a group B system on Whatman No 2 paper. Tetrahydrocortisol, allotetrahydrocortisol, and tetrahydrocortisone can then be estimated with blue tetrazolium either by scanning (Bush and Willoughby, 1957) or by elution.

The remaining three-fifths of this fraction are then available for repeating the estimation in case of accidents or bad results, or for estimating a number of other compounds such as cortols and cortolones (Fukushima et al., 1955). The methods are discussed below.

Using the scanning method for the blue tetrazolium and Zimmermann reactions this method gives results for each compound in duplicate urine samples which are within 5% of the means in 90% of estimations. It should be remembered that the overlap of zones on the chromatograms described above is considerable and should be allowed for when using elution prior to cuvette colorimetry. Brooks (1958) found for instance that mutual contamination of androsterone and DHA was of order of 1·8% with this type of paper technique. This is automatically detected and corrected for by the scanning method.

(*d*) *More Complete Fractionation of Urine*

The above scheme is designed to use the minimum amount of urine that is necessary to obtain a reasonable amount of the 11-oxygenated 17-oxosteroids that occur in fraction 2. Compounds that might and sometimes do occur in fraction 2 in addition to these and pregnanetriol are as follows:

11β-Hydroxyandrostenedione
Andrenosterone
11β- and 17α-Hydroxyprogesterone
11β-Hydroxytestosterone
Androstenediol (A^5-3β,17β-diol)
Deoxycorticosterone
Testosterone
17α-Hydroxypregnanolone (βP-3α,17α-ol-20-one)

Similarly in fraction 3:

Cortol and β-cortol (βP-3α,11β,17α,20ξ,21-ol)
Allocortols
Cortolone and β-cortolone
16α-Hydroxypregnenolone
Androstanetriol (αA-3α,16ξ,17β-ol)
Aetiocholanetriol (βA-3α,16ξ,17β-ol)
Androstenetriol (A^5-3β,16α,17β-ol) (both epimers at 16)
Pregnane-3α,16α,20α,-triol
11-Oxo-and 11β-Hydroxy-pregnanetriol (βP-3α,17α,20α-ol-11-one and βP-3α,11β,17α,20α-ol) (Finkelstein *et al.*, 1953; Fukushima and Gallagher, 1957)
Pregnanetriolone (βP-3α,17α,21-ol-20-one)

This does not exhaust the possibilities. The estimation of each individual steroid would require further pre-fractionation and there are several possible modifying reactions that may be used. The Δ^4-3-ketones in fraction 2 are separable by prolonged overrunning in systems of group A but it would be an advantage to acetylate part of this fraction after a second sub-fractionation and estimate the acetates of testosterone and deoxycorticosterone separately from the others.

11β-Hydroxytestoterone, found in many samples of urine (Bush and Mahesh, unpublished) is well separated from the other compounds and does not need to be acetylated. Both the esters and free compounds would then best be run in an acetic acid system to use the large positive ΔR_{Ms} values of 3-,17- and 20-ketone groups.

17-Hydroxypregnanolone runs in a characteristic position and gives a pink but stable colour with the Zimmermann reagent. It easily separated and estimated (Bush, Stern, Swyer and Willoughby, 1958). Androstenediol is also easily separated from other contaminants by overrunning in a group-B system. It can then be estimated by the chromic acid–Zimmermann procedure (Wilson and Fairbanks, 1955), phosphomolybdate and scanning, or ultraviolet absorption in sulphuric acid (Eik-Nes and Oertel, 1958).

Fraction 3 is a much more complicated problem. This is not to say that it is difficult in principle. It is merely difficult to devise a procedure which is both reasonably watertight and yet as economical of effort as possible. The main problem is the mixture of very polar C_{19} steroids with the moderately polar C_{21} steroids. Probably the most economical solution is the one suggested in Fig. 6.8 where the C_{21} compounds are divided into several fractions by sub-fractionation in a group C system and then oxidized to 17-oxosteroids with bismuthate. The resultant 17-oxosteroids are then separated from the C_{19} triols which are oxidized to the D-seco-16,17-dialdehydes. The latter would be more readily separated by a solvent system of group A.

The problem would be simpler if longer strips of paper than usual were used (Reineke, 1956). In view of the relative ease of this procedure (p. 174) it must be considered seriously whenever a complicated mixture needing sub-fractionation into four or more groups of compounds has to be dealt with. In cases of this sort, however, the use of partition columns with marker dyes marking the cuts becomes a definite rival to paper methods in terms of economy, and for small numbers of samples may be easier than finding the materials and space for double-length tanks. The same principles can be followed with column methods and the solvent systems can be directly transferred to column use or slightly modified if necessary (e.g. Ayres et al., 1957a; Tait and Tait, 1960).

		Urinary steroid	Oxidation product with NaBiO$_3$
Sub-fraction 4	O	2α-Hydroxycortisol	→ 2α,11β-Dihydroxyandrostenedione
		Cortols	→ 11β-Hydroxyaetiocholanolone
		Cortolones	→ 11-Oxoaetiocholanolone
		20β-Dihydrocortisol	→ 11β-Hydroxyandrostenedione
Sub-fraction 3		Tetrahydrocortisol	→ 11β-Hydroxyaetiocholanolone
		Allotetrahydrocortisol	→ 11β-Hydroxyandrosterone
		20β-Dihydrocortisone	→ Adrenosterone
		Tetrahydrocortisone	→ 11-Oxoaetiocholanolone
		Cortisol	→ 11β-Hydroxyandrostenedione
Sub-fraction 2		Aetiocholane-3α,16,17β-triol	→ D-Secodialdehyde
		βP-3α,11β,7α,20α-ol	→ 11β-Hydroxyaetiocholanolone
		Androstane-3α,16,17β-triol	→ D-Secodialdehyde
		Cortisone	→ Adrenosterone
		Δ^5-3β,11β,17β-ol	→ unchanged
		Pregnane-3α,16α,20α-triol	→ unchanged
		Aetiocholane-3α,11β,17β-triol	→ unchanged
		11-Oxopregnane-3α,17α,20α-triol	→ 11-Oxoaetiocholanolone
		Tetrahydrocorticosterone	→ Etianic acid
Sub-fraction 1		Allotetrahydrocorticosterone	→ Etianic acid
		βP-3α,11β,21-ol-20-one	→ Etianic acid
		16α-Hydroxypregnenolone	→ unchanged
		Corticosterone	→ Etianic acid
		Z front	

Fig. 6.12. A suggested further fractionation of the polar urinary steroids of fraction C. System benzene–methanol–water (2:1:1 v/v) approx. 25° C. The positions of the C$_{19}$ triols and of 16α-hydroxypregnanolone are only approximate. It is seen that by taking the suggested three sub-fractions oxidation with NaBiO$_3$ gives three or four C$_{19}$O$_3$ 17-oxosteroids from each fraction but their sources are not confused. The oxidation products of the 16,17-glycols will be separated easily from these 17-oxosteroid derivatives. The C$_{19}$ triols shown as examples by no means exhausts the list of possible urinary metabolites in this fraction. For clarity only the centres of the zones are drawn.

(e) Δ^4-3-*Ketones*

It is often useful to concentrate on one group of compounds at a time. The Δ^4-3-oxosteroids are estimated most easily by fluorimetric methods (Ayres et al., 1957b; Abelson and Bondy, 1955) and in any class of compounds usually have a different biological significance from the other types of steroid. It is often convenient to devise a method for them alone. In this case, a good general method is to sub-fractionate in an aqueous methanol system, acetylate the fractions and then run in appropriate acetic acid systems. The polar compounds of this class can then be estimated by NaOH fluorimetry and the less polar compounds by fluorimetry in t-butanol or by their reaction with Zimmermann reagent. In particular, the low R_F values of the non-polar C_{19}-3-oxo-steroids in the acetic acid systems allow one to obtain high resolutions with considerably overrun chromatograms (p. 100).

(f) Pregnane-3α,20α-diol

This important metabolite of progesterone (Venning and Browne, 1937) is difficult to measure specifically in urine because of several closely related epimers accompanying it. Klopper *et al.* (1955) used alumina columns and acetylation as modifying step but it is doubtful if these accompanying contaminants are completely eliminated. Eberlein and Bongiovanni (1958) use paper and a group-B aqueous methanol system but fail to eliminate three of the epimers: for clinical purposes this does not matter because all of them are metabolites of progesterone. Much better separation, however, would be achieved by using a group-A system (cf. also Bloch, Zimmermann and Cohen (1953) who attempted to use a group-C system) and overrunning two- or three-fold. If chromic acid oxidation were then used as modifying step the resulting diones could be completely separated in an acetic acid or butane-1,3-diol system (p. 100) since the only contaminants on the first chromatogram would be small amounts of allopregnane-3β,20β-diol and allopregnane-3α,20α-diol, both of which would give rise to allopregnane-3,20-dione. The other epimers could also be estimated specifically by taking appropriate cuts from the first chromatogram and using this modifying step.

(g) Reducing Metabolites in the Presence of Steroid Analogues

A special problem encountered by the author in some recent work (Bush and Mahesh, 1958b) illustrates the versatility of the general procedure suggested in Chapter 4. It is useful to investigate steroid metabolism in man by administering a steroid to a subject whose endogenous adrenal secretion has been partly or wholly suppressed by the previous and concurrent administration of an analogue of cortisone such as prednisolone.*
The latter compound is only separated from allotetrahydrocortisol by rather inconveniently long runs in aqueous methanol systems. The problem is

---

* *Added in proof*: Newer analogues such as Dexamethasone (Merck) avoid this problem.

easily solved by carrying out a short run and eluting tetrahydrocortisol, allotetrahydrocortisol, tetrahydrocortisone, and prednisolone as one group. The eluate is then acetylated and run on a type-B system with slight overrunning. The monoacetate of prednisolone is then separated completely from the diacetates of the other metabolites, which are themselves separated from one another sufficiently well to be estimated by the blue tetrazolium scanning method. Another solution is to oxidize the first eluate with sodium bismuthate and overrun the resultant 17-oxosteroids in a group-B system using the Zimmermann reaction for estimation. Owing to one of the minor effects of changing the mobile phase discussed in Chapter 2, the 3,17-dioxosteroid derived from prednisolone is more easily separated from its companions than the original steroids in the B/50 type system.

A similar problem arises with the metabolites of 9α-fluorocortisol. This compound overlaps the position of tetrahydrocortisone, and 9α-fluoroallotetrahydrocortisol is completely mixed with tetrahydrocortisol (Bush and Mahesh, 1958c). The first confusion is easily overcome by acetylation (Figs. 5.4, 5.5) as with prednisolone. The second is also resolved by acetylation but requires the accurate elution of this zone separately from that of allotetrahydrocortisol. This resolution depends on the different ΔR_M of acetylating equatorial and axial hydroxyl groups (p. 284); if these reducing steroids were eluted and acetylated *as a group* then the diacetate of 9α-fluorotetrahydrocortisol would overlap the zone of allotetrahydrocortisol diacetate (b, c, Fig. 5.5). Exactly the same technique was needed for the analogous group of 17-oxosteroids, namely 11β-hydroxyaetiocholanolone, 11β-hydroxyandrosterone, and their 9α-fluoro-derivatives.

(h) Non-Alcoholic Steroid Ketones

This group of compounds includes some rather non-polar diketones, such as pregnanedione (βP-3,20-one), progesterone, and androstanediones. With the exception of progesterone these compounds have not been studied extensively, partly because they are difficult to handle with partition systems of the usual type (p. 100). Reich *et al.* (1952; 1953) devised a useful method of running the 2,4-dinitrophenyl hydrazones on alumina columns and these compounds readily form thiosemicarbazones (Pearlman and Cercero, 1953; Bush, 1953). The latter derivatives behave excellently on chromatograms when in the form of monothiosemicarbazones: most of the compounds under consideration, however, form dithiosemicarbazones and these are often too insoluble to run well. In the light of recent evidence the most convenient way of obtaining a selective method for this class of compound is to sub-fractionate in a group-A system ("typical"), acetylate the mixture to give large negative R_{Mr} values to contaminating hydroxylic compounds, and then run a final chromatogram with a group-A acetic acid system. This is capable of both separating 5-epimers and of giving low enough R_F values to allow considerable overrunning of the final chromatogram. With more complicated mixtures reversed-phase methods would be needed (p. 164).

10. Conclusion

In this chapter I have tried to apply the principles suggested in earlier chapters to some practical problems which seem to be of general interest. I have omitted any examples drawn from the bile acids, sapogenins, cardiac aglycones, and steroid alkaloids, since my experience in these fields is very limited. There is every reason to suppose, however, that similar general rules can also be applied to them and it will be interesting to see whether the occasion will arise to use them. While the general theme of this chapter has been the many ways of skinning cats, it is now safe to state the corollary to this, namely that some general principles can be seen to underlie all of them.

APPENDIX I

PURIFICATION OF REAGENTS AND MATERIALS

USEFUL information and many detailed methods can be found in Vogel's (1956) *Textbook of Practical Organic Chemistry*. It should be remembered, however, that "purity" sufficient for one use is not necessarily so for another, and that the undesirable impurities in a solvent or reagent also vary with the source and mode of preparation as well as with the manufacturer and the official specification. The following list is not exhaustive but covers a widely used range of materials and may be found helpful.

1. SOLVENTS

(a) Hydrocarbons

(*i*) *Aliphatic, boiling points up to* 120 °C—E.g. light petroleum, cyclohexane, heptane, methylcyclohexane.

Wash with concentrated H_2SO_4 till wash layer is colourless; or percolate through a large (e.g. 5 × 40 cm) column of silica gel the top half of which is loaded with concentrated H_2SO_4 (0·5 ml H_2SO_4/g silica gel) the lower half dry and active (e.g. Davison's; B.D.H. "for chromatography"). Then wash, and dry with NaOH pellets. (Vogel also recommends washing with acid $KMnO_4$.) Decant, and distil through a 60–90 cm fractionating column, preferably packed with Fenske spirals. Store over freshly pressed sodium wire that has been well rinsed with the fresh distillate of the solvent itself.

(*ii*) *Aliphatic, boiling points over* 120 °C—E.g. decalin (decahydronaphthalene) "paraffin oil", "odourless kerosene".

The procedure above is sometimes adequate but if the original solvent is more than faintly coloured it is best to add 3–4 preliminary washes with concentrated H_2SO_4 before using the silica gel column.

With certain colour-reactions (e.g. alkaline *m*-dinitrobenzene) these solvents give bad background colours when used as chromatographic solvents. This is more often due to incomplete removal of the solvents from the paper than to impurities in the solvents.

(*iii*) *Aromatic*—E.g. benzene, toluene, tetralin.

The procedure (*i*) for aliphatic hydrocarbons is in our experience invariably adequate with benzene and toluene of "Laboratory Reagent" or better grades. It is cheaper to use such grades than risk the considerable losses that may occur in the purification of material of poorer quality.

Tetralin is sulphonated rapidly by concentrated H_2SO_4 (cf. Vogel, p. 949) and should be washed with dilute Fe_2SO_4 in dilute H_2SO_4. After drying with anhydrous $CaCl_2$ it is filtered through a large column of silica gel as above but omitting the H_2SO_4 layer.

(*b*) *Alcohols*

Standard solutions and solutions for absorptiometry, etc., can be prepared with some grades of ethanol without purification. Burroughs "R.R." or "A.R." ethanol and certain other makes of "rectified spirit", "S.V.R.", or "spirits of wine" can be used directly for such purposes and also for the ethanolic alkali (or organic bases) and *m*-dinitrobenzene of Zimmermann's reagent. Absolute ethanol is also usually of high quality in the U.S.A. "Absolute ethanol" made by azeotropic distillation must be avoided for spectroscopic work.

For chromatographic solvent systems commercial absolute methanol should usually be refluxed for 1 hr with 2,4-dinitrophenylhydrazine (5 g) and concentrated HCl (10 ml/l). It is then cooled, filtered rapidly, and distilled twice through a 25–60 cm fractionating column at 200–500 ml/hr using apparatus closed with $CaCl_2$ drying-tubes. Careful distillation should produce methanol with less than 0·2% by weight of water: it is not worth using the magnesium-iodine method to achieve lower water content. Various other methods using semicarbazide, or granulated zinc and solid KOH are also satisfactory.

Ethanol, n-propanol and isopropanol are purified similarly if used for chromatographic solvent systems. Shell Chemicals supply a very pure grade of isopropanol that needs no purification for most purposes (J. Swale, personal communication).

Laboratory reagent grade t-butanol (Hopkin and Williams) and "Analar" n-butanol (B.D.H.) require no purification for most chromatographic work. These and higher alcohols are conveniently purified by passing through silica gel columns (loosely packed for these viscous solvents) followed by distillation through a good fractionating column.

(c) Halogenated Hydrocarbons

E.g. chloroform, carbon tetrachloride, methylene chloride, ethylene dichloride.

After shaking the solvent with anhydrous potassium carbonate and filtering rapidly, it is percolated through a column of active silica gel and then distilled through a good fractionating column.

Methylene chloride and ethylene dichloride are far more stable than chloroform and need no preservative addition of alcohol. They and chloroform should be stored in the dark and kept out of "fluorescent" strip lighting at all times (H. L. Mason, Rochester, Minn., personal communication). Anhydrous sodium carbonate or fused calcium chloride are the best desiccants for these solvents as they also remove acidic impurities. Sodium should not be used because of the dangers of explosive reactions.

(d) Ethers

E.g. diethyl and di-isopropyl ethers, dioxan, tetrahydrofuran.

With the less polar members of the series, percolation through active alumina or silica gel usually removes peroxides completely. With the more polar ethers or with very impure samples shaking with small volumes of a solution of 50 g ferrous sulphate in 100 ml of 5% sulphuric acid (v/v) is necessary. Alternatively sodium sulphite or stannous chloride may be used (Vogel, p. 163). The ether is then dried with calcium chloride for 24 hr and distilled, or dried by percolation through silica gel and distilled.

Di-isopropyl ether and tetrahydrofuran form peroxides very rapidly. This is stopped extremely efficiently by the addition of an antioxidant of which "Topanol" (2,4-dimethyl-6-t-butylphenol; I.C.I.) has proved very effective. About 0·01% (v/v) is used and has kept tetrahydrofuran free of peroxide for over a year at room temperature: without "Topanol" the solvent showed detectable amounts of peroxide within 24 hours of purification and was grossly contaminated within a week. The antioxidant does not usually interfere with most uses of these solvents and can easily be removed by fractional distillation when necessary.

Dioxan usually needs refluxing for 6–12 hr with 1/20 vol. of 10% HCl (v/v) with a stream of nitrogen to remove acetals. The solution is cooled and treated with solid KOH until a solid residue remains and the aqueous layer is separated. Fresh KOH is added and the solvent allowed to dry for 24 hr. Good grades of solvent can then be distilled from fresh sodium but technical grades need preliminary refluxing over sodium for 6–12 hr (pure: b.p. 101·5°/760 mm).

(e) Esters

E.g. ethyl acetate, n-butyl acetate.

Even with the best grades an estimate of acid content should be made before starting to purify these solvents. Contact with acid and alkaline reagents should be kept to a minimum and a low temperature maintained.

The ester is washed with sodium carbonate or bicarbonate in small volumes until the wash is slightly alkaline. It is then washed with small volumes of saturated aqueous calcium chloride to remove ethanol (Prof. T. Reichstein, personal communication) and finally dried over calcium chloride or calcium sulphate. It is then distilled through a good fractionating column.

Solvent systems based on esters should contain the acid and alcohol of the ester at or near equilibrium concentrations, or be allowed a considerable time to reach equilibrium before use (Bate-Smith and Westall, 1950).

Vogel (p. 174) suggests adding acetic anhydride (1/10 vol.) and a trace of concentrated H_2SO_4 followed by refluxing for 4 hr and fractionation. The distillate is shaken with anhydrous potassium carbonate, filtered and redistilled.

The best rule is always to use fresh solvent for extraction, and equilibrated solvent with chromatographic systems. Strong acids and bases cannot easily be used in such systems.

(f) Glycols and Cellosolves

E.g. propylene glycol, carbitol, phenylcellosolve.

These can all be purified sufficiently for most purposes by drying for 24 hr with anhydrous potassium carbonate, filtering, and distilling carefully through a good fractionating column. Many of these solvents are extremely hygroscopic and the distillation apparatus should be carefully dried and fitted with drying tubes before use. It is usually advantageous to use reduced pressure for the distillation.

Peroxides are best removed by refluxing for 30 min with anhydrous stannous chloride (Vogel, p. 886) (10–15 g/l) but in many cases percolation through silica gel under pressure will suffice.

(g) Formamides

E.g. formamide, dimethylformamide.

This can be purified by shaking with anhydrous potassium carbonate, adding acetone and filtering. The filtrate is dried over calcium or magnesium sulphate and distilled through a fractionating column under reduced pressure (b.p. 105° C/11 mm, Vogel, p. 179).

Dr A. Izzo (Rochester, N.Y., personal communication) has found that impurities upsetting the absorptiometry of eluates from paper chromatograms may be removed by shaking with concentrated H_2SO_4: this is not necessary with all batches.

(h) Pyridine and Other Bases

Pyridine is refluxed over pellets of NaOH and distilled from fresh pellets. It can be stored over barium oxide for most purposes.

Lutidine, collidine, and other heterocyclic bases are often seriously impure and always contain a large number of isomeric forms (A. T. James, personal communication). Steam-distillation of an acidified solution removes many non-basic impurities and the base can then be separated from most of the water by adding solid NaOH and extracting with ether or by steam distillation. The base is dried with anhydrous potassium carbonate and distilled through a good fractionating column. With some examples (e.g. the picolines) fractional freezing is useful (Vogel, p. 179).

(i) Organic Acids

E.g. acetic, propionic, formic, benzoic acids.

"Analar" grades can usually be used without purification, or at most require distillation through a good fractionating column. Very wet samples are best treated with the appropriate anhydride and distilled. Aromatic acids are usually adequately purified by recrystallization from hot water or dilute aqueous alcohol.

(j) Ketones

E.g. acetone, ethyl methyl ketone.

"Analar" samples are usually pure enough for immediate use. Purification is achieved by refluxing with potassium permanganate, adding small amounts till the colour persists. After filtering, the solvent is dried with calcium sulphate or chloride and distilled.

A very quick method is to add aqueous silver nitrate (1/40 vol. of 15%) and sodium hydroxide (1/40 vol. N) and shake the mixture for 15 min. The solvent is then filtered, dried, and distilled (Vogel, p. 171).

2. Adsorbents

(a) Alumina

(i) Active, alkaline Al_2O_3—A technical grade of pure $Al(OH)_3$ is stirred for 3–4 hr at 380–400° C. This gives the most active type of preparation and the best separations. It is not suitable for many sensitive compounds and may prevent the use of certain solvents (see p. 38). It contains a variable amount of alkali (Reichstein and Shoppee, 1949).

Any less active grade can be prepared by mixing with small amounts of water and standardizing with dyes (Brockmann and Schodder, 1941); or by equilibration with the vapour over previously determined concentrations of salts or sulphuric acid (Dingemanse et al., 1952; Lakshmanan and Lieberman, 1954).

(ii) Neutral alumina—Active alumina is suspended in water and the mixture is boiled. The supernatant is made just acid to litmus with dilute nitric acid and the process continued until the supernatant remains acid after boiling for 10 min. It is then filtered off and boiled repeatedly with water until the filtrate is neutral, and finally boiled with methanol. The alumina is then activated by heating for 12–16 hr at 160–200 °C under reduced pressure (10 mm); 180 °C is usually the best temperature (Reichstein and Shoppee, 1949).

Neutral, basic, and acid aluminas of high quality are now supplied commercially, notably by Woelm (M. Woelm, Eschwege, Germany).

(iii) Acid alumina—Suspend a good commercial grade of alumina in 3–4 vol. N-HCl and stir for 10–15 min. Decant the supernatant and very small particles and repeat several times. Wash the alumina slowly with water on a Büchner filter (preferably sintered glass) until the filtrate is only slightly acid to litmus. Dry at 100° C. This material is really an anion-exchange adsorbent in chloride form (Wieland, 1942).

(iv) Regeneration—Alumina that has been used can be regenerated by boiling repeatedly with methanol and then water, with NaOH added if too much colour remains. It is then ready for one of the above procedures (Reichstein and Shoppee, 1949).

(b) Magnesium Silicate ("Florisil", etc.)

Add acetic acid to an aqueous suspension until the supernatant is acid to litmus after mixing well. Wash with hot distilled water on a Büchner funnel until the filtrate is neutral. Boil with 2–3 separate lots of pure methanol and then dry at 110 °C in air. (Rittel, Hunger and Reichstein, 1952).

For a more active preparation (Nelson and Samuels 1952) "Florisil" is dried at 120 °C and then heated for 4 hr at 600 °C. The material is stored over silica gel. If the humidity in the laboratory is high, this material only needs heating to 120 °C for 24 hr to obtain the original activity.

(c) *Silica Gel*

Use the same procedure as for magnesium silicate except that full activity is obtained by heating finally for 24 hr at 120 °C in air and allowing to cool in a desiccator.

As with alumina various grades are obtainable by deactivation with known amounts of water. This is mixed in, or obtained by mixing, "dry" and "wet" silica gel until the desired water content is obtained.

Neher (1959a, p. 108) and Cahnman (1957) obtained the best separations of polar steroids on silica gel containing large quantities of water (15–25%). This, however, is probably due to the partition type of mechanism becoming dominant (see p. 40).

Regeneration is achieved by boiling with an excess of 1% sodium hydroxide for 1 hr, adding further alkali until the supernatant is alkaline to phenolphthalein. The gel is then filtered hot, washed with hot distilled water, and boiled for 0·5 hr with an excess of "5% acid". It is again washed with hot water till neutral and finally dried at 120° C (Neher, 1959a, p. 108).

3. Kieselguhr ("Celite")

Various grades are available; "Hyflo Supercel" and "Celite 545" (Johns Manville Corp., New York) have been used most frequently. They are used either to support stationary phases in partition columns (Martin, 1949) or to increase the velocity of the mobile phase with certain adsorbents such as magnesium silicate, silica gel, or charcoal.

(a) *Ordinary Celite*

The powder is boiled with distilled water and then with methanol. After filtering it is stood with an excess of 11N-HCl for about 16 hr and then washed with water until the filtrate is free of acid, chloride ion, and metals (Butt *et al.*, 1951; Simpson *et al.*, 1954; Bauld, 1956). The powder is then washed with methanol and dried at 110–120° C. Some workers finish the acid treatment by heating for 1 hr before filtering off the acid.

(b) Hydrophobic Celite

Celite purified as above is exposed for 16 hr to the vapour of dimethyldichlorosilane in a vacuum desiccator. Alternatively it can be stirred in a cylinder with a stream of silane-saturated air entering the cylinder at the bottom. The treated Celite is then floated on water and gently agitated to allow any uncoated particles to sink, skimmed off the surface with a beaker and washed on a sintered glass funnel with warm methanol until the filtrate is not acid to bromothymol blue. After drying at 110 °C it is stored in a tightly stoppered jar (Howard and Martin, 1950).

4. Cellulose

(a) Powder

Powdered cellulose and cotton linters can be washed with hot water and alcohols. Acetic acid (2N) and ethanolic ammonium hydroxide (1 vol. conc. ammonia, sp. gr. 0·880, to 9 vol. 95% ethanol) can be used in the cold to advantage. Hot acids or stronger bases hydrolyze or mercerize the cellulose in a way that is difficult to control and usually deleterious.

(b) Paper Sheets

Washing with hot solvents or in a Soxhlet extractor is commonly done with benzene, ethanol, methanol, acetone, chloroform, or mixtures of these (Neher, 1959a, p. 147). Such procedures do not give an exhaustively cleaned product but one in which continuous decomposition imposes a lower and, at present, impassable limit of impurities.

The following sequence has been found useful for cold washing (cf. Isherwood and Hanes, 1953):

(a) Ethanolic ammonium hydroxide (as above);
(b) 50% aqueous ethanol;
(c) 2N-aqueous acetic acid;
(d) 95% aqueous ethanol;
(e) Ether or methylene chloride.

It has usually been done chromatographically. Washing or impregnation with 0·01% ethylenediaminetetraacetic acid disodium salt ("Versene") is carried out when metallic impurities are likely to prove troublesome (see Hanes and Isherwood, 1949; Lewbart and Mattox, 1959). Vogt (1960) found that a wash with light petroleum was needed to remove traces of machine oils.

5. Glassware

Cold chromic acid in sulphuric acid or warm concentrated nitric acid are commonly used for cleaning. The glassware is then rinsed copiously with tap water and distilled water and finally dried in a large oven. In our experience, this procedure is inadequate or even positively dangerous when working with small amounts of sensitive steroids in extracts of biological materials. The main potential sources of danger are:

(a) Chromic acid adsorbed by or impregnating the glass surface;
(b) Metallic impurities in tap and even "glass-distilled" water (see especially Lewbart and Mattox, 1959);
(c) The production of actively adsorptive points in the glass by high-temperature drying;
(d) The deposition of resinous and other materials on the glass from hot objects in the oven, e.g. from plastic wrappings or stoppers, dust or other material being burnt on exposed heating elements, etc., dust and vapours of soaps, oils and polishes in the atmosphere circulating in or through the oven.

Our own sequence of washing is as follows:

(a) "Teepol" (detergent: British Petroleum Ltd.) and hot water added to vessel immediately after use;
(b) Scrub well with brush (preferably with wooden or plastic handle) as soon as possible. Rinse copiously with hot water.
(c) Remove foam and metallic impurities by rinsing well with acetic acid–methanol (or ethanol) (approx. 1 : 4 by vol.) then glass-distilled water and finally methanol with a trace of acetic acid.
(d) Leave to drain upside down and when practically dry close with a glass stopper or with metal foil pressed over the aperture to form a cap.
(e) Before use rinse with acetic acid–methanol; or rinse with 0·1% ethylenediaminetetraacetic acid disodium salt ("Versene") in 50% aqueous methanol, followed by methanol.

It is only rarely that the conventional procedure is necessary if the apparatus is rinsed well immediately after use, and complete dryness is only needed for certain volumetric operations. If chromic acid or similar caustic reagents have to be used the glassware should be warmed finally with acetic acid–methanol for about 30 min and then thoroughly rinsed as above before use (cf. Bauld, 1956, who used ethanol to reduce traces of chromic acid).

6. REAGENTS

Only special precautions and methods will be considered.

(a) *Sulphuric Acid*

Some bottles of concentrated sulphuric acid will be found unsuitable for colour reagents and/or for obtaining spectra of sulphuric acid chromogens. It is usually easier to select a "good" bottle and keep it specially for such purposes rather than attempt to purify this acid.

(b) *Sodium and Potassium Hydroxides*

These materials are nearly always coated with a film or crust of the carbonates and may also contain brown stains (Bush and Willoughby, 1957) which are usually correlated with unsatisfactory solutions for use with Zimmermann's reagent. In our experience, the "stick" form is better for reagents giving colour and fluorescent reactions; stained sticks should be rejected. The sticks are rinsed immediately before use with approx. 50% aqueous ethanol (cooling the flask under a tap) and the solvent decanted after a few seconds' gentle shaking. The sticks are then rinsed twice with absolute ethanol and finally dissolved in ethanol or water, according to need (see p. 375). Dissolution in ethanol should be slow enough to avoid more than a slight rise in temperature and can be controlled by adjusting the violence with which the flask is shaken.

(c) *Phosphomolybdic Acid*

This should be bright yellow but commonly contains traces of reduced material recognized by a greenish tinge. This is often removable by proceeding as for the alkali hydroxides but using absolute ethanol and decanting after 10–25 sec.

(d) m-*Dinitrobenzene*

In our experience the procedure of Callow *et al.* (1938) is long and expensive and not as good as a simple sublimation in air or *in vacuo*. A reasonable commercial product is placed in a small evaporating basin and covered with a clean watch-glass. The basin is heated with a low flame and the glass cooled with a wad of cotton-wool periodically soaked in cold water. The sublimate is scraped off the watch-glass from time to time and 2–10 g are purified in one batch (originally suggested by Prof. F. T. G. Prunty).

(e) Tetrazolium Salts

Triphenyltetrazolium chloride usually requires no purification.

The various "blue" tetrazoliums are almost invariably seriously impure and for quantitative work unsatisfactory to varying extents (p. 225).

Good samples of 3,3'-dianisole-bis-4,4'-(3,5-diphenyl) tetrazolium chloride can be purified by dissolving the salt in water and extracting 3 × 1 vol. n-butanol. The base is then extracted from the organic solvent by diluting with an equal volume of benzene and shaking 3 times with ½ vol. of N·HCl. The acid is neutralized with ammonium hydroxide in methanol and the solution concentrated *in vacuo* at 50° C. The solid residue is then extracted with warm methanol and recrystallized from methanol-ethyl acetate.

Usually, however, it is better to reduce all the tetrazolium salt to the formazan with aqueous alkaline ascorbate (5% w/v ascorbic acid in 2N-NaOH) and collect the precipitate on a sintered glass funnel. After washing with water the precipitate is washed exhaustively with hot n-propanol or 50% aqueous ethanol till no purple colour appears in the filtrate. The indigo residue (formazan) is then reoxidized by the method of Rutenberg, Gofstein and Seligman (1950) (Bush and Gale, 1957).

Dr A. J. Izzo (personal communication) has found that many commercial samples can be purified to acceptable quality by recrystallization from hot water but we have not been so fortunate with this method.

APPENDIX II

MICROCHEMICAL REACTIONS FOR STEROIDS

NEARLY all the reactions given here have been tried out in the author's laboratory under the conditions described. Most of them are well known in ordinary chemistry and are modified only in scale; references are therefore usually to the example which formed the basis of the conditions given below. In certain cases special conditions have been adopted, or the modification given has been devised to achieve a selective action not envisaged in the original description; in these cases the source and essential features of the modification are referred to. The usual amount of steroid was 10–100 µg. Reagents are assumed to be good laboratory or analytical grade unless otherwise described; solvents were purified as in Appendix I.

(a) *Acetylation*

(i) *Slow*—Most primary and secondary hydroxyl groups are completely acetylated by dissolving the steroid in pyridine, adding an equal volume of acetic anhydride, and leaving overnight at room temperature. On the micro-scale (total vol 0·05–0·4 ml) the reagents can be removed by a stream of filtered air or nitrogen (Zaffaroni and Burton, 1951). These conditions have the merit of leaving tertiary and certain hindered secondary hydroxyl groups unesterified. There is also no danger of dehydrating such hydroxyl groups, or of forming enol acetates with ketones. It is probably best to start with this traditional method when identifying completely unknown compounds, but for many purposes it is unnecessarily wasteful of time. Demey and Verly (1958) showed for instance that the 21-hydroxyl group of cortisol is acetylated completely in 1 hr under these conditions.

(ii) *Fast*—Cholesterol and many other steroids can be acetylated quantitatively by refluxing for 1 hr on a boiling water-bath in a 1: 1 (v/v) mixture of acetic anhydride and pyridine (Engel and Baggett, 1954).

The 21-hydroxyl group of steroid α-ketols is completely esterified in 15 min by heating the steroid at 60 °C in a mixture of pyridine (0·1 ml) and acetic anhydride (0·4 ml). Under these conditions the 18-hydroxyl

group of aldosterone (hemiacetal form) is only esterified to about 15%.
The 3α-hydroxyl group of tetrahydrocortisone and tetrahydrocortisol is more than 90% esterified in 0·5 hr under these conditions. The 18-hydroxyl group of aldosterone is completely esterified by prolonging the reaction to 2 hr. No evidence of the formation of dehydration products or other artifacts was obtained using these conditions (Baird and Bush, 1960; Author and Miss S. Hunter, unpublished).

(*iii*) *Forcing*—With certain acid catalysts acetylation can be forced to esterify hindered or tertiary hydroxyl groups (Oliveto *et al.*, 1953). Dehydration and other artifacts are invariably produced and such conditions are rarely useful in chromatographic work.

(*iv*) *Partial*—The isolation of partially esterified polyhydroxy compounds is often useful in chromatographic, as in classical, identifications but is less easily controlled on the micro-scale. The best conditions seem to be to conduct the acetylation at 20–22 °C in acetic anhydride (0·4 ml) and pyridine (0·025 ml) and examine samples drawn off at 0·5, 1·0, 3·0 and about 8 hr. Under these conditions, the 3α-hydroxyl group (*e*) of aetiocholanolone is 75% acetylated in 1·0 hr while the same group (*a*) in androsterone is only 30% acetylated (see p. 306).

(*b*) *Benzoylation*

Dissolve the steroid in dry pyridine (0·2–0·3 ml) and cool in ice; add ice-cold benzoyl chloride (0·1 ml); mix; leave at room temperature 16–18 hr. Add ice water (2 ml) and leave 4 hr at room temperature. Extract with ether or methylene chloride (2 × 3 vol.), wash with hydrochloric acid (2 × 0·1 vol., 3N), sodium carbonate (2 × 0·1 vol., N) and water (2 × 0·1 vol.). Dry with sodium sulphate, filter, and evaporate. (Brooks, Klyne and Miller, 1953).

(*c*) *Hydrolysis of Esters*

(*i*) *Potassium bicarbonate*—Introduced by Reichstein for the gentle saponification of the 21-acetates of adrenal steroid α-ketols; the best conditions for micro-scale work seem to be those of Meyer (1953). The steroid is dissolved in methanol (0·05–0·1 ml) and an excess (about 0·5 ml) of 0·4% methanolic potassium bicarbonate is added, having been previously de-gassed and saturated with nitrogen. After 8 hr at room temperature in the dark under nitrogen the steroid is extracted with methylene chloride or ether and the solution is washed with water and dilute acetic acid.

Under these conditions only about 5% of 3α-acetoxyl groups is hydrolysed, which is useful in the characterization of 3,21-dihydroxysteroids (Bush and Willoughby, 1957).

(*ii*) *Sodium hydroxide*—The steroid is dissolved in methanol (0·2 ml); 0·4 ml of 0·1 N-NaOH in 70% methanol is added. The reagents are previously saturated with nitrogen and the mixture is left in the dark at room temperature for 18–21 hr under nitrogen. The steroid is extracted as in (*i*). (Meyer, 1953; Bush and Willoughby, 1957). This method hydrolyses 3-acetoxyl groups completely and is useful for 17-oxosteroids; it has not been recommended for 20,21-ketols but might be satisfactory.

(*iii*) *Sodium carbonate*—The steroid is dissolved in methanol (1·0 ml) saturated with nitrogen, and sodium carbonate in water (0·5 ml of 0·5 N) is added. Ethylene dichloride (0·25 ml) is added, the tube gassed with nitrogen, and put on a mechanical shaker for 48 hr. The suspension is then extracted with further ethylene dichloride or ether and washed in the usual way. These conditions are not necessarily the best but have been found to give a very clean and complete hydrolysis of tetrahydrocorticosterone 3-monoacetate (author, unpublished).

(*iv*) *Perchloric acid*—The steroid is dissolved in methanol saturated with nitrogen (0·4 ml) and aqueous perchloric acid (0·2 ml of 1·0 N) is added. The tube is gassed with nitrogen and left for 18–20 hr in the dark at room temperature. The mixture is extracted with ether and the extract washed with dilute sodium bicarbonate and water. This method is necessary for hydrolysing the 21-acetates of 9α-chloro- and 9α-bromo-20,21-ketols. (Fried and Sabo, 1957).

(*v*) *Digestive juice of* Helix pomatia—The ester is incubated with digestive juice (0·05 ml) and acetate buffer (0·5 ml of 0·5 M at pH 4·5–5·5) for 16 hr at 45 °C. (author, unpublished; see also Henry and Thevenet, 1952; Simpson and Tait, 1953, who used blood pseudocholinesterase). 21-Acetates of 20,21-ketols are completely hydrolysed.

(*vi*) *11β-Acetoxyl group*—This can be introduced by forcing conditions (Oliveto *et al.*, 1953) but is not hydrolysed by the most energetic chemical means. It has now been found possible to hydrolyse this group using the micro-organism *Flavobacterium dehydrogenans* (Charney *et al.*, quoted in Fieser and Fieser, 1959, p. 701).

(d) Hydrolysis of Ethylene Ketals

(*i*) *Acetic acid*—The steroid is dissolved in acetic acid (1–2 ml) and heated under reflux in a small tube at 100 °C for 30 min. After adding water (2 ml) the mixture is extracted with ethyl acetate or methylene chloride, and the extract washed with sodium bicarbonate and water in the usual way. (Antonucci *et al.*, 1952; Allen, Bernstein and Littel, 1954; Bush and Mahesh, 1959b).

(*ii*) *Dilute sulphuric acid*—The steroid is dissolved in methanol (2 ml) and refluxed in a small tube on a water-bath. Aqueous sulphuric acid (0·1 ml, 10%) is added and, after exactly 3 min, the reaction is stopped by adding iced water (5 ml). The steroid is extracted with ethyl acetate or methylene chloride and washed as in (*i*). This method can be used with 11β-hydroxysteroids (Bernstein and Lenhard, 1955).

(e) Hydrolysis of Ketonides

The steroid (e.g. triamcinolone-16α,17α-acetonide) is dissolved in a few drops of methanol, and 50% formic acid (1·0 ml) is added. The mixture is heated for 1 hr on a boiling water-bath, cooled in ice, and solid sodium bicarbonate added. The steroid is extracted with ethyl acetate and washed with water. 75% of the acetonide is hydrolysed to the free steroid and 25% of the original acetonide remains unchanged. (Author unpublished; based on a personal communication of J. Fried.)

(f) Formation of Hydrazones, etc.

The number of different methods available is considerable and most of them need careful study before being applied to a particular problem. The principal features of these reactions in all cases are: each type of hydrazine derivative varies both in the maximum reactivity that can be obtained and in the pH optimum for the condensation; the condensation is reversible and complete reaction must be achieved by providing a good excess of reagent; the reagents have variable tendencies to produce osazones with certain groups; with polycarbonyl compounds it is often difficult to obtain a single product because of differing reactivities of the different carbonyl groups; with most hydrazine derivatives and hydroxylamine the products of condensation are *cis/trans* isomers which are often both stable and separable on chromatograms (see e.g. Markees, 1954).

(*i*) *Dinitrophenylhydrazones*—Dissolve steroid in ethanol (0·1–0·2 ml), and add 0·3 ml of a reagent containing 2,4-dinitrophenylhydrazine (25 mg) and 6 drops concentrated hydrochloric acid in ethanol (3 ml). Leave at room temperature overnight and add Benedict's reagent (0·4 ml) and water (0·4 ml). Heat 10 min at 95 °C, and cool; then extract with chloroform (5 ml), wash extract with water (0·5 vol.), and dry with sodium sulphate. Filter, and wash sodium sulphate with chloroform until quite colourless.

Benedict's reagent oxidizes the excess reagent to *m*-dinitrobenzene and also destroys the hydrazones of formaldehyde and acetaldehyde. The final filtrate is ready for chromatography of the 2,4-dinitrophenylhydrazones after evaporation and dissolution in a suitable solvent (Reich *et al.*, 1952). On the milligram scale the hydrazones will precipitate and can be filtered off and washed with ethanol or aqueous methanol to remove most of the excess reagent; they are then recrystallized.

Using a similar reagent mixture it was found possible to get quantitative conversion of six 17-oxosteroids to their 2,4-dinitrophenylhydrazones by incubating at 37 °C for 2 hr (author, unpublished; see also references in Markees, 1954).

(*ii*) p-*Nitrophenylhydrazones*—The same conditions as above give quantitative yields of the hydrazones of 17-oxosteroids and Δ^4-3-oxosteroids (author, unpublished).

(*iii*) *Girard hydrazones*—Dissolve the dry steroid residue in glacial acetic acid (0·5 ml) and add Girard's Reagent T (approx. 100 mg.) Heat the lightly stoppered tube at 90–100 °C for 20 min. Add iced water (15 ml) and NaOH (3·5 ml, 10% w/v) and extract with ether (3 × 20 ml). Wash the combined ether extracts with water (1 × 10 ml). To recover the ketonic steroid add conc. HCl (3 ml) to the combined aqueous fractions and leave at room temperature for 2 hrs. The freed ketones are then extracted in the usual way (Pincus & Pearlman, 1941).

Partial acetylation occurs during the hydrolysis of the Girard hydrazones. To overcome this the ketonic fraction is evaporated and the residue dissolved in methanol (1·0 ml) and aqueous $KHCO_3$ (0·2 ml, 50 mg/ml) is added. The solution is left in a stoppered tube overnight at 30 °C and then extracted with ether and washed with water in the usual way (Brooks, 1958).

(*iv*) *Semicarbazones*—Grind semicarbazide hydrochloride (100 mg) with sodium acetate (150 mg NaOAc . $3H_2O$) until liquid. Add a small volume of ethanol, and filter. Make up to 10 ml with ethanol. Add 20–30 mg of steroid to 2–3 ml of this reagent and reflux for 24 hr. Add excess water and filter off the derivative. Wash the precipitate with water (Reich and Collins, 1951; Reichstein, 1936).

(*v*) *Thiosemicarbazones*—Dissolve thiosemicarbazide (9·5 mg) in water (0·5 ml), and add ethanol (5 ml) and steroid (20–30 mg). Reflux for 2 hr and dilute with water (40 ml). Extract with chloroform (3 × 0·2 vol.), wash extract with water (0·15 vol.), and dry with sodium sulphate (H. Reich, personal communication).

Various conditions can be used on the micro-scale; e.g. with a reagent of thiosemicarbazide (100 mg) in acetic acid (20 ml)—add the reagent (0·2 ml) to a solution of the steroid (0·5–100 µg) in methanol (0·5 ml) and reflux in a small tube on a boiling water-bath for 0·3–2 hr, or leave at 35° C for 12–16 hr (Bush, 1953). The reaction with Δ^4-3-ketones and saturated 3-ketones is complete in 15 min at 70° or 100° C, or in 16 hr at room temperature (see also Pearlman and Cercero, 1953).

(vi) *Formation of ethylene ketals*—Dissolve the steroid (10–20 µg) in ethylene glycol (0·2 ml) and benzene (10 ml), add *p*-toluenesulphonic acid (5 mg) and reflux for 3·5 hr with a water trap in the side-arm of the apparatus. Evaporate at 40–50° C, add 5 ml of aqueous NaOH (0·2N) and extract with methylene chloride or ethyl acetate (3 × 2 vol.). Wash the extracts with water and dry with sodium sulphate. (Antonucci *et al.*, 1952; Allen *et al.*, 1954).

(vii) *Reactivity of different positions*—The above conditions will secure the complete reaction of ketone groups at most of the common positions in the steroid nucleus or side chains. The 7-position (Haslewood, 1958) is unreactive and the 20-ketone in a dihydroxyacetone side chain, and some 12-ketones, are much less reactive than normally. The latter type of 20-ketone group will react completely with semicarbazide on prolonged refluxing (Reichstein, 1936), poorly with thiosemicarbazide, completely with ethylene glycol, hardly at all with Girard's reagent T (Morris and Williams, 1953), and probably gives osazones with phenylhydrazines (Porter and Silber, 1950; Gornall and MacDonald, 1953). The Δ^4-3-ketone group is very reactive with all the above reagents and condenses completely in 5–15 min under relatively mild conditions.

(g) *Oxidations with Chromic Acid*

The only procedures used extensively so far in chromatographic identification are mild oxidations of secondary hydroxyl groups to ketone groups and oxidation of certain side chains to give 17-ketones or etianic acids.

The steroid is dissolved in a small volume of acetic acid and an aqueous solution of chromic acid (0·5–0·7 vol. 2% w/v chromium trioxide) is added (Lieberman, Katzenellenbogen, Schneider, Studer and Dobriner, 1953). The tube is shaken gently, stoppered and left in the dark. Aldosterone diacetate is oxidized to the 18→11-lactone-21-acetate in 10 min (Mattox *et al.*, 1956). The 11β- and 11α-hydroxyl groups in typical C_{19} and C_{21} steroids are oxidized completely in 10–20 min; the 17β-hydroxyl group of testosterone and oestradiol-17β in 2 hr (author, unpublished). The 16α- and 16β-hydroxyl group in some C_{21} steroids are not oxidized under mild conditions (Neher et al., 1958). Other groups are usually oxidized after 16 hr (Lieberman et al., 1953; Zaffaroni and Burton, 1951). At the desired time, the solution is diluted with water (5–10 vol.) and extracted with methylene chloride, ether, or ethyl acetate, washed with sodium bicarbonate and water (0·1 vol.), dried, and evaporated.

It has recently been found that the adrenocortical type of side chain is oxidized surprisingly slowly under these conditions and the oxidation of 11α- and 11β-hydroxyl groups can be carried out safely without the usual protection by acetylation. The side chain is noticeably oxidized (1–5%) after 45 min and completely oxidized only after 8–16 hr at 20–22 °C (author, in press).

Other conditions remain to be explored for chromatographic work.

(h) Oppenauer Oxidations

The steroid (20–500 μg) is put into a small tube and the solvent evaporated. Two additions of absolutely dry ethanol followed by evaporation under reduced pressure are carried out to remove all traces of water. The residue is dissolved in absolutely dry benzene (0·5 ml) and aluminium t-butoxide (14–15 mg) is added, followed by 0·1 ml dry benzene–acetone (1:1 by vol.). The tube is sealed and heated at 100 °C for 75 min. After cooling the tube it is opened and the solution transferred with benzene and small volumes of N-HCl to a small separating funnel. The benzene layer is retained and washed with further N-HCl, dilute sodium bicarbonate (2%, 0·05 vol.) and water (1 $\times$ 0·05 vol.), dried over sodium sulphate, and evaporated. (Hershberg, Wolfe and Fieser, 1941). These conditions give quantitative conversion of Δ^5-3β-ols to Δ^4-3-ketones.

The use of sealed tubes is avoided by using cyclohexanone instead of acetone and an apparatus closed with drying tubes.

(i) Oxidations with Periodate

Dissolve the steroid in methanol (0·5 ml) and add aqueous periodic acid (0·5 ml of a solution of 4g $HIO_4 \cdot 2H_2O$ in 100 ml 0·2 N-H_2SO_4). Leave in the dark for 15–18 hr. Add methanol (2 ml) and concentrated ammonia solution (0·1 ml). Centrifuge, decant the supernatant, and wash the precipitate of ammonium salts with methanol. Evaporate the combined supernatants to obtain the ammonium salts of the etianic acids, or the 17-oxosteroids (see p. 292) (Zaffaroni and Burton, 1951).

(j) Oxidations with Sodium Bismuthate

Dissolve the steroid in acetic acid (0·5 ml) and add water (0·5 ml) and sodium bismuthate (25 mg). Leave 1–2 hr in the dark at room temperature. Add methylene chloride (10 ml) or ethyl acetate and filter or centrifuge. Wash the filter with the organic solvent (2 × 5 ml). Then wash the extract with aqueous sodium bicarbonate (0·1 vol., saturated) and water (0·1 vol.), dry with sodium sulphate, and evaporate. We have not found it necessary to use the treatment with metabisulphite (Appleby et al., 1954). Note the interesting degradation of glucuronosidates to formates with this reagent, followed by hydrolysis to the free steroid with alumina (Norymberski and Sermin, 1953). Some samples of bismuthate oxidize the 11β-hydroxyl group but this is *not* typical (cf. Fieser and Fieser, 1959, p. 633). Note also the complications caused by chloride ion (Norymberski and Stubbs, 1956b).

(k) Oxidations with Manganese Dioxide

(i) Dissolve the steroid in chloroform (5–10 ml) and add manganese dioxide (0·5–1·0 g). Shake the suspension for 24 hr at room temperature and wash it well with warm chloroform. Wash the filtrates with water (2 × 0·1 vol.), dry over sodium sulphate, and evaporate.

Equatorial allylic hydroxyl groups are oxidized in good yield under these conditions; little oxidation of axial groups occur. (Amendolla et al., 1954).

(ii) Using same proportions as in (i), reflux the suspension for 6 hr. This procedure brings about oxidation of axial as well as equatorial allylic hydroxyl groups (e.g. Δ^4-6β-ol $\rightarrow$ Δ^4-6-ketone) (Amendolla et al., 1954).

Dr B. Riegel (personal communication) finds this oxidation to be very dependent upon the sample of manganese dioxide and suggests that a fresh specimen is best. It is prepared as follows (Mancera, Rosenkranz and Sondheimer, 1953). Add concentrated aqueous potassium permanganate to a solution of manganese sulphate stirred at 90 °C until a faint colour persists. Stir a further 15 min and then collect the dioxide on a filter. Wash well with hot water, methanol, and ether. Dry at 120–130 °C to constant weight. Keep in a dry well-stoppered bottle. It is stable for "several months" according to these authors.

(l) Reductions with Metal Hydrides

(i) *Selective*—Ketone groups at positions 3, 17 and 20 are reduced by adding 1·5 moles of sodium borohydride in methanol at 0 °C to the steroid dissolved in methanol at 0 °C. Excess hydrochloric acid is added to decompose the reagent and the steroid is extracted with a convenient organic solvent, washed with alkali, acid, water, and evaporated. Reduction at C-20 is 50–60% complete. A Δ^4 double-bond reduces the reactivity of the 3-oxo group. The 11-oxo group is unchanged (Norymberski and Woods, 1954; Appleby and Norymberski, 1955)

Selective reduction of the 11-oxo group is achieved by ketalation (p. 363) followed by refluxing with sodium borohydride (15 mg), NaOH (1·0 ml, 2·5% w/v) and tetrahydroforan (10 ml) for 10 hr (Bush and Mahesh, 1959b).

(*ii*) *Selective reduction of 3-oxo groups*—The steroid is dissolved in methanol (9 ml) and dry nitrogen is bubbled through. To this is added a mixture of methanol (3 ml), 2·5 N-sodium hydroxide (0·1 ml), and a pyridine solution of sodium borohydride (2·8 ml, 0·18 M). Leave for 10 min and then add excess hydrochloric acid. Extract with ether, wash with dilute sodium hydroxide and water, dry over sodium sulphate, and evaporate. Good yields of the related 3-hydroxysteroid are obtained with predominance of the equatorial isomer (i.e. 3β-OH with 5α-(H)-, and 3α-OH with 5β-(H)-steroids). Ketone groups at positions, 11, 12, 17, and 20 are unchanged. 20,21-Ketol side chains are partially destroyed, probably by the alkali (Soloway, Deutsch and Gallagher, 1953).

(*iii*) *Extensive reduction*—Dissolve the steroid in t-butanol (3 ml), add sodium borohydride (50 mg), and leave at room temperature overnight. Decompose excess reagent with aqueous acetic acid, add water (10 ml), and extract with chloroform (3 × 10 ml). Wash the extracts with water (1 × 0·3 vol.) and evaporate. (Appleby and Norymberski, 1955).

(*iv*) *Catalytic hydrogenation*—The steroid (2α-methylcortisol, 3 mg) is dissolved in acetic acid (2 ml), and platinum oxide (1 mg) is added. The solution is hydrogenated for 18 hr. The solution is then diluted with methylene chloride, filtered, and evaporated at 45 °C *in vacuo*. Degradation with bismuthate gives 2α-methyl-3β,11β-dihydroxy-5α-androstan-17-one, confirming reduction of the Δ^4-3-ketone group to the saturated 3-ol. Under these conditions the equatorial isomer predominates and with steroids reduction of double-bonds is almost invariably from the α-side of the molecule.

(*v*) *Reductive elimination with zinc*

(α) Dissolve the steroid in acetic acid (2 ml) and add water (0·5 ml) and zinc dust (200 mg). Shake the mixture at room temperature for 2 hr. Filter and wash the dust with methylene chloride. Evaporate under reduced pressure to near dryness. Extract with methylene chloride

(5–10 ml) and wash the extract with water (1 × 0·5–1·0 ml), dry over sodium sulphate, and evaporate. Allylic hydroxyl groups are eliminated and replaced by hydrogen (e.g. 6β-hydroxyprogesterone → progesterone) (Fieser, 1953).

(β) Dissolve the steroid in aqueous acetic acid (2 ml, 50% v/v), and add zinc dust, 150 mg). Reflux for 1 hr. Cool, filter, and wash the precipitate with methylene chloride. Evaporate to near dryness under reduced pressure and extract and wash as in (α).

Under these conditions the dihydroxyacetone side chain of adrenal steroids loses the 17α-hydroxyl group entirely, and about 50% of the 21-hydroxyl groups are reduced. Principal products from cortisone are, therefore, 11-dehydrocorticosterone and 11-oxoprogesterone (Norymberski and Stubbs, 1956).

(m) Formation of Ethers

(i) For extracts of urinary oestrogens dissolve the extract in water (50 ml) and add boric acid (0·9 g) and aqueous sodium hydroxide (4 ml, 20% w/v). Warm to 37 °C and add dimethyl sulphate (1 ml). Shake the mixture violently till homogeneous and leave for 10–30 min at 37 °C. Then add further dimethyl sulphate (1 ml) and 20% sodium hydroxide (2 ml) and shake violently. Leave for another 20–30 min at 37 °C or cool and leave at room temperature overnight. Add 20% sodium hydroxide (10 ml) and 30% hydrogen peroxide (2·5 ml) to destroy interfering impurities. Extract with benzene (1 × 5 ml) and the extract is ready to add to a column of alumina (Brown, 1955).

(ii) For purified oestrogen fractions make a reagent of aqueous boric acid (25 ml, 1·8% w/v) and add sodium hydroxide (2 ml, 20% w/v) and dimethyl sulphate (0·5 ml). Shake this violently till homogeneous. Add the oestrogen in ethanol (0·1 ml) to the reagent (2 ml) and incubate at 37 °C for 20 min. Then add further dimethyl sulphate (0·05 ml) and 20% sodium hydroxide (0·1 ml) and shake well. Leave 10–15 min at 37 °C, cool, and add sodium hydroxide (0·5 ml, 5N). After a few minutes extract with methylene chloride (5–10 ml), wash with water (2 × 0·5 ml), add a drop of acetic acid, and evaporate (author, unpublished).

(n) Formation of Ketonides

Useful for characterizing *cis*-α-diols. The steroid (e.g. triamcinolone, 10–100 μg) is dissolved in dry acetone (2 ml) in a stoppered tube and perchloric acid (0·01 ml "60%" $HClO_4$) is added. The solution is incubated at 45 °C for 15–120 min. After cooling, half-saturated aqueous sodium bicarbonate is added, the mixture extracted (0·5 ml) with methylene chloride (10 ml), and the extract washed with water (2 × 0·5 ml). (Fried et al., 1958).

Conversion was complete for all incubation times over 15 min. No reaction occurred with α- or β-cortolone (author, unpublished; cf. von Euw and Reichstein, 1940).

(o) Dehydrohalogenation, etc.

(*i*) The steroid is dissolved in a small volume of freshly distilled collidine (0·2–0·5 ml) and refluxed for 10–15 min. The mixture is extracted with methylene chloride (10 ml) or ethyl acetate. The extract is washed with dilute HCl (2 × 0·5 ml), aqueous sodium bicarbonate (1 × 0·5 ml) approx. 5%) and water (1 × 0·5 ml), and evaporated. 9α-Halogeno-11-ketones are converted quantitatively to the $\Delta^{8(9)}$-11-ketones, (Fried and Sabo, 1957). The refluxing must not be prolonged if isomerization to $\Delta^{8(14)}$ products is to be avoided (J. Fried, personal communication).

(*ii*) *Reduction with zinc in acetic acid*—The steroid is dissolved in acetic acid (0·5–1·0 ml), and zinc dust (1–2 mg) is added. The mixture is refluxed for 15 min on a steam-bath, extracted with methylene chloride (5–10 ml), and the zinc removed by filtration or centrifuging. The extract is then evaporated. 9α-Halogeno-11-ketones are reduced to the normal 11-ketones (i.e. 9α-H). 9α-Halogeno-11β-ols undergo elimination of the elements of water to yield the 9(11)-anhydro-derivatives (Fried and Sabo, 1957).

Prolonging the reaction to 1·5 hr gives almost complete removal of 17α-hydroxyl groups from dihydroxyacetone side chains by hydrogenolysis (Norymberski and Stubbs, 1956a; see (*l*)(*v*)).

(*iii*) Conditions for bromination in the ring A have not been worked out for the micro-scale. Removal of halogen should, however, be easy on the micro-scale with suitable modification of the above conditions.

(p) Dehydrations

(*i*) The 11β-hydroxyl group is smoothly eliminated give the $\Delta^{9(11)}$-anhydro-compound by dissolving the steroid in 0·1 ml ethanol, adding hydrochloric acid (5 ml, 15% v/v), and heating on a boiling water-bath for 30 min. The solution is cooled, extracted with methylene chloride or ether (2 × 3 vol.), and the extract washed with sodium bicarbonate (1 × 0·05 vol., half-saturated), with water (1 × 0·05 vol.), and evaporated (author and Miss M. Gale, unpublished).

(*ii*) The steroid is dissolved in 1–2 ml absolutely dry pyridine and cooled in ice. Phosphorus oxychloride (0·3–0·6 ml) is added and the mixture left 16–18 hr at room temperature. It is then cooled in ice and the oxychloride decomposed with crushed ice (1–2 ml). The solution is then extracted with ether or methylene chloride (2 × 3 vol), the extracts washed with sodium bicarbonate (2 × 0·05 vol. half-saturated) and water (1 × 0·05 vol.), and evaporated (Rosenkranz, 1954).

cis-Testosterone was completely converted to the $\Delta^{4,16}$-3-one, testosterone was unchanged, and 11β-hydroxytestosterone converted to $\Delta^{4,9(11)}$-17β-ol-3-one by this method (Miss M. Gale, unpublished) using 50–100 μg.

(*iii*) Dissolve steroid in 2–5 ml anhydrous *ethanol-free* chloroform and pass in a dry stream of gaseous hydrochloric acid slowly for 90 min. Wash with ice-cold water (1 × 3 ml), ice-cold half-saturated sodium bicarbonate (2 × 0·5 ml), and water (2 × 0·5 ml). Dry with sodium sulphate and evaporate. This dehydrates 5α-hydroxy-3-oxo-steroids to Δ^4-3-ketones. In the presence of a trace of ethanol, epimerization of 6β-acetoxyl groups occurs (Herzig and Ahrenstein, 1951).

(*iv*) *Thionyl chloride*—The steroid is dissolved in anhydrous pyridine (2 ml) and cooled in ice. Thionyl chloride (0·2 ml) is added and the mixture left (stoppered) for 5 min at 0 °C. Water (5 ml) is added and the product extracted with methylene chloride or ether (2 × 3 vol.). The extracts are washed with 3N-HCl (2 × 0·1 vol.), sodium bicarbonate (1 × 0·1 vol. half-saturated) and water (2 × 0·1 vol.), dried over sodium sulphate, and evaporated. This reagent gives a high yield of 6β-hydroxy-Δ^4-3-ketones from the related 5α,6β-diols (Amendolla *et al.*, 1954).

(*q*) *Hydrolysis of "Labile" Steroid Sulphates*

Urine or an aqueous solution of the steroid sulphate is made just alkaline to litmus with sodium bicarbonate and heated on a steam-bath for 8 hr. Δ^5-3β-Hydroxysteroid sulphates are completely hydrolysed and converted to the 3,5-cyclo-6β-hydroxysteroid derivatives (Patterson and Swale, 1953; cf. Bitman and Cohen, 1951). The solution is extracted with methylene chloride or ether (2 × 3 vol.), washed with water (2 × 0·05 vol.), dried with sodium sulphate, and evaporated.

Brief treatment of the cold aqueous hydrolysate with hydrochloric acid (bring to pH < 2 for 5 min before extraction) converts 60% of the 3,5-cyclo-6β-ol to the Δ^5-3β-ol and 3β-chloroandrost-5-en-17-one (Dingemanse *et al.*, 1948; Teich, Rogers, Lieberman, Engel and Davis, 1953).

(For solvolysis see McKenna and Norymberski, 1957; Burstein and Lieberman, 1958b).

(*r*) *Hydrolyses with Succus Entericus of* Helix pomatia

This material contains a wide variety of enzymes capable of splitting glucuronosidates and sulphates (Henry and Thevenet, 1952; Jarrige and Henry, 1952), glycosides (Huber, Blindenbacher, Mohr, Speiser and Reichstein, 1951, who give references back to 1909), acetates (above), and certain organic phosphates (Miss M. Gale, unpublished).

The snails are found on chalky ground and are "re-vivified" in the laboratory by two days in a warm, lighted room, with plenty of green food (cabbage, lettuce, etc.), filter paper (Huber *et al.*, 1951), and water, of which the last is perhaps the most important. It is then best to leave them 24 hr on water alone to empty the gut of particulate matter (author, unpublished). Huber *et al.* (1951) then anaesthetize the snails in a bucket of water containing 0·8% ethylurethane for 4 hr. We drown them for 24 hr without anaesthetic which gives a larger volume of intestinal fluid. The shell is then partly removed to expose the back of the snail and a slit about 5 cm long is made in the back to one side of the midline, starting about 2 cm behind one horn. The oedema fluid is drained off and the organs "dissected" and lifted out of the slit by a jet of tap water. The gut is then hung over the mouth of a small conical flask and cut half-across with scissors so that the brown fluid in it drains into the flask. One should get 1·0–2·0 ml from each snail containing 75–95 × 10^3 units glucuronidase/ml (Talalay-Fishman units). The digestive organs can also be removed and extracted with acetone or ethanol to obtain further enzymic activity (Huber *et al.*, 1951).

Although optimum conditions vary for the different classes of compound and should be sought out in the original papers, a wide range of compounds are completely hydrolysed by adjusting the aqueous medium to a 0·5M concentration of acetic acid/sodium acetate buffer with a pH of 4·8–5·2, and incubating at 45 °C for 16 hr. Dehydroepiandrosterone and aetiocholanolone sulphates, but *not* androsterone sulphate (R. D. Bulbrook, personal communication), are hydrolysed by this material.

It is reasonable to consider this procedure as a mild enzymic version of "hot acid" hydrolysis, i.e. it is very unspecific but cleaves a wide variety of conjugates, glycosides and esters. So far no enzymic alterations of the steroid part of such conjugates has been observed with this method, but the possibility should be watched for with each batch of snails. If necessary the solution of conjugates is heated on a boiling water-bath for 20 min to destroy inhibitors of β-glucuronidase and proteolytic enzymes before adding the snail enzyme (Cohen, 1951).

(s) *Hydroxylation of Double Bonds*

(i) *Osmium tetroxide*—To 48 mg of an extract of steroid in dry ether (4 ml) add 3 drops of pyridine and osmium tetroxide in dry ether (15 mg in 1·5 ml). Stopper the tube and leave 6–7 days at room temperature. (The vapour of osmium tetroxide is toxic.) Saturate with hydrogen sulphide and filter or centrifuge. Evaporate the filtrate. (Pregn-5-en-3β-ol-20-one → 5α,6α-diol: Ruzicka and Prelog, 1943; Δ^6 → 6α,7α-diol: Zderic, Carpio and Djerassi, 1959).

(ii) *Potassium permanganate (alkaline)*—Dissolve the steroid in acetone (1–5 ml) and cool in ice. Add dropwise an ice-cold solution of potassium permanganate (N/30) in aqueous potassium hydroxide (N/40) and shake rapidly. Add the reagent until the colour persists for 2 min. Then add saturated aqueous sodium bisulphite until colour is discharged. Extract the solution with ether–ethyl acetate (2:1 by vol.) or ethyl acetate (2 × 3 vol.) and wash the extracts with water (3 × 0·05 vol.). Dry over sodium sulphate, filter or centrifuge, and evaporate the filtrate (Savary and Desnuelle, 1953).

(iii) *Potassium permanganate (acid)*—Dissolve the steroid in acetone–acetic acid (9:1 by vol.; 1 ml) and cool in ice. Add potassium permanganate (2% w/v) in ice-cold aqueous acetone (85% v/v) dropwise until colour persists for 15 min. Discharge the colour by bubbling sulphur dioxide through the solution and filter. Wash the precipitate with ethyl acetate (10 ml) and then wash the combined filtrates with half-saturated aqueous sodium bicarbonate (1 × 0·5 ml) and water (2 × 0·5 ml). Dry with sodium sulphate, filter, and evaporate the filtrate. This gives an approximately 50% yield of 16α,17α-diols from Δ^{16}-20-oxosteroids and a fair amount of the Δ^{14}-16α,17α-diol: the concentration of acetic acid is critical (Ellis, Hartley, Petrow and Wedlake, 1955).

APPENDIX III

METHODS OF DETECTION ON PAPER CHROMATOGRAMS

1. Selective Absorption of Light

(a) *Direct Observation*

With a suitable lamp it is usually possible to detect ultraviolet-absorbing zones by their contrast with the light blue fluorescence of the paper. It is not a very sensitive method ($\geqslant$ 2–3 μg/cm^2) but is useful for locating radioactive zones in which a known amount of ultraviolet-absorbing carrier (20–50 μg) has been added to all samples in order to avoid losses and to provide a "marker". The details are otherwise the same as in (b) below.

(b) *Contact Photography*

A most useful general method capable of many variations (Fisher, Parsons and Morrison, 1948; Markham and Smith, 1949; Bush, 1952; Haines, 1952). For $\alpha\beta$-unsaturated ketones (230–250 mμ) a low-pressure mercury resonance arc is the most convenient source (British Thompson-Houston) and needs no filter, although commercial models with filters are available (Hanovia Ltd., Slough; "Chromatolite", etc.). Steroid-iodine complexes (Bush, 1950) are excellently recorded using the same wavelength as for $\alpha\beta$-unsaturated ketones (author, unpublished). The chromatogram is pressed flat over a sheet of "reflex document" paper (e.g. Ilford reflex No 50) and the lamp switched on for 1–3 sec. The film is developed and fixed in a few minutes, and is used for a record, for marking zones on the chromatogram for elution, or for comparison with subsequent treatment with a colour or fluorescence reaction.

Fluorescent zones can be recorded by exposing to a suitable source (more usually a high-pressure mercury arc with Wood's glass filter giving most of its output in the 350–420 mμ range) with ultraviolet-absorbing and colour-selecting filters between the chromatogram and the reflex film. "Cobex" No 5038 (green; BX Plastics Ltd., London) absorbs ultraviolet strongly and allows blue, green and yellow fluorescence to be recorded. Ilford Micro gelatine filters (No 4—yellow) allows yellow, orange and red fluorescence to be recorded. (R. Markham, personal communication; author, unpublished).

For $\alpha\beta$-unsaturated lactones filtered ultraviolet in the range 300–340 mμ is best (Corning filter No. 9863, with preferably an interference filter in addition, can be used (Heftmann et al., 1954)).

For oestrogens (and many other phenols) a magnesium spark gives a convenient source of light of 280 mμ wavelength (Perrin, 1956).

Good contact methods can detect about 0·5 $\mu g/cm^2$ of any substance with a molecular extinction coefficient of 14,000 or above. Coloured substances can be similarly recorded using ordinary light and appropriate filters (e.g. Fisher et al., 1948).

(c) Contact Fluoroscopy

Visual detection of light-absorbing zones is improved by combining a filtered source of light with a fluorescent screen exhibiting a reasonably narrow exciting waveband. Drake, Haines, Knauff and Nielson (1956) have given details of their successful apparatus for detecting $\alpha\beta$-unsaturated oxosteroids. They use a Corning No 9863 filter with a 15-W (2537 Å) "germicidal" lamp (WL–15 "Sterilamp"; Westinghouse Corp.) and a Du Pont zinc silicate phosphor painted on glass in a polyvinyl alcohol base. The chromatogram is pressed flat against the screen between it and the filtered ultraviolet light.

Although simple and elegant the method has some disadvantages compared with (b). Firstly a much longer exposure to ultraviolet light is needed in order to make a tracing or pencil mark around the absorbing zones. Secondly such marks cannot be made except from behind the screen. The necessary manoeuvres can lead to excessive rubbing of the chromatogram against the screen and inaccurate location of zones. A photographic record is made with a few seconds' exposure and can be measured accurately at leisure.

2. Colour and Fluorescence Reactions

Considerable attention to detail is needed to ensure that the best optical and chemical conditions are obtained, e.g. most fluorescent substances are excited maximally in the longer wavelengths of the ultraviolet ($\geqslant$ 300 mμ) and a high-pressure mercury arc with a Wood's glass filter is the best general source (e.g. Hanovia, Ltd.: Model XI); blue colours are best observed in yellow light, yellow colours in daylight or blue light.

Feigl's (1956) classic on "Spot Tests" is a valuable reference work when devising new methods and is a good source of sensitive general or group methods.

(a) Alkaline Silver Nitrate (Tollens reagent)

Silver nitrate (10 ml, 0·1N in water), concentrated ammonia solution (0·5 ml), and sodium hydroxide (5 ml, 10% w/v) are freshly mixed. Dip, and hang the chromatogram in still air until maximum contrast is obtained (3–10 min). Then rinse in sodium thiosulphate (5% w/v) and water. Dry.

It is a useful but very unspecific reagent for reducing substances— α-ketols, α-bromoketones, lactones, some aldehydes, sugars, polyhydroxyphenolic acids, etc. (Zaffaroni *et al.*, 1950; Heftmann, Berner, Hayden, Miller and Mosettig, 1954; Isherwood, 1954; Smith, 1958, p. 196).

(b) Triphenyltetrazolium Chloride

Triphenyltetrazolium chloride in water (0·2% w/v, 2 vol.) and sodium hydroxide (2N, 1 vol.) are mixed immediately before use. Spray or dip paper. Reducing substances give pink zones. Strong and immediate reactions are given with 20,21-ketols and some sugars. Weaker reactions are given with cyclic α-ketol groups (Burton *et al.*, 1951).

(c) Complex Tetrazolium Salts e.g. Blue Tetrazolium (BT)

A variety of these exists of which 3,3'-dianisole-4,4'-(3,5-diphenyl) tetrazolium chloride has been most widely used (Chen, Wheeler and Tewell, 1953). In solution the reaction is usually carried out in an alcohol medium; on paper chromatograms a largely aqueous vehicle is essential.

(*i*) *Qualitative*—A final concentration of BT $\geqslant$ 0·01% is used with a final concentration of sodium hydroxide 0·2–2·0N. Mix appropriate volumes of stock solutions before use. 0·01% BT in 1·8N-sodium hydroxide is stable for about 36 hr if kept in a refrigerator.

(*ii*) *Quantitative (scanning or elution)*—Rigid conditions must be fixed and adhered to. One suitable reagent is: ethanolic BT (1·0% w/v, 2 vol.), 4N-sodium hydroxide (3 vol.), water (5 vol.). Mix immediately before use and filter off any scum (preferably avoided by the use of very pure BT). Dip the strip or sheet slowly over the surface of the reagent in a watch-glass *twice in the same direction* (i.e. not back and forth). Leave 10–20 min in still air on a glass sheet or hanging horizontally between glass rods. Then hang strips for 20 min (vertically) in a glass tank saturated with sulphur dioxide. Dry the strips in a fume cupboard at room temperature (Bush and Willoughby, 1957).

Specificity: Similar to that of triphenyltetrazolium but apparently wider, probably due largely to its greater sensitivity (approx. 0·2 μg α-ketol/cm^2).

(d) Alkaline m-*Dinitrobenzene*

An enormous number of procedures exist (Neher, 1959a, p. 153). In our experience a single dip through an alkaline reagent is more satisfactory than using two dips, one through ethanolic base and the other through a solution of *m*-dinitrobenzene (see also p. 231). The conditions used by the author are suitable for quantitative estimation by scanning (Bush and Mahesh 1959a).

Make up a fresh solution of *m*-dinitrobenzene in a good sample of absolute ethanol or "doubly rectified" ethanol (Burroughs Co.: "R.R." ethanol). At the same time, dissolve washed potassium hydroxide in similar ethanol at a rate that does not raise the temperature excessively (p. 355), filter the solution through a sintered glass funnel (porosity 3), estimate the concentration of KOH by titration with dilute hydrochloric acid, and dilute with ethanol to give a concentration of 3N-potassium hydroxide. Add the ethanolic potash (1 vol) to the ethanolic *m*-dinitrobenzene (2 vol) and mix rapidly. Pour into a watch-glass. Dip the strips rapidly over the surface of the reagent, steadying them against the rim of the watch-glass. Then hold the strip *horizontally* in the strong draught of an almost closed fume cupboard until the wet surface loses its sheen and appears matt. Then heat strongly for 1–2 min in an oven at 60–100 °C, or in front of a radiating source of heat (p. 233).

(i) *Ideal reagent*—This is recognized by a pale yellow background, a sensitivity of about 0·5 μg/cm^2 with dehydroepiandrosterone, an almost immediate pink colour (before heating) with 11,17-dioxosteroids, and an almost immediate blue colour with saturated or αβ-unsaturated 3-oxo-steroids (just before or immediately after beginning to heat).

(ii) *Common faults*—(α) A "pink flash" which fades slowly on heating and leaves a brownish, often blotchy, background giving a low sensitivity ($\geqslant 2$ μg/cm^2). Most often due to a combination of excessive water content of reagent, inadequate drying in cold air before heating, and exacerbated by the concentration of the ethanolic KOH being much over 3N.

(β) A brownish background, immediate colours with 11β-hydroxy- and 11-deoxy-17-oxosteroids (as well as the usual behaviour with 11-oxosteroids) and with 3-oxosteroids, and moderate to good sensitivity but for the background, which is probably due to excessive KOH concentration.

(γ) A good background but low sensitivity usually due to inadequate KOH concentration, inadequate heating, occasionally to a faulty sample of m-dinitrobenzene, or high water content with low concentration of KOH.

(δ) A blue or purple background colour developing during heating and steadily worsening. This is most often seen with chromatograms that have been run in systems containing decalin, tetralin or xylene. This is usually due to incomplete removal of the solvent and/or impurities therein and is avoided by drying the paper in a draught of air at 50–60 °C for 1–2 hr before applying the reagent.

(ϵ) A brown-grey background worsening during heating but not preceded by excessive "pink flash". This is usually due to impurities in solvent systems remaining on the paper and is commonly seen with chromatograms run in systems having a water content of less than 2% (or total volume). It is avoided by prolonged drying as under (d) or by dipping the dry chromatogram through water or 50% methanol and redrying it at 40–50 °C before applying the reagent. Care is needed to avoid excessive crinkling of the strips after the second drying.

Specificity: A variety of active methylene groups react to give pink, violet, or blue coloured products. The reaction with 17-oxo groups is violet, intense, and fairly long lasting (see p. 231). With 3-oxosteroids the colours are intense with less polar steroids, very weak with polar steroids such as cortisone, and are blue or grey. 5β-(H)-3-Ketones give a blue which rapidly fades to grey and distinguishes them clearly from 5α-(H)-3-ketones which give a more permanent blue colour. 20-Ketones give a pink or brown colour (Savard, 1953) which is permanent enough to allow quantitative estimation by scanning (Bush et al., 1957). 16-Ketones give a weak orange-brown colour. Δ^4-3,6-Diketones give an immediate orange colour before heating (obtainable with alkali alone) followed by a dirty brown-green colour (Savard, 1954). Rather evanescent blue colours are given by cardenolides and their glycosides (Schindler and Reichstein, 1951) and by other $\alpha\beta$-unsaturated lactones. More permanent colours are given by other compounds such as β-ionones etc. (Klyne and Wright, 1957).

(e) 3,5-*Dinitrobenzoic Acid*

This reagent is similar to the above (d) but has the advantage of giving much more durable colours with cardenolides (Bush and Taylor, 1952). One vol. N-KOH is mixed with 1 vol. 3,5-dinitrobenzoic acid (2% w/v) in methanol and applied by spraying or dipping. A final concentration of N-alkali was used by Tschesche et al. (1953). The colours appear in the cold.

(f) *Dinitrophenylhydrazine*

An enormous number of procedures have been described. The reagent is not entirely satisfactory since alkaline *m*-dinitrobenzene or the sodium hydroxide fluorescence test are more sensitive (and specific) for almost all the types of ketone groups commonly found in steroids. With saturated ketones the sensitivity is rather low even when the excess reagent is removed. As a "group reaction", however, it has a certain usefulness.

(i) Dissolve dinitrophenylhydrazine (300 mg) in ethanol (10 ml) and concentrated sulphuric acid (0·3 ml) with warming. Dilute to 200 ml with ethanol. Spray or dip (Reineke, 1956). This method is prone to result in a certain amount of oxidation giving a high background colour due presumably to acetaldehyde.

(ii) Dissolve dinitrophenylhydrazine (150 mg) in water (25 ml) and add concentrated hydrochloric acid (22 ml). Dilute to 100 ml with water (Axelrod, 1953b).

(iii) Make a saturated solution of dinitrophenylhydrazine in 3N-HCl and filter before use (author, unpublished).

All these reagents give easily seen deep orange colours with Δ^4-3-oxo-steroids after a few minutes at room temperatue. Saturated oxosteroids react best on heating for 5 min at 80–90° C (beware of charring) or by spraying on the reagent with a supporting blast of hot air (Kochakian and Stidworthy, 1952).

For detecting saturated oxosteroids the background must be removed either by washing the chromatograms with 2N-NaOH, 2N-HCl, and water (Neher, 1959a, p. 153), or by treating with alkaline silver and then water (Reineke, 1956). The author is not convinced that the silver is really an effective agent, and alkali alone seems equally effective.

(g) *Other Hydrazine Derivatives*

(i) p-*Nitrophenylhydrazine*—Can be substituted for dinitrophenylhydrazine in (f).

(*ii*) *Isonicotinic acid hydrazide*—The hydrazide (25 mg) is dissolved in ethanol (50 ml) containing a trace of hydrochloric acid (0.625 ml concentrated HCl/l.) The chromatogram is dipped or sprayed and dried at room temperature for 1 hr (Weichselbaum and Margraf, 1957). A bluish-white fluorescence is given with Δ^4-3-ketones and also with $\Delta^{1,4}$-3-ketones. Sensitivity is moderate (approx. 1 μg/cm^2).

(*iii*) *Salicyloyl hydrazide*—Reflux hydrazine hydrate (90–100%; 1·5 moles) and add methyl salicylate dropwise (1·0 mole). Reflux for a further 2 hr. Cool and collect the white hydrazide on a filter. Recrystallized from absolutely metal-free water it has a melting point of 149 °C. The reagent forms chelates with various metals. Make up a 2% (w/v) stock solution in acetic acid and dilute with ethanol (5 vol.) before use for spraying or dipping. The chromatogram is then dried "in a hot-air oven". It can then be examined under ultraviolet light, treated with a saturated aqueous solution of zinc acetate, dried, and re-examined. Saturated 3- and 17-oxosteroids give a blue fluorescence (sensitivity $\geqslant$ 3 μg/cm^2); Δ^4-3-oxosteroids give a yellow fluorescence (sensitivity $\geqslant$ 0·3 μg/cm^2); $\Delta^{1,4}$-3-oxosteroids give a faint blue fluorescence (sensitivity $\geqslant$ 13 μg/cm^2). Subsequent treatment with alkali intensifies the fluorescence of the first group and destroys that of the last two groups. A wide variety of non-steroidal carbonyl compounds gives fluorescent products. Various combinations with other secondary reagents are described (Camber, 1957).

(*h*) *Phosphomolybdic Acid*

A solution of the acid in ethanol (Kritchevsky and Kirk, 1952) or acetone (Kuhn and Löw, 1954) is usually used in a concentration of 5–10% (w/v). The chromatogram is dipped or sprayed and then heated at 60–100 °C until the background begins to turn green. R. P. Martin (1957) found that the greatest sensitivity with various 3-hydroxysteroids was obtained by using a 10% (w/v) solution in n-propanol and drying the chromatogram slowly for several hours at 37 °C.

Nearly all substances with reactive groups of any sort will give blue spots with this reagent with a few exceptions. Δ^5-3β-Hydroxysteroids react most rapidly and intensely ($\geqslant 0\cdot5 - 1$ μg/cm^2), followed closely by oestrogens (0·5–2 μg/cm^2), and saturated 3-hydroxysteroids (2–4μg/cm^2). Non-hydroxylic ketones give weak reactions (5–10 μg/cm^2); steroid sulphates give weak reactions; glucuronosidates give strong reactions (Lewbart and Schneider, 1955; author, unpublished). 7-Oxo-Δ^5-3β-hydroxysteroids unfortunately give negligible colours (Haslewood, 1958). The limit is set by the eventual reaction with the paper as heating is continued.

(i) Iodine

(i) *In potassium iodide* (0·3% iodine w/v in 5% w/v aqueous potassium iodide). This gives an immediate blue colour with cortisone, dehydroepiandrosterone, and some other steroids, notably several steroid carboxylic acids (Barger and Field, 1912; Zaffaroni et al., 1950; Zaffaroni and Burton, 1951; Bassil and Boscott, 1951).

(ii) *As the vapour or in light petroleum solution (saturated)*—This gives brown colours with many nitrogenous compounds (Brante, 1949) and orange or yellow colours with many steroids (Bush, 1950; Kritchevsky and Kirk, 1952; Prochazka, 1954). A blue colour is given with cortisone as with (i) and also with 11-deoxy-17-hydroxycorticosterone. The blue colour reaction is dependent on pH: with an apparent pH of the dry paper in the range 4·0–6·5 (indicators) the typical blue is given. At pH 7·0 the colour is olive green; at pH 7·5 the colour is orange-yellow (Bush, 1952). The blue colours are sometimes useful in confirming identifications, but the more usual yellow, orange and brown colours must be reckoned as practically valueless, being quite non-specific.

(j) Alkaline Fluorescence Reagents

The simplest is aqueous approx. 2N-NaOH (carbonate free) which should not be more than three weeks old under usual laboratory conditions. The original reagent in aqueous methanol (Bush, 1952) gave a lower sensitivity because the reaction only lasted 10–15 min on drying at 70–90° C. The sensitivity is improved by adding a small amount of triphenyltetrazolium chloride (Bush, 1953a) or preferably "blue tetrazolium" (Ayres et al., 1957b) and using an aqueous reagent; this quenches the fluorescence of the paper and allows the detection of α-ketol groups in the few minutes before starting to heat the chromatogram (Bush, 1953a).

Brooks (1960) obtained the best sensitivity with a light spraying but the author and others (e.g. Gowenlock, 1960) preferred to soak the chromatogram very thoroughly with the reagent. For visual estimation by matching with standards the following procedure is now recommended together with the precaution of randomizing and coding the chromatograms to be assessed (Bush and Mahesh, 1959a).

Dip the strip over the surface of the reagent (5 vol. 4N-NaOH and 4 vol. water freshly mixed) in a watch-glass. Illuminate the watch-glass with a desk lamp giving a bright reflection from the surface so that the degree of wetness of the paper can be assessed. Dip at a speed such that the top surface of the paper acquires a definite "sheen" and looks really wet. (The paper will break if it is not well dried before dipping, if the wetness is much greater than just "really wet", or if the dipping movement is jerky). Then "roll" the wet strip on to a glass plate so as to avoid trapping bubbles of air. Clip the ends of each strip to the plate and place in a large completely closed oven for 45–55 min at 63–67 °C. Then open the ports of the oven to allow air through and raise the temperature to 80–85 °C. Examine the strips frequently and remove them by peeling them off the glass when absolutely dry and just beginning to turn pale yellow (a further 5–15 min).

Mount the strips with known amounts of a suitable standard, vertically on a glass plate in front of a powerful high-pressure mercury arc *with a Wood's glass filter* (e.g. Hanovia, Model XI). Place a sheet of Ilford gelatine filter No 4 (Micro "Minus Blue") or Cobex (Yellow, No. 5013) between the glass plate and the observer and match each test strip with the two "nearest" standards (either side of it).

Dr W. Balfour (Cambridge, personal communication) found that the reaction proceeded well at room temperature under excitation with the 2537 Å mercury line for 1 hr (low-pressure resonance arc, B.T.H. Ltd.). The chromatogram was then dried with a hair drier. Dr J. F. Tait (personal communication), however, has found recently that excellent sensitivity is obtained by leaving the soaked chromatograms at room temperature in ordinary light for about 1 hr, followed by drying with a hair drier.

The most important considerations for the best results appear to be:

(*i*) The chromatogram must not be dried in less than 50–60 min whatever conditions are used.

(*ii*) It must be absolutely dry when examined, as shown by a light blue background and bright yellow spots. When damp the spots are still orange or red, and the background is deep violet.

(*iii*) The alkali must be free of carbonate (allowable limits not determined).

(*iv*) The chromatogram must not be over-soaked or the fluorescent zone will move if the strips dry unevenly, or are left in any position other than strictly horizontal.

Equally good results are given with choline (10% w/v) or tetraethylammonium hydroxide (15% w/v) (author, unpublished), and Abelson and Bondy's (1955) potassium t-butoxide reagent can also be used.

Specificity: It is almost specific for Δ^4-3-oxosteroids. A large number of other $\alpha\beta$-unsaturated ketones do not give the reaction (Bush, 1954a; see p. 297).

Sensitivity: This is maximal with very polar steroids ($\geqslant$ 0·02–0·05 μg/cm^2); considerable with polar steroids such as cortisol ($\geqslant$ 0·05–0·1 μg/cm^2); weaker with 11-ketones and 2α-methyl analogues of adrenal steroids ($\geqslant$ 0·2–0·4 μg/cm^2); and very weak with non-polar compounds such as progesterone ($\geqslant$ 1–2μg/cm^2) (see also p. 298). The sensitivity with non-polar compounds is improved to a small extent by adding 2% (v/v) commercial "Teepol" (detergent: British Petroleum Ltd) to the reagent (author, unpublished).

(k) Acid Fluorescence Reagents

A very useful group although their specificity is indeterminate.

(i) Phosphoric acid—A 15–20% (w/v) solution in water (for dipping) or ethanol (for spraying) is the most reliable reagent mixture (Neher and Wettstein, 1951; von Arx and Neher, 1956). The wet chromatogram is dried at 80–100 °C and examined under ultraviolet light (Wood's glass filter generally best). The fluorescence is improved by traces of glycols, and diminished by formamide (Neher, 1959a, p. 155). Sensitivities are usually in the range 0·5–2·0 μg/cm^2.

(ii) Trichloroacetic acid: This was first used for cardiac glycosides by Svendsen and Jensen (1950); subsequently used for sapogenins (Heftmann and Hayden, 1952) and pregnanetriol (De Courcy, 1956).

The usual reagent is a 20–33% (w/v) solution in ethanol or chloroform. After spraying or dipping the chromatogram it is dried at 100–120 °C for 5–15 min and examined under ultraviolet light (Wood's glass filter).

(iii) Sulphuric acid: This is rarely used now because the paper is charred and the results are not as useful as with phosphoric acid.

(l) Folin-Ciocalteu Reagent

This reagent is useful for oestrogens (Mitchell, 1952) but is very non-specific. It is about as sensitive as phosphomolybdic acid. The B.D.H. reagent is generally quite satisfactory but is best kept in the refrigerator.

(m) Ferric Chloride Reagents

As a general reagent for enols and some types of phenol these reagents are very useful. Blue or violet colours are given instantaneously at room temperature.

(*i*) Ferric chloride (2% w/v) in absolute or 95% methanol (Axelrod, 1953a).

(*ii*) Mix immediately before use equal volumes of an aqueous solution of ferric chloride (1% w/v) and one of potassium ferricyanide (1% w/v). To obtain permanence of the colour wash the paper in dilute hydrochloric acid and then water (Barton, Evans and Gardner, 1952).

(n) Other Reagents

Among these may be mentioned aromatic aldehydes in concentrated anhydrous acid solutions; various metal chlorides in anhydrous solution; other strong acids, e.g. *p*-toluenesulphonic, phenolsulphonic and phosphoric; sodium nitroprusside (Legal reagent); permanganate; bromine, alone or followed by fluorescein. A very useful list is given by Neher (1959a, p. 157) but in the author's experience they are rarely, if ever, an improvement on one or other of the above methods, and are for the most part of unknown mechanism and low specificity.

3. Detection of Radioactive Material

The techniques are similar to those used in other fields of work and are reviewed well in Block, Durrum and Zweig (1958) and Lederer and Lederer (1957). One new technique is, however, worth describing because of its particular connexion with steroid work.

(a) Autoradiography of Chromatograms (tritium).

The usual method involves close contact between an X-ray film and the paper chromatogram and is insensitive with very weak radiations such as the β-emission of tritium. To overcome this, Rogers (1959) impregnates the chromatogram itself with photographic emulsion as follows.

Melt Ilford K2 emulsion at 50 °C in the dark-room and add an equal volume of warm water. The chromatogram is dried, dipped through the diluted emulsion and hung in the dark-room to drain and dry for 2 hr. The strip is then put in a lightproof box containing silica gel and the box kept at 5 °C for 1–4 weeks. Next the strip is placed in standard X-ray film developer for 5–6 min and fixed similarly for 10–15 min.

Four weeks' exposure will detect typical spots of tritiated progesterone in amounts of about 0·05 μc.

APPENDIX IV

CALCULATIONS WITH R_F AND R_M VALUES

The conversion of all the published figures of the values for steroids to a consistent set of values for one temperature of operation is wellnigh impossible. The following section therefore provides the approximate ΔR_M values characteristic of various systems which can be calculated with reasonable confidence from the literature and from the author's own figures. It is hoped in the future to provide a comprehensive and more accurate set of figures.

How to Use the Tables

(a) Interconversion of R_F and R_M

Use the graph given in Fig. 2.6 (p. 76). Values of R_M less than -1.0 ($R_F \geqslant 0.9$) are of little use but may be obtained when necessary by remembering that the graph is symmetrical about the point 0.5, 0.0. Thus the R_M for an R_F of 0.93 is obtained by reading off the value for R_F $(1-0.93) = 0.07$ and changing the sign, i.e. -1.09.

(b) Calculation of the R_M Value of a New Steroid

(i) By using a known derivative—The most exact method is to start with the R_M value for a known steroid differing from the unknown by the minimum of structural features and work out the probable ΔR_M between the two.

Suppose we know that the R_M value of a C_{21} 11β-hydroxysteroid (I) in a "typical" system (p. 83) is 0.15 and we want to obtain the R_F value of the related 15α-hydroxy-11-oxosteroid (II). The differences between the two compounds are tabulated as follows:

ΔR_{Mr} (11β-OH → 11:0) = -0.4
ΔR_{Mg} (15α-H → 15α-OH) = $\underline{1.8}$
ΔR_M (I → II) = $+1.4$
$\therefore R_{MII} = 1.4 + 0.15$
$= 1.55$
$\therefore R_{FII} = 0.03 \qquad$ (from Fig. 2.6)

II will therefore move 1·2 cm on a single-length run of 40 cm which will not give a clear-cut separation from material on the origin, and will give a large error in the measurement of its R_F value. To obtain a convenient R_F value of about 0·2 ($R_M = 0.60$) we therefore need a change of solvent systems giving a ΔR_{Ms} of -0.95. Since this will usually be largely made up of a change in the ΔR_{Mg} for the hydrocarbon skeleton (if we keep within the range of typical systems) we can select such a system simply by referring to Table 2.13 which is repeated below for convenience. If the original R_M value of I was obtained with the system B_1 (Table 2.13), then changing to system C will give a ΔR_{Ms} of about -1.2 for the C_{19} skeleton. If we allow about -0.1 for the side chain we can calculate for II as follows:

$$
\begin{aligned}
&\text{Probable } R_M \text{ in system } B_1 \text{ (calc.)} = 1{\cdot}55 \\
&\Delta R_{Ms} \text{ (system } B_1 \rightarrow C_1\text{)} = -1{\cdot}3 \\
\hline
&R_{M_{II}} \text{ in system } C_1 = +0{\cdot}25 \\
\end{aligned}
$$

$\therefore R_{F_{II}}$ in $C_1 = 0.355$

(from Fig. 2.6)

This is quite convenient for accurate measurement.

(*ii*) *Ab initio*—This method is much less accurate at present and consists of adding up the ΔR_{Mg} values of all the constituents. Any solvent system will do for starting the calculation but it is best to begin with a "typical" one and to separate the hydrocarbon contribution from that of the polar groups. Thus for 21-deoxycortisol (P^4-11β,17α-ol-3,20-one), we find from Tables 2.13 and App IV.1, using system C_2:

	ΔR_{Mg} (C_2)	
C_{19}		-4.5
Side chain (17α,20-ketol)	$+1.7$	
Δ^4-3-ketone	$+1.2$	
11β–OH	$+1.1$	
	$+4.0$	-4.5
R_M (C_2)	-0.5	

$\therefore R_F$ in $C_2 = 0.76$ (From Fig. 2.6)

CALCULATIONS WITH R_F AND R_M VALUES

Table App. 1. Approximate ΔR_{Mg} Values known for Polar Substituents in C_{19} and C_{21} Steroids.

Substituent	Other features	D/85 Group A	LT21/85 Group B	DX21/85 Group B	T/75 Group C	B/50 Group C	LT11/70 Group B 34°C	LB21/80 Group B 34°C
$_{19}$-skeleton		−2·3	−2·7	−2·3	−3·9	−4·5	—	—
: O	Δ^4	1·75	1·35	—	—	1·2	—	—
: O	5α	1·2	0·9	—	—	—	—	—
: O	5β	1·2	0·9	—	—	—	—	—
3—OH	5β	1·60	1·30	—	—	—	—	—
α—OH	5α	1·70	1·40	—	—	—	—	—
α—OH	5β	1·85	1·55	—	—	—	—	—
β—OH	5α	1·92	1·62	—	—	—	—	—
3—OH	Δ^5	2·00	1·9	—	—	—	—	—
yl-3—OH	—	—	—	2·0	—	—	—	—
: O	$\Delta^{4,3}$: O	—	·31	·40	—	—	0·30	0·51
3—OH	$\Delta^{4,3}$: O	—	1·2	1·3	—	1·6	1·1	1·3
α—OH	$\Delta^{4,3}$: O	—	(1·8)	—	—	—	—	—
α—OH	—	—	1·70	—	1·3	—	—	—
: O	—	1·0	0·61	0·67	0·65	0·82	0·72	0·81
β—OH	—	1·4	1·0	1·1	0·92	1·0	1·0	1·2
α—OH	—	—	1·6	1·6	1·5	1·6	1·6	1·8
α—OH	—	—	—	—	—	—	1·2	1·4
β—OH	—	1·55	1·40	—	—	—	1·45	2·05
α—OH	—	1·75	1·55	—	—	—	1·60	1·85
β—OH	—	—	—	—	—	—	—	—
α—OH	—	—	—	—	—	—	1·4	2·1
: O	—	1·2	—	—	—	—	—	—
α—OH	—	1·6	—	—	—	—	—	—
β—OH	—	1·7	—	—	—	—	—	—
—OH	Δ^4-3:0	0·75	—	—	—	—	—	—
—OH	—	—	—	—	—	2·0	—	—
: O	—	—	—	—	—	0·60	—	—
-17β side chains								
: O	—	0·9	0·7	—	—	0·7	0·6	0·7
β—ol	—	1·60	1·5	—	—	—	—	—
α-ol	—	1·75	1·7	—	—	—	—	—
,21-ketol	—	1·9	1·5	—	—	—	1·6	—
α,20-ketol	—	2·1	1·6	—	—	1·7	1·8	2·0
hydroxyacetone	—	3·1	(2·4)	—	—	(2·5)	2·8	3·0
α,20β-diol	—	—	2·15	—	—	—	—	—
α,20α-diol	—	—	(2·3)	—	—	—	—	—
β,21-ol	—	—	—	—	—	2·7	—	—
α,21-ol	—	—	—	—	—	2·9	—	—
ycerol-20β	—	—	—	—	—	3·55	—	—
ycerol-20α	—	—	—	—	—	3·60	—	—
rboxyl	—	—	—	—	—	∼2·6	—	—

Values in brackets are of limited value because too few substances have been examined, or because errors are larger than usual due to various causes.

The last two columns are derived entirely by calculation from figures given by Neher (1959a). The other figures are drawn from data of Reineke (1956), Neher (1959a) and the author. Slight differences between the three authors have been eliminated as far as possible by making corrections, as described in the appendix, based upon the figures for 11α- and 11β-hydroxyl groups for which the data are most numerous.

Table 2.13 (Appendix IV). Approximate ΔR_{Mg} Values for the 5α–H–C_{19} Steroid Hydrocarbon Skeleton in Various Systems (22–28°C).

	Group A				Group B			Group C	
	1 L/96	2 L/85	3 D/85	4 D/A90	1 LT21/85	2 DT21/85	3 DT21/A70	1 T/75	2 B/50
Skeleton	−2·2	−2·2	−2·3	−2·5	−2·7	−2·5	−3·0	−3·9	−4·5
Methylene	−0·08	−0·12	—	—	—	—	—	—	—
Methyl	−0·10	−0·15	—	—	—	—	—	—	—
Double-bond	+0·03	+0·05	—	—	+0·03	—	—	—	—

These values are in doubt because of insufficient data but are good enough for deciding which solvent systems will be best for untried compounds.

Values for double-bonds, methyl, and methylene groups are for *isolated* groups. They will be considerably larger if substituted close to polar groups due to steric hindrance by methyl groups, and to conjugation with double-bonds.

If it is thought that a lower R_F value would be desirable we use the ΔR_{Mg} values for the C_{19}-skeleton as above. Thus the probable R_F of 21-deoxy-cortisol in aqueous methanol systems would be calculated from Table 2.13 as:

	Solvent Systems			
	C_2	C_1	B_1	A_1
ΔR_{Mg} of C_{19}-skeleton	−4·5	−3·9	−2·7	−2·2
ΔR_{Ms} of C_{19}-skeleton relative to C_2	0·0	+0·6	+1·8	+2·3
R_M of 21-deoxycortisol in C_2	−0·5	−0·5	−0·5	−0·5
Calc. R_M of 21- deoxycortisol	−0·5	+0·1	+1·3	+1·8
Probable R_F of 21-deoxycortisol	0·76	0·44	0·045	0·02

(c) *Sources of Error*

(i) ΔR_{Mg} *values*—The figures given are the best available but are by no means as good as could be desired. With most isolated groups the values will be the same in different runs of the same compounds to within ±0·02 R_M units on the same chromatogram. With any one solvent system the range of ΔR_{Mg} values for a particular substituent in different compounds is usually ⩽ ±15% of the mean value. The ΔR_M between two compounds differing in only one or two features can usually be predicted to within ±5–10% of the calculated value.

(ii) *Large and small values of R_M*—The geometrical error of measuring R_F values should be small and show an inverse relation to the R_F which is compensated by the increasing sharpness of zones with diminishing R_F value. The error in R_M per unit error in R_F is greater for extreme values and rises sharply outside the range 0·2–0·8 R_F units (see Fig. 2.6).

(iii) *Systematic variations*—Within any one family of solvent systems (largely determined by the nature of the more polar phase, p. 83) some factors produce systematic changes in ΔR_{Mg} values. These can be estimated by measuring a relatively small number of ΔR_{Mg} values; the results can then be extrapolated with reasonable confidence to the ΔR_{Mg} values of other substituents. If for instance it is found to be preferable to use a solvent system containing a higher proportion of methanol than one of the published systems for which ΔR_M data are available, then a reduction in the magnitude of all ΔR_{Mg} values will be expected, irrespective of their sign (p. 125). If the ΔR_{Mg} is measured carefully for one type of polar group and found to be x% of the value known for the original system, then the ΔR_{Mg} values of other polar groups (in the new system) will also be approximately x% of their values in the original system. This rule breaks down near the point of complete miscibility of the two phases, with systems containing high concentrations of organic acids, and with complicated quaternary systems such as those containing both methanol and t-butanol.

The same rule can be applied to variations due to differences in the temperature of operation. Changes of paper, and, on the other hand, changes in the relative cross-section of the phases in partition columns must be allowed for by *adding* a constant factor to all R_M values.

(d) *Complex Functions and Vicinal Effects*

Most of these lead to a diminution in ΔR_{Mg} value and are of steric origin. While few exact figures are available (p. 120) the following rough rules may be used as a guide:

(i) *Hydroxyl groups*—Subtract 0·2–0·5 from the sum of the expected ΔR_{Mg} values for α-diols except the *trans*-diaxial conformation. This rule will not hold for highly hindered positions. For example, the figures tend to be of the order of 0·5–1·0 for 11,12-diols (Brooks *et al.*, 1957) and one would expect difficulties with 6,7-diols.

Subtract 0·4–0·6 from the expected ΔR_{Mg} value of the hydroxyl group for each non-polar group substituted α to a hydroxyl group with the conformations *trans-e,e'*, and *cis-e,a*.

(ii) *Ketone groups*—Subtract 0·3–0·6 for α-substituted non-polar equatorial groups.

(iii) *α-Ketol groups*—Subtract 0·3–0·5 from the expected sum of the ΔR_{Mg} values when the hydroxyl is axial, and 0·6–0·8 when it is equatorial.

(iv) *Halogen*—Add 0·1–0·20 for a fluorine atom substituted α to a hydroxyl group. Subtract 0·05 – 0·1 for a fluorine α to a ketone group.

Subtract 0·05–0·1 for a chlorine or bromine α to a hydroxyl group; and 0·1–0·20 when α to a ketone group.

(v) *Conjugated double bonds*—Add 0·25–0·4 for each double-bond conjugated with an existing αβ-unsaturated ketone group.

It should be emphasized that the above figures are only the *deficits* to be applied. Thus the calculation for a non-polar substituent must also include its probable ΔR_{Mg} as an isolated group. This can be taken at present as approximately −0·15 for each methyl or methylene group in "typical" systems. Since the main factor in the chemical potential of such groups is their steric effect on the co-ordination of the solvent molecules of the polar phase, it is probable that the ΔR_M will vary with the partial molar volume of such groups (p. 12). This is born out by a few values for isolated chloro- and bromo-substituents (ΔR_{Mg} −0·1 to −0·2).

Example: calculate the probable ΔR_{Mg} value of the complex function 1β-methyl-2α-hydroxyl:

	ΔR_{Mg}
1β-methyl (e)	−0·15
2α-hydroxyl (e′)	+1·8
Vicinal deficit of hydroxyl group	−0·5
Probable ΔR_{Mg}	+1·15

Models should be inspected whenever groups at unfamiliar positions have to be considered. Pairs of hydroxyl groups in positions *sterically* allowing hydrogen-bonding will probably show a drop of 0·5–0·8 in the sum of the $\triangle R_{Mg}$ values (e.g. 11β,19-diols). The behaviour of highly hindered positions and changes in B,C- and C,D-ring junctions cannot be predicted at present.

REFERENCES

ABELSON, D. and BONDY, P. K. (1955), *Arch. Biochem. Biophys.* **57**, 208.
ABELSON, D. and BORCHERDS, D. (1957), *Nature, Lond.* **179**, 1135.
ACKLIN, W., PRELOG, V. and PRIETO, A. P. (1958) *Helv. chim. Acta* **41**, 1416.
ACKLIN, W. and PRELOG, V. (1959), *Helv. chim. Acta* **42**, 1239.
ACKERMAN, B. J. and CASSIDY, H. G. (1954), *Analyt. Chem.* **26**, 1874.
AEBI, A. and REICHSTEIN, T. R. (1950), *Helv. chim. Acta* **33**, 1013.
AHRENS, E. H. and CRAIG, L. C. (1952), *J. biol. Chem.* **195**, 763.
AITKEN, E. H. and PREEDY, J. R. K. (1956), *Biochem. J.* **62**, 15P.
AITKEN, E. H. and PREEDY, J. R. K. (1957), *Ciba Found. Colloq. Endocrin.* **11**, 331.
ALLEN, W. M. (1930), *Amer. J. Physiol.* **92**, 174, 612.
ALLEN, W. M. (1932), *Amer. J. Physiol.* **100**, 650.
ALLEN, W. S., BERNSTEIN, S. and LITTEL, R. (1954), *J. Amer. chem. Soc.* **76**, 6116.
ALM, R. S., WILLIAMS, R. J. P. and TISELIUS, A. (1952), *Acta chem. scand.* **6**, 826.
AMENDOLLA, C., ROSENKRANZ, G. and SONDHEIMER, F. (1954), *J. chem. Soc.* 1226.
ANDERSON, E. J. and WARREN, F. L. (1951), *J. Endocrin.* **7**, 65P.
ANTONUCCI, R., BERNSTEIN, S., HELLER, M., LENHARD, R. H., LITTEL, R. and WILLIAMS, J. H. (1953), *J. org. Chem.* **18**, 70.
APPLEBY, J. I., GIBSON, G., NORYMBERSKI, J. K. and STUBBS, R. D. (1954), *Biochem. J.* **57**, 14P.
APPLEBY, J. I. and NORYMBERSKI, J. K. (1955), *Biochem. J.* **60,** 460.
ARROYAVE, G. and AXELROD, L. R. (1954), *J. biol. Chem.* **208**, 579.
ARBOUZOV, B. A. and ISAERA, Z. G. (1953), *Izv. Akad. Nauk. SSSR Otdel. Khim. Nauk.*, 843.
AUTERHOFF, H. (1954), *Arch. Pharm.* **287**, 380.
AVIVI, P., SIMPSON, S. A., TAIT, J. F. and WHITEHEAD, J. K. (1954), *Radioisotope Conference*, London, Butterworth, **1**, 313.
AXELROD, L. R. (1953a), *J. biol. Chem.* **201**, 59.
AXELROD, L. R. (1953b), *J. biol. Chem.* **205,** 173.
AXELROD, L. R. (1954), *Recent Progr. Hormone. Res.* **9**, 69.
AXELROD, L. R. (1955), *Anal. Chemyt.* **27**, 1308.
AYRES, P. J., GARROD, O., PEARLMAN, W. H., TAIT, J. F., TAIT, S. A. S. and WALKER, G. (1957c), *Ciba Found. Colloq. Endocrinol.* **11**, 309.
AYRES, P. J., GARROD, O., SIMPSON, S. A. and TAIT, J. F. (1957a), *Biochem. J.* **65**, 639.
AYRES, P. J., PEARLMAN, W. H., TAIT, J. F. and TAIT, S. A. S. (1958), *Biochem. J.* **70**, 230.
AYRES, P. J., SIMPSON, S. A. and TAIT, J. F. (1957b), *Biochem. J.* **65**, 647.
BAGGETT, B. and ENGEL, L. L. (1957), *J. biol. Chem.* **229**, 443.
BAGGETT, B., ENGEL, L. L. and FIELDING, L. L. (1955), *J. biol. Chem.* **213**, 87.
BAIRD, C. W. and BUSH, I. E. (1960), *Acta endocrin., Copenhagen* **34**, 97.

BAKER, L. R. (1955), *J. sci. Instrum.* **32**, 418.
BAKER, P. B., DOBSON, F. and STROUD, S. W. (1951), *Nature, Lond.* **168**, 114.
BALANT, C. P. and EHRENSTEIN, M. (1952), *J. org. Chem.* **17**, 1587.
BALSTON, J. N. and TALBOT, B. E. (1952). *A Guide to Filter Paper and Cellulose Powder Chromatography.* Edited by Jones, T. S. G.; Reeve Angel, London.
BANES, D., CAROL, J. and HAENNI, E. O. (1950a), *J. biol. Chem.* **185**, 267.
BANES, D., CAROL, J. and HAENNI, E. O. (1950b), *J. biol. Chem.* **187**, 557.
BARGER, G. and FIELD, E. (1912), *J. chem. Soc.* **101**, 1404.
BARLOW, J. J. (1957), *Biochem. J.* **65**, 34P.
BARRER, R. M. (1957), *Experientia*, (Suppl. VII), 113.
BARROLLIER, J., HEILMAN, J. and WATZKE, E. (1958), *J. Chromatog.* **1**, 434.
BARTON, D. H. R. (1950), *Experientia*, **6**, 316.
BARTON, D. H. R. (1955), *Experientia* (Suppl. II), 121.
BARTON, D. H. R. and COOKSON, R. C. (1956), *Quart. Rev.* **10**, 44.
BARTON, D. H. R. and COX, J. D. (1948), *J. chem. Soc.* 783.
BARTON, D. H. R. and MILLER, E. (1950), *J. Amer. chem. Soc.* **72**, 370.
BARTON, D. H. R. and SCHMEIDLER, G. A. (1948), *J. chem. Soc.* 1197.
BARTON, D. H. R. and SCHMEIDLER, G. A. (1949), *J. chem. Soc.* 5232.
BARTON, G. M., EVANS, R. S. and GARDNER, J. A. F. (1952), *Nature, Lond.* **170**, 249.
BASSIL, G. T. and BOSCOTT, R. J. (1951), *Biochem. J.* **48**, xlviii.
BATE-SMITH, E. C. and WESTALL, R. G. (1950), *Biochim. biophys. Acta* **4**, 427.
BAULD, W. S. (1955), *Biochem. J.* **59**, 294.
BAULD, W. S. (1956), *Biochem. J.* **63**, 488.
BAUMANN, P. and PRELOG, V. (1958a), *Helv. chim. Acta* **41**, 2362.
BAUMANN, P. and PRELOG, V. (1958b), *Helv. chim. Acta* **41**, 2379.
BAUMANN, P. and PRELOG, V. (1959), *Helv. chim. Acta* **42**, 736.
BEATTY, J. (1949), private communication.
BERGNER, K. G. and SPERLICH, H. (1953), *in Chem. Abstr.* (1954) **48**, 2273.
BERGSTROM, S. and SJÖVALL, J. (1951), *Acta chem. scand.* **5**, 1267.
BERLINER, D. L. (1957), *Proc. Soc. exper. Biol.*, N.Y. **94**, 126.
BERLINER, D. L., DOMINGUEZ, O. V. and WESTENSKOW, G. (1957), *Analyt. Chem.* **29**, 1797.
BERLINER, D. L., JONES, J. E. and SALHANICK, H. A. (1956), *J. biol. Chem.* **223** 1043.
BERLINER, D. L. and SALHANICK, H. A. (1946), *Analyt. Chem.* **28**, 1608.
BERNAL, J. D. and FOWLER, R. H. (1933), *J. chem. Phys.* **1**, 515.
BERNASCONI, R., SIGG, H. P. and REICHSTEIN, T. (1955), *Helv. chim. Acta* **38**, 1767.
BERNSTEIN, S., ALLEN, W. S., LINDEN, C. E. and CLEMENTE, J. (1955), *J. Amer. chem. Soc.* **77**, 6612.
BERNSTEIN, S. and LENHARD, R. H. (1953), *J. org. Chem.* **18**, 1146.
BERNSTEIN, S. and LENHARD, R. H. (1954), *J. org. Chem.* **19**, 1269.
BERNSTEIN, S. and LENHARD, R. H. (1955), *J. Amer. chem. Soc.* **77**, 2331.
BEROZA, M. (1956), *Analyt. Chem.* **28**, 1550.
BEYREDER, J. and RETTENBACHER-DÄUBNER, H. (1953), *Monatsh. Chem.* **84**, 99.
BIBILE, S. W. (1953), *J. Endocrin* **9**, 357.
BITMAN, J. and COHEN, S. L. (1951), *J. biol. Chem.* **191**, 35.

REFERENCES

BITMAN, J. and SYKES, I. F. (1953), *Science* **117**, 356.
BLOCH, H. S., ZIMMERMAN, B. and COHEN, S. L. (1953), *J. clin. Endocrin.* **13**, 1206.
BLOCK, R. J. (1948), *Science* **108**, 608.
BLOCK, R. J. (1950), *Analyt. Chem.* **22**, 1327.
BLOCK, R. J., DURRUM, E. L. and ZWEIG, G. (1958), *A Manual of Paper Chromatography and Paper Electrophoresis.* New York, Acad. Press.
BOGGS, L. A. (1952), *Analyt. Chem.* **24**, 1673.
BOJESON, E. (1956), *Scand. J. clin. Lab. Invest.* **8**, 55.
BOJESON, E. (1958), *Proc. Int. Congr. Biochem.* 118.
BOLDINGH, J. (1948), *Experientia*, **4**, 270.
BOLDINGH, J. (1950), *Rec. Trav. chim. Pays-Bas.* **69**, 247.
BONDY, P. K., ABELSON, D., SCHEUER, J., TSEU, T. K. L. and UPTON, V. (1957), *J. biol. Chem.* **224**, 47.
BONDY, P. K. and UPTON, G. V. (1956), *Proc. Soc. exper. Biol., N.Y.* **94**, 585.
BONGIOVANNI, A. M. (1953), *Bull. Johns Hopkins Hosp.* **92**, 244.
BONGIOVANNI, A. M., and EBERLEIN, W. R. (1955), *Proc. Soc. exper. Biol., N.Y.* **89**, 281.
BOSCH, L. (1953), *Biochem. biophys. Acta* **11**, 301.
BOSCOTT, R. J. (1952), *Biochem. J.* **51**, xlv.
BOWERS, A. and RINGOLD, H. J. (1958), *Tetrahedron* **3**, 14.
BRADFIELD, A. E. and BATE-SMITH, E. C. (1950), *Biochim. biophys. Acta* **4**, 441.
BRANTE, G. (1949), *Nature, Lond.* **163**, 651.
BRAUNSBERG, H., STERN, M. I. and SWYER, G. I. M. (1954), *J. Endocrin.* **11**, 189.
BRAY, H. G., THORPE, W. V. and WHITE, K., (1950) *Biochem. J.* **46**, 271.
BREMNER, J. M., and KENTON, R. H. (1951), *Biochem. J.* **49**, 651.
BRIEGLEB, G. and KAMBEITZ, J. (1934), *Z. phys. Chem.* **27B**, 161.
BROCKMANN, H. and SCHODDER, H. (1941), *Ber. dtsch. chem. Ges.* **74**, 73.
BROOKS, R. V. (1958), *Biochem. J.* **68**, 50.
BROOKS, R. V. (1960), *Mem. Soc. Endocrin.* **8**, 9.
BROOKS, R. V., KLYNE, W. and MILLER, E. (1953), *Biochem. J.* **54**, 212.
BROOKS, S. G., HUNT, J. S., LONG, A. G. and MOONEY, B. (1957), *J. chem. Soc.* 1175.
BROWN, J. A. and MARSH, M. M. (1953), *Analyt. Chem.* **25**, 1865.
BROWN, J. B. (1955), *Biochem. J.* **60**, 185.
BURSTEIN, S. (1956a), *J. Amer. chem. Soc.* **78**, 1769.
BURSTEIN, S. (1956b), *Science* **124**, 1030.
BURSTEIN, S. and DORFMAN, R. I. (1954), *J. biol. Chem.* **206**, 607.
BURSTEIN, S. and DORFMAN, R. I. (1955), *J. biol. Chem.* **213**, 581.
BURSTEIN, S., DORFMAN, R. I., and NADEL, E. M. (1954), *Arch. Biochem. Biophys.* **53**, 307.
BURSTEIN, S. and LIEBERMAN, S. (1958a), *J. biol. Chem.* **233**, 331.
BURSTEIN, S. and LIEBERMAN, S. (1958b), *J. Amer. chem. Soc.* **80**, 5235.
BURTNER, H. J., BAHN, R. C. and LONGLEY, J. B. (1957), *J. Histochem. Cytochem.* **5**, 127.
BURTON, R. B., ZAFFARONI, A. and KEUTMANN, E. H. (1948), *J. clin. Endocrin.* **8**, 618.
BURTON, R. B., ZAFFARONI, A. and KEUTMANN, E. H. (1951a), *J. biol. Chem.*, **188**, 763.

BURTON, R. B., ZAFFARONI, A. and KEUTMANN, E. H. (1951b), *J. biol. Chem.* **193**, 763.
BUSH, I. E. (1950), *Nature, Lond.* **166**, 445.
BUSH, I. E. (1951), *J. Physiol.* **115**, 128P.
BUSH, I. E. (1952), *Biochem. J.* **50**, 370.
BUSH, I. E. (1953a), *Ciba Found. Colloq. Endocrinol.* **5**, 203.
BUSH, I. E. (1953b), *Fed. Proc.* **12**, 186.
BUSH, I. E. (1954a), *Recent Progr. Hormone Res.* **9**, 321.
BUSH, I. E. (1954b), *Brit. Med. Bull.* **10**, 229.
BUSH, I. E. (1955), *Biochem. J.* **59**, 14P.
BUSH, I. E. (1957a), *Biochem. J.* **67**, 23P.
BUSH, I. E. (1957b), *Ciba Found. Colloq. Endocrinol.* **11**, 263.
BUSH, I. E. (1960a), *Mem. Soc. Endocrin.* **8**, 24.
BUSH, I. E. (1960b), *Biochem. Soc. Symposia*, **18**, 1.
BUSH, I. E. and GALE, M. M. (1957), *Biochem. J.* **67**, 29P.
BUSH, I. E. and GALE, M. M. (1958), *Analyst* **83**, 532.
BUSH, I. E. and GALE, M. M. (1960), *Biochem. J.*, **76**, 10P.
BUSH, I. E. and HOCKADAY, T. D. R. (1960), *Biochem. J.*, **77**, 7P.
BUSH, I. E. and MAHESH, V. B. (1958a), *Biochem. J.* **69**, 9P.
BUSH, I. E. and MAHESH, V. B. (1958b), *Biochem. J.* **69**, 21P.
BUSH, I. E. and MAHESH, V. B. (1959a), *J. Endocrin.* **18**, 1.
BUSH, I. E. and MAHESH, V. B. (1959b), *Biochem. J.* **71**, 705.
BUSH, I. E. and MAHESH, V. B. (1959c), *Biochem. J.* **71**, 718.
BUSH, I. E. and SANDBERG, A. A. (1953), *J. biol. Chem.* **205**, 783.
BUSH, I. E., STERN, M. I., SWYER, G. and WILLOUGHBY, M. L. N. (1957), *J. Endocrin.* **15**, 430.
BUSH, I. E. and TAYLOR, D. A. H. (1952), *Biochem. J.* **52**, 643.
BUSH, I. E. and WILLOUGHBY, M. L. N. (1957). *Biochem. J.* **67**, 689.
BUTLER, J. A. V. (1937), *Trans. Faraday Soc.* **33**, 229.
BUTLER, J. A. V. and HARROWER, P. (1937), *Trans. Faraday Soc.* **33**, 171.
BUTLER, J. A. V. and RAMCHANDANI, C. N. (1935), *J. chem. Soc.* 952.
BUTLER, J. A. V., RAMCHANDANI, C. N. and THOMSON, D. W. (1935), *J. chem. Soc.* 280.
BUTT, W. R., MORRIS, P. and MORRIS, C. J. O. R. (1948), *Abstr. Int. Congr. Biochem.* 405.
BUTT, W. R., MORRIS, P. and MORRIS, C. J. O. R. (1951), *Biochem. J.* **49**, 434.
CAHNMANN, H. J. (1957), *Analyt. Chem.* **29**, 1307.
CALLOW, N. H. (1939), *Biochem. J.* **33**, 559.
CALLOW, N. H., CALLOW, R. K. and EMMENS, C. W. (1938), *Biochem. J.* **32**, 1312.
CALLOW, N. H. and CALLOW, R. K. (1939), *Biochem. J.* **33**, 931.
CALLOW, R. K., DICKSON, D. H. W., ELKS, J., EVANS, R. H., JAMES, V. H. T., LONG, A. G., OUGHTON, J. F. and PAGE, J. E. (1955), *J. chem. Soc.* 1966.
CAMBER, B. (1957), *Clin. chim. Acta* **2**, 188.
CARSTENSEN, H. (1956), *Acta Soc. Med. Uppsalien.* **61**, 26.
CASSIDY, H. G. (1952), *J. biol. Chem.* **24**, 1415.
CASSIDY, H. G. (1957), *Fundamentals of Chromatography*, 10, "Techniques of Organic Chemistry", Edited by Weissberger, New York, Interscience.

CAVINA, G. (1955a), *Bull. Soc. Ital. Biol. Sper.* **31**, 1666.
CAVINA, G. (1955b), *Bull. Soc. Ital. Biol. Sper.* **31**, 1668.
CAVINA, G. (1957), *R.C. Ist. sup. Sanita* **20**, 923.
CHAROLLAIS, E. J. (1955), *Bull. Soc. Chim. biol., Paris* **37**, 299.
CHEN, C., WHEELER, J. and TEWELL, M. E. (1953), *J. Lab. clin. Med.* **42**, 749.
CHEVREUL, M. E. (1823) quoted (1956) LEICESTER, H. M., *Historical Background of Chemistry*, New York, Wiley.
CHIBNALL, A. C. (1954), *Brit. med Bull.* **10**, 183.
CLAESSON, S. (1947), *Arkiv. Kemi. Mineral. Geol.* **24A**, No. 16.
CLARK, I. (1955), *Nature, Lond.* **175**, 123.
CLAYTON, R. A. (1950), *Analyt. Chem.* **28**, 904.
COHEN, S. L. (1950), *J. Lab. clin. Med.* **36**, 769.
COHEN, S. L. (1951) ,*J. biol. Chem.* **192**, 147.
CONSDEN, R., GORDON, A. H. and MARTIN, A. J. P. (1944), *Biochem. J.* **38**, 224.
CONSDEN, R., GORDON, A. H. and MARTIN, A. J. P. (1947), *Biochem. J.* **41**, 590.
CONWAY, E. J. (1947), *Microdiffusion Analysis and Volumetric Error*, London, Lockwood.
COOK, E. R., DELL, B. and WAREHAM, D. J. (1955), *Analyst* **80**, 215.
COPE, C. L. (1960), *Mem. Soc. Endocrin.* **8**, 18.
COPE, C. L. and HURLOCK, B. (1954), *Clin. Sci.* **13**, 69.
COSGROVE, J., BASTIAN, R. P. and MORRISON, G. H. (1958), *Anal. Chem.* **30**, 1872.
COX, R. I. (1952), *Biochem. J.* **52**, 339.
COX, R. I. and MARRIAN, G. F. (1953), *Biochem. J.* **54**, 353.
COX, R. I. and FINKELSTEIN, M. (1957), *J. clin. Invest.* **36**, 1726.
COX, J. S. G., HIGH, L. B. and JONES, E. R. H. (1958), *Proc. chem. Soc.* 234.
CRAIG, L. C. and GRAIG, D. (1956), *Techniques of Organic Chemistry*, Vol. 3, 2nd. Edn. Edited by Weissberger. New York, Interscience.
CRÉPY, O., JAYLE, M. F. and MESLIN, F. (1957a), *Acta endocr., Copenhagen* **24**, 233.
CRÉPY, O., MALASSIS, D., MESLIN, F. and JAYLE, M. F. (1957b), *Acta endocr. Copenhagen* **26**, 43.
CROOK, E. R., HARRIS, M., HASSAN, F. and WARREN, F. L. (1954), *Biochem. J.* **56**, 434.
CROOK, E. R., HARRIS, M. and WARREN, F. L. (1952), *Biochem. J.* **51**, xxvi.
CUENDET, R. S., MONTGOMERY, R. and SMITH, F. (1953), *J. Amer chem. Soc.* **75**, 2764.
DAO, T. L. (1957), *Endocrinology* **61**, 242.
DASLER, W. and BAUER, C. D. (1946), *Industr. Engng. Chem. (Anal.)* **18**, 52.
DATTA, S. P., OVERELL, B. G. and STACK-DUNNE, M. (1949), *Nature, Lond.* **164**, 673.
DAVID, K., DINGEMANSE, E., FREUD, J. and LAQUEUR, E. (1935), *Hoppe-Seyl. Z.* **233**, 281.
DAVIS, R. B., MCMAHON, J. M. and KALNITSKY, G. (1952), *J. Amer. chem. Soc.* **74**, 4483.
DE COURCY, C. (1956), *J. Endocrin.* **14**, 164.
DE COURCY, C., BUSH, I. E., GRAY, C. H. and LUNNON, B. (1953), *J. Endocrin.* **9**, 401.
DEMY, E. and VERLY, W. G. (1958), *Arch. int. Physiol.* **66**, 62.

DESTY, D. H., GOLDUP, A. and SWANTON, W. T. (1959), *Nature, Lond.* **183**, 107.
DIAZ, G., ZAFFARONI, A., ROSENKRANZ, G. and DJERASSI, C. (1952), *J. org. Chem.* **17**, 747.
DINGEMANSE, E. and HUIS IN 'T VELD, L. G. (1946), *Acta brev. neerl. Physiol.* **14**, 34.
DINGEMANSE, E., HUIS IN 'T VELD, L. G. and DE LAAT, B. M. (1946), *J. clin. Endocrin.* **6**, 535.
DINGEMANSE, E., HUIS IN 'T VELD, L. G. and HARTOGH-KATZ, S. L. U. (1948), *Nature, Lond.* **162**, 492.
DINGEMANSE, E., HUIS IN 'T VELD, L. G. and HARTOGH-KATZ, S. L. U. (1952), *J. clin. Endocrin.* **12**, 66.
DIXON, H. B. F. and STACK-DUNNE, M. (1955), *Biochem. J.* **61**, 483.
DJERASSI, C., MILLS, J. S. and VILOTTI, R. (1958), *J. Amer. chem. Soc.* **80**, 1005.
DOBRINER, K., LIEBERMAN, S. and RHOADS, C. P. (1948), *J. biol. Chem.* **172**, 241.
DODGSON, K. S. and SPENCER, B. (1953), *Biochem. J.* **55**, 315.
DOHAN, F. C., TOUCHSTONE, J. C. and RICHARDSON, E. M. (1955), *J. clin. Invest.* **34**, 485.
DONALDSON, K. D., TULANE, V. J. and MARSHALL, L. M. (1952), *Analyt. Chem.* **24**, 185.
DORFMAN, L. (1953), *Chem. Rev.* **53**, 47.
DOWLATABADI, H. (1958). Private communication.
DRAKE, N. A., HAINES, W. J., KNAUFF, R. E. and NIELSON, E. D. (1956), *Analyt. Chem.* **28**, 2036.
DURRUM, E. L. and GILFORD, S. R. (1955), *Rev. sci. Instrum.* **26**, 51.
DUTTA, N. L. (1955), *Nature, Lond.* **175**, 85.
EBERLEIN, W. and BONGIOVANNI, A. M. (1955), *Arch. Biochem. Biophys.* **59**, 90.
EBERLEIN, W. and BONGIOVANNI, A. M. (1958), *J. clin. Endocrin.* **18**, 300.
EDGAR, D. G. (1953a), *Biochem. J.* **54**, 50.
EDGAR, D. G. (1953b), *J. Endocrin.* **10**, 54.
EDWARDS, R. H. W. and KELLIE, A. E. (1956), *Chem. and Ind.* 250.
EDWARDS, R. H. W., KELLIE, A. E. and WADE, A. P. (1953), *Mem. Soc. Endocrin.* **2**, 53.
EIK-NES, K. and OERTEL, G. (1958), *J. biol. Chem.* **232**, 543.
ELLIS, B., HARTLEY, F., PETROW, V. and WEDLAKE, D. (1955), *J. chem. Soc.* 4383
ENGEL, L. L. and BAGGETT, B. (1954), *Recent Progr. Hormone Res.* **9**, 251.
ENGEL, L. L., CARTER, P. and FIELDING, L. L. (1955), *J. biol. Chem.* **213**, 99.
ENGEL, L. L., CARTER, P. and SPRINGER, M. J. (1954), *Fed. Proc.* **13**, 204.
ENGEL, L. L., SLAUNWHITE, R., CARTER, P. and NATHANSON, E. (1950), *J. biol. Chem.* **185**, 255.
EPPSTEIN, S. H., MEISTER, P. D., MURRAY, H. C. and PETERSON, D. H. (1956) *Vitam. and Horm.* **14**, 359.
EVANS, R. M., HAMLET, J. C., HUNT, J. S., JONES, P. G., LONG, A. G., OUGHTON, J. F., STEPHENSON, L., WALKER, T. and WILSON, B. M. (1956), *J. chem. Soc.* 4356.
FEIGL, F. (1956), *Spot Tests*, Vol. 2, 5th Edn., Elsevier, Amsterdam.
FIESER, L. F. (1953), *J. Amer. chem. Soc.* **75**, 4377.
FIESER, L. F. and FIESER, M. (1959), *Steroids*, Reinhold, New York.

FINK, K. and FINK, R. M. (1949), *Proc. Soc. exp. Biol. N.Y.* **70**, 654.
FINKELSTEIN, M., VON EUW, J. and REICHSTEIN, T. (1953), *Helv. chim. Acta* **36**, 1286.
FISHER, R. B., PARSONS, D. S. and MORRISON, G. A. (1948), *Nature, Lond.* **161**, 764.
FOLKERTSMA, J. P., SOKOLSKI, W. T. and SNYDER, J. W. (1957–58), *Antibiot. Ann.* 114.
FOTHERBY, K., COLAS, A., ATHERDEN, S. M. and MARRIAN, G. F. (1957), *Biochem. J.* **66**, 664.
FRESCO, J. R. and WARNER, R. C. (1955), *J. biol. Chem.* **215**, 751.
FRIED, J., BORMAN, A., KESSLER, W. B., GRABOWICH, P. and SABO, E. F. (1958), *J. Amer. chem. Soc.* **80**, 2338.
FRIED, J. and SABO, E. F. (1957), *J. Amer. chem. Soc.* **79**, 1130.
FRIERSON, W. J., THOMASON, P. F. and RAAEN, H. (1954), *Analyt. Chem.* **26**, 1210.
FUKUSHIMA, D. K. and GALLAGHER, T. F. (1952), *J. biol. Chem.* **198**, 871.
FUKUSHIMA, D. K. and GALLAGHER, T. F. (1957), *J. biol. Chem.* **229**, 85.
FUKUSHIMA, D. K., LEEDS, N. S., BRADLOW, H. L., KRITCHEVSKY, T. H., STOKEM, M. B. and GALLAGHER, T. F. (1955), *J. biol. Chem.* **210**, 129.
FUJIMOTO, G. I. (1951), *J. Amer. chem. Soc.* **73**, 1856.
FUJITA, H. (1952), *J. phys. Chem.* **56**, 625.
GALLAGHER, T. F., BRADLOW, H. L., FUKUSHIMA, D. K., BEER, C. T., KRITCHEVSKY, T. H., STOKEM, M. C., EIDINOFF, M. L., HELLMAN, L. and DOBRINER, K. (1954a) *Recent Progr. Hormone Res.* **9**, 411.
GALLAGHER, T. F., FUKUSHIMA, D. K., BARRY, M. C. and DOBRINER, K. (1954b), *Recent Progr. Hormone Res.* **9**, 131.
GALLAGHER, T. F. and KOCH, F. C. (1934), *Endocrinology* **18**, 107.
GHILLAIN, A. (1956), *Ann. Endocr., Paris* **16**, 477.
GLASSTONE, S. (1947), *Textbook of Physical Chemistry* (2nd Ed.) p. 508. Van Nostrand. New York.
GLENN, E. M. and NELSON, D. H. (1953), *J. clin. Endocrin.* **13**, 911.
GLUECKAUF, E. (1949a), *J. chem. Soc.* 3280.
GLUECKAUF, E. (1949b), *Disc. Faraday Soc.* No. 7, 12.
GLUECKAUF, E. (1949c), *Disc. Faraday Soc.* No. 7, 239.
GLUECKAUF, E. (1955), *Trans. Faraday Soc.* **51**, 34.
GLUECKAUF, E. (1958), *Trans. Faraday Soc.* **54**, 1203.
GOPPELSROEDER, F. (1909), *Z. Chem. Ind. Kolloide* **4**, 23, 94, 191, 236, 312.
GORDON, B. E., WOPAT, F., BURNHAM, D. H. and JONES, L. C. (1951), *Analyt. Chem.* **23**, 1754.
GORNALL, A. and MACDONALD, M. P. (1953), *J. biol. Chem.* **201**, 279.
GOWENLOCK, A. (1960), *Mem. Soc. Endocrin.* **8**, 77.
GRASSMAN, W. and KANNIG, K. (1954), *Klin. Wschr.* **32**, 838.
GRAY, C. H., HELLMAN, K., LUNNON, J. B. and WEINER, J. S. (1955), *J. Endocrin.* **12**, 6P.
GRAY, C. H., GREEN, M. A. S., HOLNESS, M. J. and LUNNON, J. B. (1958), *J. Endocrin.* **14**, 146.
GREEN, F. C. and KAY, L. M. (1952), *Analyt. Chem.* **24**, 726.
GREGORY, G. F. (1955), *Science* **121**, 169.
GROLLMAN, A. (1937), *Cold Spr. Harb. Symp. Quant. Biol.* **5**, 313.

GROLLMAN, A., FIROR, W. H. and GROLLMAN, E. (1935), *J. biol. Chem.* **109**, 189.
GRUNDY, H. M., SIMPSON, S. A. and TAIT, J. P. (1952), *Nature, Lond.* **169**, 795.
GUNZEL, C. and WEISS, F. (1953), *Z. anal. Chem.* **140**, 189.
HABERMANN, E., MÜLLER, W. and SCHREGLMANN, A. (1953), *Arzneimittel-Forsch.* **3**, 30.
HAENNI, E. D., CAROL, J. and BANES, D. (1953), *J. Amer. pharm. Assoc.* Sci. Ed., **42**, 162.
HAINES, W. J. (1952), *Recent Progr. Hormone Res.* **7**, 255.
HAINES, W. J. and KARNEMAAT, J. N. (1954), *Methods of Biochemical Analysis*, Vol. I, p. 171. Edited by Glick, D.
HAMILTON, J. G. and HOLMAN, R. T. (1952), *Arch. Biochem. Biophys.* **36**, 456.
HANES, C. S. and ISHERWOOD, F. A. (1949), *Nature, Lond.* **164**, 1107.
HARMAN, R. E., HAM, E. A., DE YOUNG, D. D., BRINK, N. G. and SARETT, L. H. (1954), *J. Amer. chem. Soc.* **76**, 5035.
HASLEWOOD, G. A. D. (1954), *Biochem. J.* **56**, 581.
HASLEWOOD, G. A. D. (1958), *Biochem. J.* **70**, 551.
HASLEWOOD, G. A. D. and SJÖVALL, J. (1954), *Biochem. J.* **57**, 126.
HASSALL, C. H. and MARTIN, S. L. (1951), *J. chem. Soc.* 2766.
HECHTER, O. (1949), *Fed. Proc.* **8**, 70.
HEFTMANN, E. (1950), *Science* **111**, 571.
HEFTMANN, E. (1951), *J. Amer. chem. Soc.* **73**, 851.
HEFTMANN, E. (1955), *Chem. Rev.* **55**, 679.
HEFTMANN, E. and LEVANT, A. J. (1952), *J. biol. Chem.* **194**, 703.
HEFTMANN, E. and HAYDEN, A. L. (1952), *J. biol. Chem.* **197**, 47.
HEFTMANN, E., BERNER, P., HAYDEN, A. L., MILLER, H. K. and MOSETTIG, E. (1954), *Arch. Biochem. Biophys.* **51**, 329.
HEFTMANN, E. and JOHNSON, D. F. (1954), *Analyt. Chem.* **26**, 519.
HEGEDÜS, H. M. (1943), *Helv. chim. Acta* **36**, 357.
HEGEDÜS, H. M., TAMM, CH. and REICHSTEIN, T. (1953), *Helv. chim. Acta* **36**, 357.
HENRY, R. and THEVENET, M. (1952), *Bull. Soc. Chim. biol., Paris* **34**, 886.
HENRY, R. and THEVENET, M. (1953), *Ann. Endocr., Paris.* **14**, 628.
HERMANS, P. H. (1949), *Physics and Chemistry of Cellulose Fibres*, Amsterdam, Elsevier.
HERRMANN, W. R. and SILVERMAN, L. (1957), *Proc. Soc. exp. Biol., N.Y.* **94**, 426.
HERSHBERG, E. B., WOLFE, J. K., and FIESER, L. F. (1941), *J. biol. Chem.* **140**, 215.
HERZIG, P. G. and EHRENSTEIN, M. (1951), *J. org. Chem.* **16**, 1050.
HEUSGHEM, C. (1953), *Nature, Lond.* **171**, 42.
HILDEBRAND, J. H. and SCOTT, R. L. (1950), *Solubility of Non-Electrolytes*; A.C.S. Monograph 17, Reinhold, New York.
HILL, B. R. (1952), *Arch. Biochem.* **40**, 479.
HIRSCHMANN, H. (1952), *J. Amer. chem. Soc.* **74**, 5357.
HÖBER, R. (1947), *The Physical Chemistry of Cells and Tissues*; London, Churchill.
HODGSON, H. H., and MARSDEN, E. (1944), *J. Soc. Dy. Col.* **60**, 210.
HOFMANN, H. and STAUDINGER, Hj. (1951), *Biochem. Z.* **322**, 230.
HOFMANN, H. and STAUDINGER, Hj. (1951), *Naturwissenschaften* **38**, 213.
HOLLANDER, N. and HOLLANDER, V. P. (1958), *J. clin. Endocrin.* **18**, 966.
HOWARD, G. A. and MARTIN, A. J. P. (1950), *Biochem. J.* **46**, 532.

HUBER, H., BLINDENBACHER, F., MOHR, K., SPEISER, P. and REICHSTEIN, T. (1951), *Helv. chim. Acta* **34**, 46.
HUFFMAN, M. N. and DARBY, H. H. (1944), *J. Amer. chem. Soc.* **66**, 150.
HUFFMAN, M. N. and LOTT, M. H. (1947), *J. Amer. chem. Soc.* **69**, 1835.
HUNTER, L. (1954), *Progress in Stereochemistry*, Vol. 1, Ed. W. Klyne, London, Butterworth, p. 223.
HURLOCK, B. and TALALAY, P. (1956), *Proc. Soc. exp. Biol., N.Y.* **93**, 560.
INOUYE, Y. and NODA, M. (1955), *Bull. agric. chem. Soc., Japan* **19**, 214.
INOUYE, Y., NODA, M. and HIRAYAMA, O. (1955), *J. Amer. Oil Chem. Soc.* **32**, 132.
IRIARTE, J. and RINGOLD, H. J. (1958), *Tetrahedron* **3**, 28.
ISHERWOOD, F. A. (1946), *Biochem. J.* **40**, 688.
ISHERWOOD, F. A. (1954), *Brit. med. Bull.* **10**, 202.
ISHERWOOD, F. A. and HANES, C. S. (1953), *Biochem. J.* **55**, 824.
ISHERWOOD, F. A. and JERMYN, M. A. (1951), *Biochem. J.* **48**, 515.
JAMES, A. T. (1952), *Biochem. J.*, **52**, 242.
JAMES, A. T. and MARTIN, A. J. P. (1954), *Brit. med. Bull.* **10**, 170.
JAMES, A. T., MARTIN, A. J. P. and RANDALL, S. A. (1951), *Biochem. J.* **49**, 293.
JAMINET, E. (1950), *J. Pharm. Belg.* **5**, 297.
JAMINET, E. (1951), *J. Pharm. Belg.* **6**, 90.
JAMINET, E. (1952), *J. Pharm. Belg.* **7**, 169.
JARRIGE, P. and HENRY, R. (1952), *Bull. Soc. Chim. biol., Paris* **34**, 872.
JEANES, A., WISE, C. S. and DIMLER, R. (1951), *Analyt. Chem.* **23**, 415.
JEFFRIES, H. (1957), *Scientific Inference* (2nd Ed) London, Cambridge University Press.
JENSEN, J. B. (1956), *Dansk. Tidsskr. Farm.* **30**, 293.
JENSEN, J. P. and BOJESON, E. (1958), *Proc. Int. Congr. Biochem.* 118.
JENSEN, K. B. (1953), *Acta pharm. tox., Kbh.* **9**, 99.
JERMYN, M. A. and ISHERWOOD, F. A. (1949), *Biochem. J.* **44**, 402.
JOHNSON, D. F., HEFTMANN, E. and HAYDEN, A. L. (1956), *Acta endocr. Copenhagen*, **23**, 341.
JONES, R. N. and HERLING, F. (1954), *J. org. Chem.* **19**, 1252.
JONES, R. N. and HERLING, F. (1956), *J. Amer. chem. Soc.* **78**, 1152.
JONES, R. N., HERLING, F. and KATZENELLENBOGEN, E. (1955), *J. Amer. chem. Soc.* **77**, 651.
JONES, R. N., NOLIN, B. and ROBERTS, G. (1955), *J. Amer. chem. Soc.* **77**, 6331.
JONES, J. K. N. and STITCH, G. R. (1953), *Biochem. J.* **53**, 679.
KABASAKALIAN, P. and BASCH, A. (1960), *Analyt. Chem.* **32**, 458.
KAISER, F. (1955), *Chem. Ber.* **88**, 556.
KAMIL, I. A., SMITH, J. N. and WILLIAMS, R. T. (1953), *Biochem. J.* **53**, 129.
KANDRAC, M. S. (1958), *Proc. Int. Congr. Biochem.* (Abstr. 9–73, 9–74) p. 112.
KATZENELLENBOGEN, E. R., KRITCHEVSKY, T. H. and DOBRINER, K. (1952), *Fed. Proc.* **11**, 238.
KATZENELLENBOGEN, E. R., DOBRINER, K. and KRITCHEVSKY, T. H. (1954), *J. biol. Chem.* **207**, 315.
KELLIE, A. E. and SMITH, E. R. (1956), *Nature, Lond.* **178**, 323.
KELLIE, A. E., SMITH, E. R. and WADE, A. P. (1953), *Biochem. J.* **53**, 578.
KELLIE, A. E., SMITH, E. R. and WADE, A. P. (1953), *Biochem. J.* **53**, 582.

KELLIE, A. E. and WADE, A. P. (1957), *Biochem. J.* **66**, 196.
KENDALL, E. C. and LLOYD, D. G. (1955), *P.S.G. Bull.* No. 9. 216.
KENDALL, E. C. (1937), *Cold Spr. Harb. Symp. quant. Biol.* **5**, 299.
KESTON, A. S., UDENFRIEND, S. and LEVY, M. (1947), *J. Amer. chem. Soc.* **69**, 3151.
KESTON, A. S., UDENFRIEND, S. and LEVY, M. (1950), *J. Amer. chem. Soc.* **72**, 748.
KEULEMANNS, A. I. M. (1957), *Gas Chromatography*, London, Reinhold, p. 112.
KEUTMANN, E. H., BURTON, R. B., and IZZO, A. J. (1954), U.S. Publ. Health Service rept., Grant C-1003(C4); Personal Communication.
KIRCHNER, J. G. and KELLER, G. J. (1950), *J. Amer. chem. Soc.* **72**, 1867.
KIRKWOOD, J. G. (1939), *J. phys. Chem.* **43**, 97.
KISS, P. and SZELL, T. (1956), *Naturwissenschaften* **43**, 448.
KLOPPER, A., MICHIE, E. A. and BROWN, J. B. (1955), *J. Endocrin.* **12**, 209.
KLYNE, W. (1957), *Chemistry of the Steroids*, London, Methuen.
KLYNE, W. and WRIGHT, A. A. (1957), *Biochem. J.* **66**, 92.
KOCHAKIAN, C. D. and STIDWORTHY, G. (1952), *J. biol. Chem.* **199**, 607.
KODICEK, E. and ASHBY, D. R. (1954), *Biochem. J.* **57**, xii.
KOŠTÍR, J. V. and SLAVÍK, K. (1950), *Coll. Czech. Chem. Comm.* **15**, 17.
KOWKABANY, G. N. and CASSIDY, H. G. (1950), *Anal. Chem.* **22**, 817.
KOWKABANY, G. N. and CASSIDY, H. G. (1952), *Anal. Chem.* **24**, 643.
KRAYBILL, H. R., THORNTON, M. H. and ELDRIDGE, K. E. (1940), *Ind. Engng. Chem.* **32**, 1138.
KRESS, O. and BIALKOWSKI, H. (1931), *Paper Trade J.* **93**, 35.
KRITCHEVSKY, T. (1954), *Recent Progr. Hormone Res.* **9**, 177.
KRITCHEVSKY, D. and CALVIN, M. (1950), *J. Amer. chem. Soc.* **72**, 4330.
KRITCHEVSKY, D. and KIRK, M. R. (1952a), *Arch. Biochem. Biophys.* **35**, 346.
KRITCHEVSKY, D. and KIRK, M. R. (1952b), *J. Amer. chem. Soc.* **74**, 4484.
KRITCHEVSKY, T. H. and TISELIUS, A. (1951), *Science* **114**, 299.
KUCERA, J., PROCHAZKA, Z. and VERES, K. (1957), *Coll. Czech. Chem. Comm.* **22**, 1185.
KUHN, R. and LÖW, J. (1954), *Angew. Chem.* **66**, 639.
KUMLER, W. D. (1945), *J. Amer. chem. Soc.* **67**, 1904.
LAKSHMANAN, T. K. and LIEBERMAN, S. (1954), *Arch. Biochem. Biophys.* **53**, 258.
LANGMUIR, I. (1916), *J. Amer. chem. Soc.* **38**, 2221.
LAUER, K. (1944), *Kolloid-zschr.* **107**, 86.
LAURENCE, D. J. R. (1954), *J. sci. Instrum.* **31**, 137.
LAYNE, D. S. and MARRIAN, G. F. (1958), *Biochem. J.* **70**, 244.
LE ROSEN, A. L., MONAGHAN, P. H., RIVET, C. A. and SMITH, E. D. (1951), *Anal. Chem.* **23**, 730.
LEDERER, E. and LEDERER, M. (1957), *Chromatography* (2nd Ed.) Amsterdam, Elsevier.
LEDERER, M. (1959), *J. Chromatog.* **1**, 3P.
LEEGWATER, D. C. (1956), *Nature, Lond.* **178**, 916.
LEVY, H. and KUSHINSKY, S. (1954), *Recent Progr. Hormone Res.* **9**, 357.
LEWBART, M. L. and MATTOX, V. R. (1959), *Nature, Lond.* **183**, 820.
LEWBART, M. L. and SCHNEIDER, J. J. (1955), *Nature, Lond.* **176**, 1175.
LEWIS, B. (1956), *Biochem. biophys. Acta* **20**, 396.

REFERENCES

LEWIS, B. (1957), *J. clin. Path.* **10**, 148.
LEWIS, J. L. and THORN, G. W. (1955), *J. clin. Endocrin.* **15**, 829.
LIBERMAN, L. A., ZAFFARONI, A. and STOTZ, E. (1951), *J. Amer. chem. Soc.* **73**, 1387.
LICHTI, H., TAMM, C. and REICHSTEIN, T. (1956), *Helv. chim. Acta* **39**, 1914, 1933.
LIEBERMAN, S. and DOBRINER, K. (1948), *Recent Progr. Hormone Res.* **3**, 71.
LIEBERMAN, S., DOBRINER, K., HILL, B. R., FIESER, L. F. and RHOADS, C. P. (1948), *J. biol. Chem.* **172**, 263.
LIEBERMAN, S., FUKUSHIMA, D. K. and DOBRINER, K. (1950), *J. biol. Chem.* **182**, 299.
LIEBERMAN, S., KATZENELLENBOGEN, E. R., SCHNEIDER, R., STUDER, P. E. and DOBRINER, K. (1953), *J. biol. Chem.* **205**, 87.
LIEBERMAN, S. and TEICH S. (1953), *Pharmacol. Rev.* **5**, 285.
LINDNER, H. R. (1960), *Biochim. biophys. Acta* **38**, 362.
LLAURADO, J. G., NEHER, R. and WETTSTEIN, A. (1956), *Clin. chim. Acta* **1**, 236.
LOKE, K. H., MARRIAN, G. F. and WATSON, E. J. D. (1959), *Biochem. J.* **71**, 43.
LONDON, F. (1942), *J. phys. Chem.* **46**, 305.
LOOMEIJER, F. J. and LUINGE, G. M. (1958), *J. Chromatog.* **1**, 179.
LOVELOCK, J. (1958), *J. Chromatog* **1**, 35.
LUETSCHER, J. A., NEHER, R. and WETTSTEIN, A. (1955), *J. biol. Chem.* **217**, 505.
LUGG, J. W. H. and OVERELL, B. T. (1947), *Nature, Lond.*, **160**, 87.
LUGG, J. W. H. and OVERELL, B. T. (1948), *Austral. J. sci. Res.* **1**, 98.
MCALEER, W. J. and KOLLOWSKI, M. A. (1957), *Arch. Biochem. Biophys.* **66**, 120, 125.
MCFARREN, E. F. (1951), *Anal. Chem.* **23**, 168.
MCFARREN, E. F. (1952), *Anal. Chem.* **24**, 650.
MACPHILLAMY, H. A. and SCHOLZ, C. R. (1949), *J. biol. Chem.* **178**, 37.
MCKENNA, J. and NORYMBERSKI, J. K. (1957), *J. chem. Soc.* 3889.
MADER, W. J. and BUCK, R. R. (1952), *Anal. Chem.* **24**, 666.
MANCERA, O., ROSENKRANZ, G. and SONDHEIMER, F. (1953), *J. chem. Soc.* 2189.
MARKEES, S. (1954), *Biochem. J.* **56**, 703.
MARKHAM, R., and SMITH, J. D. (1949), *Biochem. J.* **45**, 294.
MARKWARDT, F. (1954), *Naturwissenschaften* **41**, 139.
MARKWARDT, F. (1955), *Arch. Pharm.* **288**, 82.
MARRIAN, G. F., LOKE, K. H., WATSON, E. J. D. and PANATTONI, M. (1957), *Biochem. J.* **66**, 60.
MARTIN, A. J. P. (1949), *Biochem. Soc. Symposia*, **3**, 4.
MARTIN, A. J. P. (1950), *Annu. Rev. Biochem.* **19**, 517.
MARTIN, A. J. P. and JAMES, A. T. (1956), *Biochem. J.* **63**, 138.
MARTIN, A. J. P. and SYNGE, R. L. M. (1941), *Biochem. J.* **35**, 1358.
MARTIN, R. P. (1957), *Biochim. biophys. Acta* **25**, 408.
MATTOX, V. R. and MASON, H. L. (1956), *J. biol. Chem.* **223**, 215.
MATTOX, V. R. and LEWBART, M. L. (1958), *Arch. Biochem. Biophys.* **76**, 362.
MATTOX, V. R. and LEWBART, M. L. (1959), *J. clin. Endocrin.* **19**, 1151.
MATTOX, V. R., MASON, H. L. and ALBERT, A. (1956), *J. biol. Chem.* **218**, 359.
MAYER, S. W. and TOMKINS, E. R. (1947), *J. Amer. chem. Soc.* **69**, 2866.

MEIGH, D. F. (1952a), *Nature, Lond.* **169**, 706.
MEIGH, D. F. (1952b), *Nature, Lond.* **170**, 579.
MEYER, A. S. (1953), *J. biol. Chem.* **203**, 469.
MEYER, A. S. (1955), *J. org. Chem.* **20**, 1240.
MEYERE, A. S. and LINDBERG, M. C. (1955), *Analyt. Chem.* **27**, 813.
MEYER, K. (1949), *Pharm. Acta Helv.* **24**, 222.
MEYERHEIM, G. and HUBENER, H. J. (1952), *Naturwissenschaften* **39**, 482.
MICHALEC, C. (1955), *Naturwissenschaften* **42**, 509.
MICHALEC, C., JIRGL., V. and PODZIMEK, J. (1957), *Experientia* **13**, 242.
MICHEEL, F. and SCHWEPPE, H. (1954a), *Mikrochim. Acta*, 53.
MICHEEL, F. and SCHWEPPE, H. (1954b), *Angew. Chem.* **66**, 136.
MIETHINEN, J. K. and MOISIO, T. (1953), *Acta chem. scand.* **7**, 1225.
MIGEON, C. (1957), *Ciba Found. Colloq. Endocrinol.* **11**, 282.
MILLS, J. A. and KLYNE, W. (1954), *Progress in Stereochemistry*, Vol. 1, 177, London, Butterworth.
MILLS, J. S. and WERNER, A. E. A. (1952), *Nature, Lond.* **169**, 1064.
MILLS, J. S. and WERNER, A. E. A. (1955), *J. chem. Soc.* 3132.
MITCHELL, F. L. (1952), *Nature, Lond.* **170**, 621.
MITCHELL, F. L. and DAVIES, R. E. (1954), *Biochem. J.* **56**, 690.
MITCHELL, H. K. and HASKINS, F. A. (1949), *Science* **110**, 278.
MOELWYN-HUGHES, E. A. (1957), *Physical Chemistry*. London, Pergamon.
MOELWYN-HUGHES, E. A. and SHERMAN, A. (1936), *Trans. chem. Soc.* 101.
MOOLENAAR, A. J. (1957), *Acta endocr.*, Copenhagen **25**, 161.
MOORE, A. N. and BOYLEN, J. B. (1953), *Science* **118**, 119.
MOORE, S. and STEIN, W. H. (1948), *Ann. N.Y. Acad. Science* **49**, 265.
MORRIS, C. J. O. R. (1955), *Ciba Found. Colloq. Endocrinol.* **8**, 158.
MORRIS, C. J. O. R. (1959), *Mem. Soc. Endocrin.* **8**, 1.
MORRIS, C. J. O. R. and WILLIAMS, D. C. (1953a), *Ciba Found. Colloq. Endocrin.* **7**, 261.
MORRIS, C. J. O. R. and WILLIAMS, D. C. (1953b), *Biochem. J.* **54**, 470.
MORRIS, C. J. O. R. and WILLIAMS, D. C. (1955), *Ciba Found. Colloq. Endocrin.* **8**, 157.
MOSBACH, E. H., ZOMZELY, C. and KENDALL, F. E. (1953a), *Fed. Proc.* **12**, 397.
MOSBACH, E. H., NIERENBERG, M. and KENDALL, F. E. (1953b), *J. Amer. chem. Soc.* **75**, 2358.
MOSBACH, E. H., ZOMZELY, C. and KENDALL, F. E. (1954), *Arch. Biochem. Biophys.* **48**, 95.
MOSS, P. and THOMAS, A. J. (1959), *Nature, Lond.* **183**, 1161.
MUELLER, J. H. (1950), *Science* **112**, 405.
MUNIER, R. (1952), *Bull Soc. Chim. France* 852.
MUNSON, P. L., JONES, M. E., MCCALL, P. J. and GALLAGHER, T. F. (1948), *J. biol. Chem.* **176**, 73.
NACE, H. R. and TURNER, R. B. (1953), *J. Amer. chem. Soc.* **75**, 4063.
NADEL, E. M., YOUNG, B. G., HILGAR, A. C. and BURSTEIN, S. (1958), *Acta endocr. Copenhagen* **28**, 283.
NAKAGI, S. (1958), *J. Sci. of Labour (Japan)*, **34**, 681.
NASH, H. A. and BROOKER, R. M. (1953), *J. Amer. chem. Soc.* **175**, 1942.

REFERENCES

NEHER, R. (1958a), *Chromatographie von Sterinen, Steroiden und verwandten Verbindungen;* Elsevier, Amsterdam, also *J. Chromatog* **1**, 122, 205.
NEHER, R. (1958b), *Adv. clin. Chem.* **1**, 127.
NEHER, R. (1959a), *Chromatographic Reviews*, **1**, 99. Edited by Lederer, M. Amsterdam, Elsevier. (Translation of Neher (1958a)).
NEHER, R. (1959b), *Mem. Soc. Endocrin.*, **8**, 96.
NEHER, R., DESAULLES, P., VISCHER, E., WIELAND, P. and WETTSTEIN, A. (1958), *Helv. chim. Acta* **41**, 1667.
NEHER, R., MEYSTRE, C. and WETTSTEIN, A. (1959), *Helv. chim. Acta* **42**, 132.
NEHER, R. and WETTSTEIN, A. (1951) *Helv. chim. Acta* **34**, 2278.
NEHER, R. and WETTSTEIN, A. (1952), *Helv. chim. Acta* **35**, 276.
NEHER, R. and WETTSTEIN, A. (1955), *Acta endocr. Copenhagen* **18**, 386.
NEHER, R. and WETTSTEIN, A. (1956a), *J. clin. Invest.* **35**, 800.
NEHER, R. and WETTSTEIN, A. (1956b), *Helv. chim. Acta.* **39**, 2062.
NELSON, D. H. and SAMUELS, L. T. (1952), *J. clin. Endocrin.* **12**, 519.
NICHOLAS, R. E. H. and RIMINGTON, C. R. (1951), *Biochem. J.* **48**, 306.
NORMAN, A. (1953), *Acta chem. scand.* **7**, 413.
NORMAN, A. (1954), *Acta physiol. scand.* **32**, 1.
NORMAN, A. and SJÖVALL, J. (1959), *J. biol. Chem.* **233**, 872.
NORYMBERSKI, J. K. (1952), *Nature, Lond.* **170**, 1074.
NORYMBERSKI, J. K. (1953), *Mem. Soc. Endocrin.* **2**, 50.
NORYMBERSKI, J. K. and SERMIN, A. I. (1953), *Biochem. J.* **55**, 876.
NORYMBERSKI, J. K. and STUBBS, R. D. (1956a), *Biochem. J.* **64**, 168.
NORYMBERSKI, J. K. and STUBBS, R. D. (1956b), *Biochem. J.* **64**, 176.
NORYMBERSKI, J. K. and STUBBS, R. D., *Lancet*, Vol. II, 386.
NORYMBERSKI, J. K. and WOODS, G. F. (1954), *Chem. and Ind.* **5**, 18.
NOWACZYNSKI, W. J. and KOIW, E. (1957), *J. Lab. clin. Med.* **49**, 815.
NOWACZYNSKI, W. J., STEYERMARK, P. R., KOIW, E., GENEST, J. and JONES, R. N. (1956), *Canad. J. Biochem.* **34**, 1023.
NYC, T. F., MARON, D. M., GARST, J. B. and FRIEDGOOD, M. B. (1951), *Proc. Soc. exp. Biol.*, N.Y. **77**, 466.
OERTEL, G. (1954), *Acta endocr. Copenhagen* **16**, 263, 267.
OERTEL, G. and EIK-NES, K. (1959), *Acta endocr. Copenhagen* **30**, 93.
OLIVETO, E. P., GEROLD, C. and HERSHBERG, E. B. (1955), *Arch. Biochem. Biophys.* **43**, 234.
OTT, E., SPURLIN, H. M. and GRAFFLIN, M. W. (1954), *Cellulose* (Pt. I High Polymers, Vol. 5. New York, Interscience, p. 234.
PARTINGTON, J. R. (1937), *A Short History of Chemistry.* p. 16, Macmillan, London.
PARTRIDGE, S. M. (1948), *Biochem. J.* **42**, 238, 251.
PATTERSON, J. and SWALE, J. (1953), *Mem. Soc. Endocrin.* **2**, 41.
PEARLMAN, W. H. (1957a), *Biochem. J.* **66**, 17.
PEARLMAN, W. H. (1957b), *Biochem. J.* **67**, 1.
PEARLMAN, W. H. and CERCERO, E. (1953), *J. biol. Chem.* **198**, 79.
PEARLMAN, W. H., PEARLMAN, M. R. J. and RAKOFF, A. E. (1954), *J. biol. Chem.* **209**, 803.
PECHET, M. M. (1953), *J. clin. Endocrin.* **13**, 1542.
PECHET, M. M. (1955), *Science* **121**, 39.

PERRIN, D. D. (1956), *Nature, Lond.* **178**, 1244.
PETERSON, R. E. (1957), *J. biol. Chem.* **225**, 25.
PETERSON, R. E. KANER, A. and GUERRA, S. L. (1957a), *Analyt. Chem.* **29**, 144.
PETERSON, R. E., KLIMAN, B. and BOLLIER, M. E. (1957b), *Abstracts of the* 131*st Meeting American Chemical Society*, 59C.
PFIFFNER, J. J. and VARS, H. M. (1934), *J. biol. Chem.* **106**, 645.
PFIFFNER, J. J., VARS, H. M. and TAYLOR, A. R. (1934), *J. biol. Chem.* **106**, 625.
PFIFFNER, J. J., WINTERSTEINER, O. and VARS, H. M. (1935), *J. biol. Chem.* **111**, 585.
PHILIP, J. E. and SCHENK, J. R. (1957), *Analyt. Chem.* **29**, 170.
PIEROTTI, G. J., DEAL, C. A., DERR, E. L. and PORTER, P. E. (1956), *J. Amer. chem. Soc.* **78**, 2989.
PIEROTTI, G. J., DEAL, C. A. and DERR, E. L. (1959), *Ind. Engng. Chem.* **51**, 95.
PINCUS, G. and PEARLMAN, W. H. (1941), *Endocrinology* **29**, 413.
PINCUS, G. and ROMANOFF, E. B. (1950), *Fed. Proc.* **9**, 101.
PIRIE, N. W. (1940), *Biol. Rev.* **15**, 377.
PLAGER, J. E. and SAMUELS, L. T. (1953), *Arch. Biochem. Biophys.* **42**, 477.
POLSON, A., VAN ROOY, P. J., and MAVAIS, E. J. (1952), see *Chem. Abstr.* **46**, 3905.
PORTER, C. C. and SILBER, R. H. (1950), *J. biol. Chem.* **185**, 201.
PRELOG, V. and ZÄCH, D. (1959). Private communication.
PRICE, D. and ASHMAN, P. B. (1955), *Nature, Lond.* **175**, 45.
PROCHAZKA, Z. (1954), *Coll. Czech. Chem. Comm.* **19**, 98.
PROCHAZKA, Z., LABLER, L. and KOTASEK, Z. (1954) *Coll. Czech. Chem. Comm.* **19**, 1258.
PUCK, A. (1955), *Klin. Wschr.* **33**, 865.
QUAIFE, M. L., GEYER, R. P. and BOLLIGER, H. R. (1956), quoted R. NEHER (1959a), ref. no. 192.
RAMSEY, L. L. and PATTERSON, W. I. (1946), *J. Assoc. Off. Agr. Chem.* **29**, 337.
RAMSEY, L. L. and PATTERSON, W. I. (1948), *J. Assoc. Off. Agr. Chem.* **31**, 139.
REDDY, W. J., JENKINS, D. and THORN, G. W. (1952), *Metabolism*. **1**, 511.
REDFIELD, R. R. and BARRON, E. S. G. (1952), *Arch. Biochem. Biophys.* **35**, 443.
REES, V. H. and LAURENCE, D. J. R. (1955), *Clin. Chem.* **1**, 320.
REEVES, R. E. (1951), *Adv. carbohyd. Chem.* **6**, 108.
REICH, H. and COLLINS, R. W. (1951), *J. Amer. chem. Soc.* **73**, 137.
REICH, H., CRANE, K. F. and SANFILIPPO, S. J. (1953), *J. org. Chem.* **18**, 822.
REICH, H., NELSON, D. H. and ZAFFARONI, A. (1950), *J. biol. Chem.* **187**, 411.
REICH, H., SANFILIPPO, S. J. and CRANE, K. F. (1952), *J. biol. Chem.* **198**, 713.
REICH, M. (1958), *Austral. J. exp. Biol. med. Sci.* **36**, 555.
REICHL, E. R. (1955), *Monatsch. Chem.* **86**, 69.
REICHL, E. R. (1956), *Mikrochim. Acta* 958.
REICHSTEIN, T. (1936), *Helv. chim. Acta* **19**, 29.
REICHSTEIN, T. (1938), *Ergebn. Vitam. u. Hormonforsch.* **1**, 334.
REICHSTEIN, T. and SHOPPEE, C. W. (1949), *Disc. Faraday Soc.* **7**, 305.
REINEKE, L. M. (1956), *Analyt. Chem.* **28**, 1853.
RICHARDSON, E. M., TOUCHSTONE, J. C. and DOHAN, F. C. (1955), *J. clin. Invest.* **34**, 285.
RIDDELL, C. and COOK, R. P. (1955), *Biochem. J.* **61**, 657.

RIEGEL, B. and PROUT, F. S. (1948), *J. org. Chem.* **13**, 933.
RITTEL, W., HUNGER, A. and REICHSTEIN, T. (1952), *Helv. chim. Acta* **35**, 434.
RITTER, J. J. and HARTEL, J. (1958a), *J. Chromatog.* **1**, 461.
RITTER, J. J. and HARTEL, J. (1958b), *Nature. Lond.* **181**, 765.
RITTER, J. J. and LOVER, M. J. (1952), *J. Amer. chem. Soc.* **74**, 5576.
ROBERTS, G. (1957), *Analyt. Chem.* **29**, 911.
ROCKLAND, L. B., BLATT, J. L. and DUNN, M. S. (1951), *Analyt. Chem.* **23**, 1142.
ROCKLAND, L. B. and UNDERWOOD, J. C. (1956), *Analyt. Chem.* **28**, 1679.
ROGERS, A. W. (1959), *Nature, Lond.* **184**, 721.
ROLAND, J. F. and GROSS, A. M. (1954), *Analyt. Chem.* **26**, 502.
ROMANOFF, E. B. and HUNT, C. A. (1955), *Endocrinology* **57**, 499.
ROMANOFF, L. P. and WOLF, R. S. (1954), *Recent Progr. Hormone Res.* **9**, 337.
ROMANOFF, L. P., WOLF, R. S., CONSTANDSE, M. and PINCUS, G. (1953), *J. clin. Endocrin.* **13**, 928.
ROSENFIELD, R. S. (1957), *J. Amer. chem. Soc.* **79**, 5540.
ROSENKRANTZ, H. (1953), *Arch. Biochem. Biophys.* **44**, 1.
ROSENKRANTZ, H. and GUT, M. (1954), *Science* **120**, 3129.
ROSENKRANTZ, H., POTVIN, P. and SKOGSTROM, P. (1958), *Analyt. Chem.* **30**, 975.
ROSENKRANTZ, H. and SKOGSTROM, P. (1956), *Analyt. Chem.* **28**, 31.
ROSENKRANTZ, H. and ZABLOW, L. (1953), *Analyt. Chem.* **25**, 1025.
ROSENKRANTZ, H. and ZABLOW, L. (1953), *J. Amer. chem. Soc.* **75**, 903.
ROUX, I. G. and EVELYN, S. R. (1958), *J. Chromatog.* **1**, 537.
ROY, A. B. (1956), *Biochem. J.* **62**, 41.
RUBIN, B. L., DORFMAN, R. I. and PINCUS, G. (1953a), *J. biol. Chem.* **203**, 629.
RUBIN, B. L., DORFMAN, R. I. and PINCUS, G. (1954), *Recent Progr. Hormone Res.* **9**, 213.
RUBIN, B. L., ROSENKRANTZ, H., DORFMAN, R. I. and PINCUS, G. (1953b), *J. clin. Endocrin.* **13**, 568.
RUCKSTUHL, J. P. and MEYER, K. (1957), *Helv. chim. Acta* **40**, 1270.
RULIFFSON, W. S., LANG, H. M. and HUMMEL, J. P. (1953), *J. biol. Chem.* **201**, 839.
RUSHBROOKE, G. S. (1938), *Proc. Roy. Soc. A.* **166**, 296.
RUTENBERG, A. M., GOFSTEIN, R. and SELIGMAN, A. M. (1950), *Cancer Res.* **10**, 113.
RUZICKA, L. and PRELOG, V. (1943), *Helv. chim. Acta* **26**, 975.
SAIFER, A. and ORESKES, I. (1953), *Analyt. Chem.* **23**, 1539.
SAKAL, E. H. and MERRILL, E. J. (1953), *Science* **117**, 451.
SANDBERG, A. A., SLAUNWHITE, Jr., W. R. and ANTONIADES, H. N. (1957), *Recent Progr. Hormone Res.* **13**, 209.
SANGER, F. (1945), *Biochem. J.* **39**, 507.
SANGER, F. (1952), *Advanc. Protein Chem.* **7**, 1.
SANGER, F. (1955), *Bull. Soc. Chim. biol., Paris.* **37**, 23.
SANNIÉ, C., HEITZ, S. and LAPIN, H. (1951), *C.R. Acad. Sci., Paris.* **233**, 1670.
SANNIÉ, C. and LAPIN, H. (1952a), *Bull. Soc. chim., France*, 1080.
SANNIÉ, C. and LAPIN, H. (1952b), *C.R. Acad. Sci., Paris* **235**, 581.
SAVARD, K. (1953), *J. biol. Chem.* **202**, 457.
SAVARD, K. (1954), *Recent. Progr. Hormone Res.* **9**, 185.
SAVARD, K. (1959), *Ciba Found. Colloq. Endocrin.* **11**, 252.

SAVARD, K., WOTIZ, H. W., MARCUS, P. and LEMON, H. M. (1953), *J. Amer. chem. Soc.* **75**, 6327.
SAVARY, P. and DESNUELLE, P. (1953), *Bull. Soc. chim., France*, 939.
SCATCHARD, G., WOOD, S. E. and MOCHEL, J. M. (1939), *J. phys. Chem.* **43**, 119.
SCHAUER, M. K. and BURLISCH, R. (1955), *Z. Naturf.* **10b**, 683.
SCHENKER, E., HUNGER, A. and REICHSTEIN, T. (1954), *Helv. chim. Acta* **37**, 680.
SCHINDLER, O. and REICHSTEIN, T. (1951), *Helv. chim. Acta* **34**, 108.
SCHLOGL, K. and SIEGEL, A. (1953), *Hoppe-Seyl. Z.* **292**, 263.
SCHMIDLIN, J., ANNER, G., BILLETER, J. R., HEUSLER, K., UEBERWASSER, H., WIELAND, P. and WETTSTEIN, A. (1957), *Helv. chim., Acta* **40**, 1438.
SCHMIDT, H., STAUDINGER, HJ. and BAUER, V. (1953), *Biochem. Z.* **324**, 128.
SCHMIDT, H. and STAUDINGER, HJ. (1954), *Biochem. Z.* **325**, 148.
SCHNEIDER, J. J. and LEWBART, M. L. (1956), *J. biol. Chem.* **222**, 787.
SCHNEIDER, J. J. and LEWBART, M. L. (1959), *Recent Progr. Hormone Res.* **15**, 201.
SCHROEDER, W. and VOIGHT, K. K. (1955), *Rec. Trav. chim. Pays-Bas.* **74**, 603.
SCHROTER, H., TAMM, C., REICHSTEIN, T. and DEULOFEU, V. (1958), *Helv. chim. Acta* **41**, 140.
SCHWARZ, V. (1953), *Biochem. J.* **53**, 148.
SEEBECK, E. and REICHSTEIN, T. (1943), *Helv. chim. Acta* **26**, 536.
SEKI, T. (1958), *Nature, Lond.* **181**, 768.
SEKI, T. (1959), *J. Chromatog.* **2**, 667.
SHOPPEE, C. W. and SUMMERS, G. H. R. (1952), *J. chem. Soc.* 3361.
SHORT, R. V. (1957), *Ciba Found. Colloq. Endocrin.* **11**, 362.
SHORT, R. V. (1960), *Mem. Soc. Endocrin.* **8**, 92.
SHULL, G. M., SARDINAS, J. L. and NUBEL, R. C. (1952), *Arch. Biochem. Biophys.* **37**, 186.
SHULL, G. M. (1954), quoted by NEHER, R. (1959a), ref. no. 345.
SHUTE, J. B. (1953), quoted by CASSIDY, A. G. (1957), ref. no. 831.
SILBER, R. H. (1955a), *Clin. Chem.* **4**, 234.
SILBER, R. H. (1955b), *Clin. Chem.* **4**, 278.
SILBER, R. H. and BUSCH, R. D. (1956), *J. clin. Endocrin.* **16**, 1333.
SILBERMAN, H. and THORP, R. H. (1953), *J. Pharm., Lond.* **5**, 438.
SILBERMAN, H. and THORP, R. H. (1954), *J. Pharm., Lond.* **6**, 546.
SIMPSON, S. A. and TAIT, J. F. (1953), *Mem. Soc. Endocrin.* **2**, 9.
SIMPSON, S. A., TAIT, J. F., WETTSTEIN, A., NEHER, R., VON EUW, J., SCHINDLER, O. and REICHSTEIN, T. (1954), *Helv. chim. Acta* **37**, 1163.
SIPERSTEIN, M. D., HAROLD, F. M., CHAIKOFF, I. L. and DAUBEN, W. G. (1954), *J. biol. Chem.* **210**, 181.
SJÖVALL, J. (1952), *Acta chem. scand.* **6**, 1552.
SJÖVALL, J. (1954), *Acta chem. scand.* **8**, 339.
SJÖVALL, J. (1955), *Arkiv. Kemi* **8**, 299, 317.
SMITH, C. G. (1958), *J. Bact.* **75**, 577.
SMITH, E. D. and LE ROSEN, A. L. (1951), *Analyt. Chem.* **23**, 732.
SMITH, I. (1958), *Chromatographic Techniques*, London, Heinemann.
SOKOLSKI, W. T., EILERS, N. J. and SIMONOFF, P. (1957–58), *Antibiot. Ann.* 119.
SOLOMON, S., LANMAN, J. T., LIND, J., and LIEBERMAN, S. (1958), *J. biol. Chem.* **233**, 1084.

SOLOWAY, A. H., DEUTSCH, A. S. and GALLAGHER, T. F. (1953), *J. Amer. chem. Soc.* **75**, 2356.
SOLOWAY, A. H., CONSIDINE, W. J., FUKUSHIMA, D. K., and GALLAGHER, T. F. (1954), *J. Amer. chem. Soc.* **76**, 2941.
SOMMERVILLE, I. F. and DESHPANDE, G. N. (1958), *J. clin. Endocrin.* **18**, 1223.
SPITERI, J. (1954), *Bull. Soc. Chim. biol., Paris* **36**, 1355.
STEIGER, M. and REICHSTEIN, T. (1938), *Helv. chim. Acta* **21**, 546.
STEIN, W. H. and MOORE, S. (1948), *J. biol. Chem.* **176**, 337.
STERN, M. I. and SWYER, G. I. M. (1952), *Nature, Lond.* **169**, 796.
STERNBERG, J. C. and CARSON, L. M. (1959), *J. Chromatog.* **2**, 53.
STIMMEL, B. F. (1946), *J. biol. Chem.* **162**, 99.
STIMMEL, B. F., RANDOLPH, J. D. and CONN, W. M. (1952), *J. clin. Endocrin.* **12**, 371.
STOLL, A., ANGLIKER, E., BARFUSS, F., KUSSMAUL, W. and RENZ, J. (1951), *Helv. chim. Acta* **34**, 1460.
STOLL, A. and KREIS, W. (1951), *Helv. chim. Acta* **34**, 1431.
STRAIN, H. H. (1942), *Chromatographic Adsorption Analysis*, New York, Interscience.
STRAIN, H. H. (1948), *J. Amer. chem. Soc.* **70**, 588.
SVENDSEN, A. B. and JENSEN, K. B. (1950), *Pharm. Acta Helv.* **25**, 241.
SWEET, M. H. (1946), *Electronics*, November, 105.
SWYER, G. I. M. and BRAUNSBERG, H. (1951), *J. Endocrin.* **7**, lx.
SYKORA, V. and PROCHAZKA, Z. (1953), *Chem. Abstr.* (1954), **48**, 3852e.
SYNGE, R. L. M. (1946), *Analyst* **71**, 266.
SZEGO, C. M. (1956), *Ciba Found. Colloq. Endocrin.* **11**, 286.
TAIT, S. A. and TAIT, J. F. (1960), *Mem. Soc. Endocrin.* **8**, 40.
TAKAMATSU, M. and NAKAGI, S. (1958), *Reports of Inst. for Science of Labour*, p. 1.
TAYLOR, W. (1954), *Biochem. J.* **56**, 463.
TEICH, S., ROGERS, J., LIEBERMAN, S., ENGEL, L. L. and DAVIS, J. W. (1953), *J. Amer. chem. Soc.* **75**, 2523.
TEITELBAUM, C. L. (1958), *J. org. Chem.* **23**, 646.
TENNENT, D. M., WHITLA, J. B. and FLOREY, K. (1951), *Analyt. Chem.* **23**, 1748.
THOMPSON, A. R. (1951), *Austral. J. sci. Res.* **B4**, 180.
THORNTON, M. H., KRAYBILL, M. R. and BROOME, F. K. (1941), *J. Amer. chem. Soc.* **63**, 2079.
TOKUICHIVO SEK (1958), *Nature, Lond.* **181**, 768.
TOUCHSTONE, J. C. and CHIEN TIEN HSU (1955), *Analyt. Chem.* **27**, 1517.
TRAPPE, W. (1940a), *Biochem. Z.* **305**, 50.
TRAPPE, W. (1940b), *Biochem. Z.* **306**, 316.
TRUEBLOOD, K. N. and MALMBERG, E. (1950), *J. Amer. chem. Soc.* **72**, 4112.
TSCHESCHE, R., GRIMMER, G. and SEEHOFER, F. (1953), *Chem. Ber.* **86**, 1235.
TSCHESCHE, R. and PETERSON, R. (1954), *Chem. Ber.* **87**, 1719.
TSCHESCHE, R. and SEEHOFER, F. (1954), *Chem. Ber.* **87**, 1108.
TSUKAMOTO, T., KAWASAKI, T., NARAKI, A. and YAMAUCHI, T. (1954), *J. Pharmacol. Soc. Japan* **74**, 1097.
TURNER, R. B. (1950), *J. Amer. chem. Soc.* **72**, 579.

UDENFRIEND, S. (1950), *J. biol. Chem.* **187**, 65.
URSCHELER, H. R., TAMM, CH. and REICHSTEIN, T. (1955), *Helv. chim. Acta* **38**, 883.
VAN DUIN, H. (1952), *Biochim. biophys. Acta* **9**, 580.
VAN DUIN, H. (1953a), *Biochim. biophys. Acta* **10**, 198.
VAN DUIN, H. (1953b), *Biochim. biophys. Acta* **12**, 489, 490.
VAN LAAR, J. J. (1894), *Z. physik. Chem.* **15**, 457.
VEALL, N. and VETTER, H. (1958), *Radioisotope Techniques in Clinical Research and Diagnosis*, London, Butterworth.
VENNING, E. H. (1937), *J. biol. Chem.* **119**, 473.
VENNING, E. H. (1938), *J. biol. Chem.* **126**, 595.
VENNING, E. H. and BROWNE, J. S. L. (1937), *Endocrinology* **21**, 711.
VESTERGAARD, P. (1960a), *J. Chromatog.* **3**, 554.
VESTERGAARD, P. (1960b), *J. Chromatog.* **3**, 560.
VLITOS, A. J. (1953), *Contr. Boyce Thompson Inst.* **17**, 127.
VOGEL, A. I. (1956), *A Textbook of Practical Organic Chemistry* (3rd Ed.), London, Longmans and Green.
VOGT, M. (1955), *J. Physiol.* **130**, 601.
VOGT, M. (1960), *Mem. Soc. Endocrin.* **8**, 95.
VON ARX, E. and NEHER, R. (1956), *Helv. chim. Acta* **39**, 1664.
VON EUW, J. and REICHSTEIN, T. (1940), *Helv. chim. Acta*, **23**, 1258.
WACHTMEISTER, C. A. (1951), *Acta chem. scand.* **5**, 976.
WALENS, H. A., TURNER, A. and WALL, M. E. (1954), *Analyt. Chem.* **26**, 325.
WALLIS, E. S., FERNHOLZ, E., GEBHART, F. T., CHAKRAVORTY, P. and FORD, E. G. (1938), *J. Amer. chem. Soc.* **60**, 413.
WEICHSELBAUM, T. E. and MARGRAF, M. W. (1957), *J. clin. Endocrin.* **17**, 959.
WEICHSELBAUM, T. E., MARGRAF, M. W. and ELMAN, R. (1953), *Fed. Proc.* **12**, 287.
WEST, C. D., HOLLANDER, V. P., KRITCHEVSKY, T. H. and DOBRINER, K. (1952), *J. clin. Endocrin.* **12**, 915.
WIELAND, T. (1942), *Hoppe-Seyl. Z.* **273**, 24.
WIELAND, T. and KRACHT, W. (1957), *Angew. Chem.* **69**, 172.
WIEME, R. J. (1958), *J. Chromatog.* **1**, 166.
WHITEHEAD, J. K. (1958), *Biochem. J.* **68**, 653.
WILSON, H. (1954), *Arch. Biochem. Biophys.* **52**, 217.
WILSON, H., BORRIS, J. J. and GARRISON, M. M. (1958), *J. clin. Endocrin.* **18**, 643.
WILSON, H. and FAIRBANKS, R. (1954), *Arch. Biochem. Biophys.* **54**, 440.
WINTERINGHAM, F. P. W., HARRISON, A. and BRIDGES, R. G. (1950), *Nature, Lond.* **166**, 999.
WOLFF, K. (1955), *Pharmazie* **10**, 371.
WOOD, S. E. and STRAIN, H. H. (1954), *Analyt. Chem.* **26**, 260.
WOOLF, L. I. (1958), Private communication.
WOOLF, L. I. (1959), *J. Chromatog.* **2**, 431.
WOOTTON, I. D. P. and OSBORN, E. C. (1960), *Mem. Soc. Endocrin.* **8**, 70.
ZAFFARONI, A. (1950), *J. Amer. chem. Soc.* **72**, 3828.
ZAFFARONI, A. (1953), *Recent Progr. Hormone Res.* **8**, 51.
ZAFFARONI, A. and BURTON, R. B. (1951), *J. biol. Chem.* **193**, 749.
ZAFFARONI, A., BURTON, R. B. and KEUTMANN, E. H. (1950), *Science* **111**, 6.

REFERENCES

ZAFFARONI, A., RINGOLD, H. J., ROSENKRANZ, G., SONDHEIMER, F., THOMAS, G. H. and DJERASSI, C. (1958), *J. Amer. chem. Soc.* **80**, 6110.
ZANDER, J., FORBES, T. R., NEHER, R. and DESAULLES, P. (1957), *Klin. Wschr.* **35**, 143.
ZANDER, J., ZIMMER, H., MÜNSTERMAN, A. M. and MARX, E. (1954), *Klin. Wschr.* **32**, 529.
ZIJP, J. W. H. (1955), *Chem. Weekblad.* **51**, 547.
ZDERIC, J. A., CARPIO, H. and DJERASSI, C. (1959), *J. org. Chem.* **24**, 909.
ZONDEK, B. and FINKELSTEIN, M. (1954), *Acta endocr., Copenhagen* **17**, 452.

NAME INDEX

Those names followed by *r* are only referred to in Table 2,1 and/or the Reference List

Abelson, D., 186, 213, 319, 343, 381
Ackerman, B. J., 70, 71
Acklin, W., 96
Aebi, A., 39
Ahrens, E. H., 99
Aitken, E. H., 208
Albert, A., *r*
Allen, W. M., 192
Allen, W. S., 361, 363
Alm, R. S., 41, 153
Amendolla, C., 272, 273, 365, 369
Anderson, E. J., 336
Angliker, E., *r*
Anner, G., 43
Antoniades, H. N., *r*
Antonucci, R., 269, 361, 363
Appleby, J. I., 112, 203, 365, 366
Arbouzov, B. A., 39
Arroyave, G., 297
Ashby, D. R., 64
Ashman, P. B., 234
Atherden, S. M., 286
Auterhoff, H., *r*
Avivi, D., 206, 252
Axelrod, L. R., 30, 66, 70, 100, 148, 177, 231, 294, 297, 299, 314, 377, 382
Ayres, P. J., 37, 39, 205, 207, 224, 236, 248, 253, 312, 317, 319, 320, 321, 323, 343, 379

Bahn, R. C., 225
Baggett, B., 42, 247, 251, 358
Baird, C. W., 312, 340, 359
Baker, L. R., 236
Baker, P. B., *r*
Balant, C. P., 297
Balfour, W., 380
Balston, J. N., 158

Banes, D., *r*
Barfuss, F., *r*
Barger, G., 379
Barlow, J. J., 43, 329
Barrer, R. M., 13
Barrollier, J., 241
Barron, E. S. G., 236, 239
Barry, M. C., 247
Barton, D. H. R., 19, 25, 34, 92, 95, 96, 115, 122, 304, 382
Barton, G. M., *r*
Basch, A., 84
Bassil, G. T., 379
Bastien, R. D., *r*
Bate-Smith, E. C., 11, 19, 24, 25, 44, 74, 75, 83, 84, 102, 115, 119, 281, 308, 350
Bauer, C. D., 38
Bauer, V., *r*
Bauld, W. S., 312, 353, 353
Baumann, D., 96
Beatty, J., 213
Beer, C. T., *r*
Bergner, K. G., 94
Bergstrom, S., *r*
Berliner, D. L., 249, 250, 251, 252, 254
Bernal, J. D., 12, 17, 119
Bernasconi, R., *r*
Berner, P., 374
Bernstein, S., 269, 297, 326, 361
Beroza, M., *r*
Beyreder, J., *r*
Bialkowski, H., 30
Bibile, S. W., 194
Billeter, J. R., 43
Bitman, J., 369
Blatt, J. L., 236
Blindenbacher, F., 370
Bloch, H. S., 344

NAME INDEX

Block, R. J., 222, 229, 232, 233, 236, 241, 382
Boggs, L. A., 185
Bojeson, E., 251, 252
Boldingh, J., 61, 63, 162
Bollier, M. E., 65
Bolliger, H. R., r
Bondy, P. K., 186, 194, 312, 319, 343, 381
Bongiovanni, A. M., 60, 96, 143, 320, 331, 340, 344
Borcherds, D., 213
Boris, J. J., r
Borman, A., 291
Bosch, L., r
Boscott, R. J., 379
Bowers, A., r
Boylen, J. B., 185
Brachi, M., 192
Bradfield, A. E., 25
Bradlow, H. L., 247, 337
Brante, G., 379
Braunsberg, H., r
Bray, H. G., r
Bremner, J. M., 26
Bridges, R. G., 61
Briegleb, G., 21
Brink, N. G., r
Brockmann, H., 352
Brooker, R. M., r
Brooks, R. V., 117n, 139, 199, 224, 323, 337, 339, 341, 359, 362, 367, 379
Brooks, S. G., 79, 84, 94, 95, 121, 122, 123, 131, 295
Broome, F. K., 191
Brown, J. A., 224, 235
Brown, J. B., 37, 196, 204, 217, 286, 312, 314
Browne, J. S. L., 344
Buck, R. R., 225
Bulbrook, R. D., 370
Burlisch, R., 177
Burnham, H. D., 25
Burstein, S., 167, 193, 198, 199, 272, 331, 337 370
Burtner, H. J., 225
Burton, R. B., 42, 60, 62, 113, 170, 259, 283, 358, 364, 374, 379

Busch, R. D., 311
Bush, I. E., 1, 3, 4, 30, 37, 39, 42, 44, 60, 61, 66, 84, 99, 101, 113, 115, 125, 131, 139, 142, 143, 158, 162, 163, 175, 180, 182, 194, 198, 199, 208, 214, 216, 218, 220, 223, 224, 225, 228, 229, 231, 233, 234, 235, 237, 239, 249, 256, 260, 262, 268, 269, 271, 275, 283, 286, 291, 294, 297, 298, 299, 312, 329, 331, 332, 333, 335, 336, 337, 339, 340, 342, 344, 345, 356, 357, 359, 360, 361, 363, 366, 372, 375, 376, 377, 379, 381
Butler, J. A. V., 11, 12, 13, 17, 22
Butt, W. R., 4, 25, 60, 68, 74, 137, 150, 192, 353

Cahnmann, H. J., 353
Callow, N. H., 37, 297, 326, 356
Callow, R. K., 37, 100, 297
Calvin, M., 63
Camber, B., 378
Campbell, P. N., 146, 147
Carol, J., r
Carpio, H., 371
Carson, L. M., 157, 220
Carstensen, H., 31, 127
Carter, P., 32, 337
Cassidy, A. G., r
Cassidy, H. G., 1, 43, 70, 71, 73, 158, 289
Cavina, G., 329
Cercero, E., 215, 345, 363
Chaikof, I. L., r
Chakravorty, P., r
Charney, W., 360
Charollais, E. J., r
Chen, C., r
Chevreul, M. E., r
Chibnall, A. C., 201, 220
Chien Tien Hsu, 230
Claeson, S., 153
Clark, I., r
Clayton, R. A., 73
Clemente, J., r
Cohen, S. L., 198, 344, 369, 371
Colas, A., 286

NAME INDEX

Collins, R. W., 362
Conn, W. M., 37
Consden, R., 4, 60, 68, 124, 162, 183, 185
Considine, W. J., 39
Constandse, M., 37
Conway, E. J., 134
Cook, E. R., 61
Cook, R. P., *r*
Cookson, R. C., 25, 92, 96
Cope, C. L., 171, 224, 225
Cosgrove, J., *r*
Cox, J. D., 95
Cox, J. S. G., 288
Cox, R. I., 173, 340
Craig, D., 6
Craig, L. C., 6, 99
Crane, K. F., 215
Crépy, O., 43, 329
Crook, E. R., 235
Cuendet, R. S., *r*

Dao, T. L., *r*
Darby, H. H., 314
Dasler, W., 38
Datta, S. P., 162
Dauben, W. G., *r*
David, K., 192
Davies, R. E., *r*
Davis, J. W., 369
Davis, R. B., 63
Deal, C. A., 12, 157
De Courcy, C., 216, 262, 299, 320, 346, 381
Dell, B., 61
Demy, E., 358
Derr, E. L., 12
Desaulles, P., *r*
Deshpande, G. N., 194, 316
Desnuelle, P., 371
Desty, D. H., 117, 328
Deulofeun, V., *r*
Deutsch, A. S., 366
De Young, D. D., *r*
Diaz, G., 297
Dickson, D. H. W., 100
Dimler, R., 26

Dingemanse, E., 37, 352, 369
Dixon, H. B. F., 80, 82
Djerassi, C., 268, 288, 297, 371
Dobriner, K., 36, 37, 39, 40, 43, 151, 247, 324, 363
Dobson, F., *r*
Dodgson, K. S., 335, 337
Dohan, F. C., 337
Dominguez, O. V., 249
Donaldson, K. D., 153
Dorfman, L., 297
Dorfman, R. I., 30, 272
Dowlatabadi, H., 168
Drake, N. A., 373
Dunn, M. S., 236
Durrum, E. L., 241, 382
Dutta, G. L., *r*

Eberlein, W., 60, 96, 320, 331, 344
Edgar, D. G., *r*
Edwards, R. H. W., 43, 170, 171, 192, 330, 335, 336
Ehrenstein, M., 297, 369
Eidinoff, M. L., *r*
Eik-Nes, K., 196, 342
Eilers, N. J., 213
Eldridge, K. E., 191
Elks, J., 100
Ellis, B., 371
Elman, R., 208
Emmens, C. W., 297
Engel, L. L., 32, 42, 247, 251, 337, 358, 369
Eppstein, S. H., 283
von Euw, J., 61, 291, 368
Evans, R. H., 100
Evans, R. M., *r*
Evans, R. S., 382
Evelyn, S. R., *r*

Fairbanks, R., 342
Feigl, F., 374
Fernholz, E., *r*
Field, E., 379
Fielding, L. L., *r*
Fieser, L. F., 36, 273, 360, 364, 365, 367

Fieser, M., 360, 365
Fink, K., 22
Fink, R. M., 22
Finkelstein, M., 173, 336
Firor, W. H., r
Fisher, R. B., 372, 373
Florey, K., 62
Folkertsma, J. P., 213
Forbes, T. R., r
Ford, E. G., r
Fotherby, K., 286
Fowler, R. H., 12, 17, 119
Fresco, J. R., 206
Freud, J., r
Fried, J., 268, 276, 286, 291, 360, 361, 368
Friedgood, M. B., 61
Frierson, W. J., 188, 190
Fujimoto, G. I., 248
Fujita, H., 70
Fukushima, D. K., 37, 39, 247, 248, 337, 340

Gale, M. M., 196, 225, 294, 329, 332, 334, 336, 357, 368, 369, 370
Gallagher, T. F., 39, 174, 192, 247, 248, 337, 366
Gardner, J. A. F., 382
Garrison, M. M., r
Garrod, O., 37
Garst, J. B., 61
Gebhart, F. T., r
Genest, J., 207
Gerold, C., r
Geyer, R. P., r
Ghillain, A., r
Gibson, G., 203
Gilford, S. R., 241
Glasstone, S., 118
Glenn, E. M., 207
Glueckauf, E., 1, 6, 70, 80, 81, 250
Gofstein, R., 357
Goldup, A., 117
Goppelsroeder, F., 3
Gordon, A. H., 4, 60
Gordon, B. E., 25

Gornall, A., 363
Gowenlock, A., 224, 379
Grabowich, P., 291
Grafflin, M. W., 131
Grassman, W., 234
Gray, C. H., 262, 320
Green, F. C., 40, 41
Green, M. A. S., r
Gregory, G. F., 185
Grimmer, G., 62
Grollman, A., 192
Grollman, E., r
Gross, A. M., 236
Grundy, H. M., 27
Guerra, S. L., r
Gunzel, C., r
Gut, M., r

Habermann, E., r
Haenni, E. D., r
Haenni, E. O., r
Haines, W. J., 153, 185, 372, 373
Ham, E. A., r
Hamilton, J. G., 42
Hamlet, J. C., r
Hanes, C. S., 4, 32, 160, 162, 173, 331, 354
Harold, F. M., r
Harman, R. E., r
Harris, M., 235
Harrison, A., 61
Harrower, D., 11
Hartel, J., 63, 64, 162
Hartley, F., 371
Hartogh-Katz, S. L. U., 37
Haskins, F. A., 137, 186
Haslewood, G. A. D., 92, 170, 283, 300, 363, 379
Hassall, C. H., 187
Hassan, F., 235
Hayden, A. L., 60, 374, 381
Hechter, O., r
Heftmann, E., 37, 44, 60, 62, 153, 373, 374, 381
Hegedüs, H. M., 136, 149, 171
Heilman, J., 241

NAME INDEX

Heitz, S., r
Heller, M., 26
Hellman, K., 262
Hellman, L., r
Helmreich, M. L., 186
Henry, R., 37, 337, 360, 370
Herling, F., 297
Hermans, P. H., 3
Hermann, W. R., 340
Hershberg, E. B., 364
Herzig, P. G., 369
Heusghem, C., r
Heusler, K., 43
High, L. B., 288
Hildebrand, J. H., 11, 17, 18
Hilgar, A. C., 337
Hill, B. R., 36, 198
Hirayama, O., 21
Hirschmann, H., r
Höber, R., 6
Hockaday, T. D. R., r
Hodgson, H. H., 131
Hofman, H. 62, 222, 225, 314
Hollander, N., 324
Hollander, V. P., 324
Holman, R. T., 42
Holness, M. J., r
Howard, G. A., 61, 63, 68, 100, 146, 157, 354
Hubener, H. J., 62
Huber, H., 370
Huffman, M. N., r
Hughes, I., 319
Huis in 'T Veld, L. G., 37
Hummel, J. P., 252
Hunger, A., 352
Hunt, C. A., 231
Hunt, J. S., 79
Hunter, L., 17
Hunter, S., 359
Hurlock, B., 225, 283

Inouye, Y., 21, 65, 164, 165
Iriarte, J., r
Isaera, Z. G., 39
Isherwood, F. A., 4, 26, 32, 60, 61, 124, 160, 162, 173, 219, 295, 331, 354, 374

Izzo, A. J., 351, 357

James, A. T., 23, 129, 157, 201, 220, 225, 288, 289, 290
James, V. H. T., 100
Jaminet, E., r
Jarrige, P., 370
Jayle, M. F., 43, 329
Jeanes, A., 26, 173
Jeffries, A., 260
Jenkins, D., 335
Jensen, J. B., 252
Jensen, J. P., 251, 252
Jensen, K. B., 381
Jermyn, M. A., 26, 60, 124, 173, 219
Jirgl, V., r
Johnson, D. F., 37, 153
Jones, E. R. H., 288
Jones, J. E., r
Jones, J. K. N., 62
Jones, L. C., 25
Jones, M. E., r
Jones, P. G., r
Jones, R. N., 207, 297

Kabasakalian, P., 84
Kaiser, F., r
Kalnitsky, G., 63
Kambeitz, J., 21
Kamil, I. A., 335
Kandrac, M. S., 340
Kannig, K., 234
Karnemaat, J. N., r
Katzenellenbogen, E., 40, 191, 363
Kawasaki, T., r
Kay, L. M., 40, 41
Keller, G. J., 162
Kellie, A. E., 38, 43, 153, 199, 216, 231, 330, 335
Kendall, E. C., 192
Kendall, F. E., 92, 99
Kenton, R. H., 26
Kessler, W. B., 291
Keston, A. S., 206, 251
Keulemanns, A. I. M., 81
Keutmann, E. H., 42, 60, 62
Kirchner, J. G., 162

Kirk, M. R., 63, 162, 299, 378, 379
Kirkwood, J. G., 11, 13
Kiss, P., r
Klaner, A., r
Kliman, B., 65, 252
Klopper, A., 204, 205, 217, 344
Klyne, W., 120, 207, 261, 264, 309, 359, 376
Knauff, R. E., 373
Koch, F. C., 192
Kochakian, C. D., 232, 377
Kodicek, E., 64
Koiw, E., 207
Kollowski, M. A., r
Koštír, J. V., 63, 162
Kotasek, Z., r
Kowkabany, G. N., 73, 158
Kracht, W., 300
Kraybill, H. R., r
Kraybill, M. R., 191
Kreis, W., r
Kress, O., 30
Kritchevsky, D., 63, 162, 299, 378, 379
Kritchevsky, T. H., 40, 61, 63, 162, 324, 337
Kucera, J., r
Kuhn, R., 378
Kumler, W. D., 34
Kushinsky, S., 34
Kussmaul, W., r

Laat, 37
Labler, L., r
Lakshmanan, T. K., 37, 38, 41, 153, 158, 352
Lang, H. M., 252
Langmuir, I., 2
Lanman, J. T., 250
Lapin, H., r
Laqueur, E., r
Lauer, K., 71
Laurence, D. J. R., 234, 235, 241
Layne, D. S., 312, 314, 327
Lederer, E., 33, 39, 151, 157, 173, 382
Lederer, M., 33, 39, 79, 151, 157, 173, 382

Leeds, N. S., 337
Leegwater, D. C., 206
Lemon, H. M., r
Lenhard, R. H., 269, 297, 326, 361
Le Rosen, A. L., 10, 11
Levant, A. J., r
Levy, H., 34, 39
Levy, M., 206
Lewbart, M. L., 60, 131, 317, 320, 322, 323, 328, 329, 330, 331, 332, 354, 355
Lewis, B., 143, 160, 186, 312
Lewis, J. L., 198
Liberman, L. A., 164
Liberti, A., 157
Lichti, H., 39, 138
Lieberman, S., 36, 37, 38, 39, 41, 43, 151, 153, 158, 198, 199, 250, 328, 331, 352, 363, 364, 369, 370
Lind, J., 250
Linberg, M. C., 269
Linden, C. E., r
Lindner, H. R., 200
Littel, R., 269, 361
Llaurado, J. G., r
Lloyd, D. G., r
Loke, K. H., 94, 276, 312
London, F., 119
Long, A. G., 79, 100
Longley, J. B., 225
Loomeijer, F. J., 73, 163, 327
Lott, M. H., 314
Lovelock, J., 201
Lover, M. J., r
Löw, J., 378
Luetscher, J. A., r
Lugg, J. W. H., 29, 32
Luinge, G. M., 73, 163, 327
Lunnon, J. B., 262

MacDonald, M. P., 363
MacPhillamy, H. A., 248
Mader, W. J., 225
Mahesh, V. B., 131, 139, 142, 218, 224, 228, 233, 260, 271, 275, 283, 286, 291, 298, 336, 337, 339, 340, 341, 344, 345, 361, 366, 375, 379

NAME INDEX

Malassis, D., 43, 329
Malmberg, E., *r*
Mancera, O., 365
Marcus, P., *r*
Margraf, M. W., 208, 378
Markees, S., 361, 362
Markham, R., 372, 373
Markwardt, F., *r*
Maron, D. M., 61
Marrian, G. F., 94, 276, 286, 312, 314, 327, 340
Marsden, E., 131
Marsh, M. M., 224, 235
Marshall, L. M., 153
Martin, A. J. P., 1, 3, 4, 6, 7, 8, 9, 10, 15, 25, 30, 42, 43, 44, 60, 61, 62, 68, 81, 82, 83, 95, 100, 102, 129, 145, 146, 157, 173, 201, 281, 288, 289, 290, 328, 353, 354
Martin, R. P., 61, 64, 163, 164, 165, 192, 299, 316, 327, 378
Martin, S. L., 187
Marx, E., *r*
Mason, H. L., 349
Mason, W., 185, 261
Mattox, V. R., 206, 317, 320, 322, 323, 354, 355, 363
Mavais, E. J., *r*
Mayer, S. W., 80, 82
McAlfer, W. J., *r*
McCall, P. J., *r*
McFarren, E. F., 32, 236
McKenna J., 370
McMahon, J. M., 63
Meigh, D. F., 21
Meister, P. D., 283
Merrill, E. J., 62
Meslin, F., 43, 329
Meyer, A. S., 34, 153, 269, 272, 273, 297, 359, 360
Meyer, K., 39
Meyerheim, G., 62
Meystre, C., 174
Michalec, C., *r*
Micheel, F., *r*
Michie, E. A., 204
Miethinen, J. K., 246

Migeon, C., 194
Miller, E., 359
Miller, H. K., 374
Mills, J. A., *r*
Mills, J. S., 61, 64, 163, 164, 165, 236, 288, 327
Mitchell, F. L., 127, 299, 381
Mitchell, H. K., 137, 186
Mochel, J. M., 16
Moelwyn-Hughes, E. A., 11, 18, 128
Mohr, K., 370
Moisio, T., 246
Monaghan, P. H., 10
Montgomery, R., *r*
Moolenaar, A. J., 207, 323
Mooney, B., 79
Moore, A. N., 185
Moore, S., 43, 157, 220
Morris, C. J. O. R., 4, 25, 60, 63, 153, 192, 219, 229, 230, 312, 363
Morris, P., 4, 25, 60
Morrison, G. A., 372
Morrison, G. H., *r*
Mosbach, E. H., 92, 99
Mosettig, E., 374
Moss, P., *r*
Mueller, J. H., 190
Müller, W., *r*
Munier, R., 32
Munson, P. L., *r*
Münsterman, A. M., *r*
Murray, H. C., 283

Nace, H. R., *r*
Nadel, E. M., 272, 337
Nakagi, S., 198
Naraki, A., *r*
Nash, H. A., *r*
Nathanson, E., 32
Neher, R., 31, 33, 34, 37, 38, 39, 44, 61, 62, 65, 66, 68n, 73, 86, 92, 96, 115, 125, 140, 151, 159, 160, 163, 174, 176, 177, 180, 182, 187, 188, 190, 192, 207, 216, 220, 249, 253, 294, 297, 299, 309n, 311, 312, 317, 323, 327, 337, 353, 354, 364, 375, 377, 381, 382
Nelson, D. H., 37, 207, 259, 353

Nicholas, R. E. H., 26
Nielson, E. D., 373
Nierenberg, M., 92
Noda, M., 21
Nolin, B., r
Norman, A., r
Norymberski, J. K., 39, 112, 203, 260, 291, 365, 366, 367, 368, 370
Nowaczynski, W. J., 207, 320, 323
Nubel, R. C., 162
Nyc, T. F., 61

Oertel, G., 185, 196, 342
Oliveto, E. P., 268, 359
Oreskes, I., 73
Osborne, E. C., 196
Ott, E., 131
Oughton, J. F., 100
Overell, B. G., 162
Overell, B. T., 29, 32

Page, J. E., 100
Panattoni, M., 312
Parsons, D. S., 372
Partington, J. R., 257
Partridge, S. M., 61
Patterson, J., 369
Patterson, W. I., 60, 62
Pearlman, M. R. J., 248
Pearlman, W. H., 205, 215, 248, 345, 362, 363
Pechet, M. M., r
Perrin, D. D., 373
Peterson, D. H., 283
Peterson, R., 62, 141, 186
Peterson, R. E., 40, 65, 98, 203, 205, 206, 209, 216, 220, 252, 253, 254, 255, 294, 309, 311, 312, 317, 319, 321, 322, 323
Petrow, V., 371
Pfiffner, J. J., 192
Philip, J. E., 153
Pierotti, G. J., 12, 13, 14, 16, 123, 125, 127, 281, 290
Pincus, G., 30, 37, 362
Pirie, N. W., 259
Plager, J. E., 248

Podzimek, J., r
Polson, A., 236
Porter, C. C., 363
Porter, P. E., r
Potvin, P., r
Preedy, J. R. K., 208
Prelog, V., 96, 371
Price, D., 234
Prieto, A. P., 96
Prochazka, Z., 27, 64, 65, 92, 379
Prout, F. S., 248
Prunty, F. T. G., 356
Puck, A., r

Quaife, M. L., r

Raaen, H., 188
Rakoff, A. E., 248
Ramchandani, C. N., 11, 12, 17
Ramsey, L. L., 60, 62
Randall, S. A., 157
Randolph, J. D., 37
Reddy, W. J., 335
Redfield, R. R., 236, 239
Rees, V. H., 234, 235
Reeves, R. E., 121, 122
Reich, H., 215, 259, 345, 362
Reich, M., 319
Reichl, E. R., 23, 110, 214
Reichstein, T., 34, 36, 38, 39, 43, 61, 144, 151, 171, 174, 192, 221, 286, 291, 299, 300, 352, 359, 362, 363, 368, 370, 376
Reineke, L. M., 31, 62, 66, 67, 68, 73, 89, 92, 174, 180, 215, 220, 249, 283, 290, 298, 327, 342, 377
Renz, J., r
Rettenbacher-Däubner, H., r
Rhoads, C. P., 36, 39
Richardson, E. M., 337
Riddell, C., r
Riegel, B., 248, 365
Rimington, C. R., 26
Ringold, H. J., 268
Rittel, W., 352
Ritter, J. J., 63, 64, 162
Rivet, C. A., 10

Roberts, G., *r*
Rockland, L. B., 236
Rogers, A. W., 382
Rogers, J., 369
Roland, J. F., 236
Romanoff, E. B., 37, 158, 231
Romanoff, L. D., 37, 39
Rosenfield, R. S., *r*
Rosenkrantz, H., *r*
Rosenkranz, G., 30, 268, 272, 297, 365, 369
Roux, I. G., *r*
Roy, A. B., 329
Rubin, B. L., 30, 43, 73, 286, 327, 337
Ruckstuhl, J. D., *r*
Ruliffson, W. S., 252
Rushbrooke, G. S., 11
Rutenberg, A. M., 357
Ruzicka, L., 371

Sabo, E. F., 268, 276, 286, 291, 360, 368
Saifer, A., 73
Sakal, E. H., 62
Salhanick, H. A., 249, 250, 251, 254
Samuels, L. T., 37, 207, 248, 353
Sandberg, A. A., 37, 39, 113, 208, 260, 291, 312
Sanfilippo, S. J., 215
Sanger, F., 281, 286
Sannié, C., *r*
Sardinas, J. L., 162
Sarett, L. H., *r*
Savard, K., 30, 37, 39, 44, 66, 84, 88, 116, 153, 273, 376
Savary, P., 371
Scatchard, G., 16
Schauer, M. K., 117
Schenk, J. R., 153
Schenker, E., *r*
Scheuer, J., 186
Schindler, O., 61, 171, 174, 299, 376
Schlogl, K., *r*
Schmeidler, G. A., 304
Schmidlin, J., 43
Schmidt, H., 62

Schneider, J. J., 60, 131, 328, 329, 330, 331, 332, 335, 363
Schodder, H., 352
Scholz, C. R., 248
Schreglmann, A., *r*
Schroeder, W., 329
Schröter, H., 174
Schwarz, V., 173, 222
Schweppe, H., *r*
Scott, R. L., 11, 17, 18
Seebeck, E., 43, 286
Seehoffer, F., 62
Seki, T., *r*
Seligman, A. M., 357
Sermin, A. I., 39, 365
Sherman, A., 128
Shoppee, C. W., 34, 36, 38, 39, 144, 221, 352
Short, R. V., 127, 173, 174, 194, 213, 312
Shull, G. M., 162
Shute, J. B., *r*
Siegel, A., *r*
Sigg, H. P., *r*
Silber, R. H., 194, 311, 363
Silberman, H., 44
Silverman, L., 340
Simonoff, D., 213
Simpson, S. A., 27, 37, 61, 147, 206, 224, 252, 294, 323, 353, 360
Siperstein, M. D., *r*
Sjövall, J., 29, 99, 163, 164
Skogstrom, P., *r*
Slaunwhite, Jr., W. R., 32
Slavík, K., 63, 162
Smith, C. G., 213
Smith, E. D., 10, 11
Smith, E. R., 216, 231
Smith, F., *r*
Smith, I., 23
Smith, J. D., 372
Smith, J. N., 335
Smith, P., 170, 171, 175, 177, 192, 374
Snyder, J. W., 213
Sokolski, W. T., 213
Solomon, S., 250
Soloway, A. H., 39, 366

NAME INDEX

Sommerville, I. F., 194
Sondheimer, F., 268, 272, 326, 365
Spackman, 157
Speiser, P., 370
Spencer, B., 335, 337
Sperlich, H., 94
Spiteri, J., 61, 65
Springer, M. J., 337
Spurlin, H. M., 131
Stack-Dunne, M., 80, 82, 162
Stark, 157
Staudinger, Hj., 62, 222, 225
Steiger, M., 300
Stein, W. H., 43, 157, 220
Stephenson, L., r
Stern, M. I., 342
Sternberg, J. C., 157, 220
Steyermark, P. R., 207
Stidworthy, G., 232, 377
Stimmel, B. F., 37
Stitch, G. R., 62
Stokem, M. B., 337
Stoll, A., r
Stotz, E., r
Strain, H. H., 37, 38, 41, 71, 328
Stroud, S. W., r
Stubbs, R. D., 203, 365, 367, 368
Studer, P. E., 363
Summers, G. H. R., r
Svendsen, A. B., 381
Svensson, H., 189
Swale, J., 232, 369
Swanton, W. T., 117
Sweet, M. H., 241, 242
Swyer, G. I. M., 342
Sykes, I. F., r
Sykora, V., r
Synge, R. L. M., 1, 3, 4, 6, 15, 25, 29, 32, 43, 44, 61, 81, 82, 145, 289
Szego, C. M., 196
Szell, T.,

Tait, J. F., 27, 37, 42, 43, 61, 80, 82, 129, 131, 150, 205, 206, 219, 220, 223, 224, 226, 248, 249, 251, 252, 254, 298, 320, 322, 323, 360, 380

Tait, S. A., 42, 43, 80, 82, 129, 131, 150, 205, 219, 220, 223, 226, 249, 251, 252, 254, 320, 322
Takamatsu, M., r
Talalay, P., 283
Talbot, B. E., 158
Tamm, Ch., 39
Taylor, A. R., 192
Taylor, D. A. H., 299, 377
Taylor, W., r
Teich, S., 198, 328, 369
Teitelbaum, C. L., 200
Tennent, D. M., 62
Tewell, M. E., r
Thevenet, M., 37, 337, 360, 370
Thomas, A. J., r
Thomas, G. H., 268
Thomason, P. F., 188
Thompson, A. R., 22
Thomson, D. W., 11
Thorn, G. W., 198, 335
Thornton, M. H., 191
Thorp, R. H., 44
Thorpe, W. V., r
Tiselius, A., 41, 61, 63, 162
Tokuichvo Sek, r
Tompkins, E. R., 80, 82
Touchstone, J. C., 230, 337
Trappe, W., 38
Trueblood, K. N., r
Tschesche, R., 62, 377
Tseu, T. K. L., 186
Tsukamoto, T., r
Tulane, V. J., 153
Turner, A., r
Turner, R. B., 248

Undenfriend, S., 42, 206, 257
Ueberwasser, H., 43
Underwood, J. C., 236
Upton, V., 143, 186, 194
Urscheler, H. R., r

Vallee, L., 194
Van Duin, H., 23, 25, 68, 74
Van Laar, J. J., 11
Van Rooy, P. J., r

Vars, H. M., 192
Veall, N., 248
Venning, E. H., 335, 344
Veres, K., r
Verly, W. G., 358
Vestergaard, P., 157, 158, 221
Vetter, H., 248
Vilotti, R., 288
Vischer, E., r
Vlitos, A. J., 329
Vogel, A. I., 347, 348, 349, 350, 351
Vogt, M., 160, 183, 354
Voight, K. K., 329
Von Arx, E., 140, 187, 188, 190, 381

Wachtmeister, C. A., 295
Wade, A. P., 38, 153, 199, 335
Walens, H. A., r
Walker, G., r
Walker, T., r
Wall, M. E., r
Wallis, E. S., r
Wareham, D. J., 61
Warner, R. C., 206
Warren, F. L., 235, 336
Watson, E. J. D., 94, 276, 312
Watzke, E., 241
Wedlake, D., 371
Weichselbaum, T. E., 208, 378
Weiner, J. S., 262
Weiss, F., r
Werner, A. E. A., 61, 64, 163, 164, 165, 327
West, C. D., 324
Westall, R. G., 11, 19, 24, 25, 44, 74, 75, 83, 84, 102, 115, 119, 281, 308, 350
Westenskow, G., 249
Wettstein, A., 43, 61, 62, 66, 68n, 73, 92, 96, 125, 163, 174, 207, 299, 317, 323, 327, 381
Wheeler, J., r
White, K., r
Whitehead, J. K., 206, 252
Whitla, J. B., 62

Wieland, P., 43
Wieland, T., 300, 352
Wieme, R. J., 237, 242
Williams, D. C., 25, 60, 63, 153, 192, 230, 312, 363
Williams, J. H., 269
Williams, R. J. P., 41
Williams, R. T., 335
Willoughby, M. L. N., 61, 113, 115, 162, 180, 199, 216, 220, 223, 224, 225, 228, 229, 233, 239, 260, 268, 269, 291, 299, 337, 340, 342, 356, 360, 375
Wilson, B. M., r
Wilson, H., 269, 298, 342
Winteringham, F. P. W., 61, 64, 162, 163, 164
Wintersteiner, O., 192
Wise, C. S., 26
Wolf, R. S., 37
Wolfe, J. K., 364
Wolff, K., r
Wood, S. E., 16, 71
Woods, G. F., 366
Woolf, L. I., 30, 127
Wootton, I. D. P., 196
Wopat, F., 25
Wotiz, H. W., r
Wright, A. A., 207, 376

Yamauchi, T., r
Young, B. G., 337

Zablow, L., r
Zäch, D., 96
Zaffaroni, A., 42, 60, 62, 140, 183, 186, 199, 259, 261, 265, 268, 283, 297, 299, 300, 301, 358, 364, 374, 379
Zander, J., 316
Zderic, J. A., 371
Zijp, J. W. H., 63
Zimmer, H., r
Zimmerman, B., 344
Zomzely, C., 99
Zondek, B., 336
Zweig, G., 241, 382

Svensson, H., Agrell, C.-E., Dehlén, S.-O. & Hagdahl, L. (1955) *Science Tools*, **2**, A.

SUBJECT INDEX

Absorptiometry
  specificity of, 203
[^{131}I] Acethydrazide-3-iodopyridinium bromide, 252
Acetic acid, 17, 66, 107, 163, 213, 360, 371
  useful properties of, as stationary phase, 167
Acetic anhydride, 291
  tritium-labelled, 205, 215, 252, 254, 319, 321, 326
Acetone, 351
  use with alumina, 38
Acetyl-glycosides, 39
Acetylation, 113, 205, 209, 218, 268, 283, 309, 316, 344, 345, 358
  fast, 358
  forcing, 359
  partial, 291, 359
  slow, 358
Acids, 6, 31
  swamping, 32
Activity, 28
  constant specific, 42
  coefficient, 15
Acylating agents, 63
Adhesive surgical tape, 170
Adsorbent, xv
  "active point" on the, 41
  standardization of, 311
  swelling of, 41
Adsorption, xiv, 2, 15, 31, 32, 125, 191, 336
  isotherm, 10
  mechanism of adsorption systems, 41
  monomolecular layers, 41
  of steroids, in warm air, 140
  on cellulose, 4, 131
  quasi-partition behaviour, 40, 41
  systems, 36
    displacement in, 42

Adsorption—*contd.*
  to paper, 30
  to supporting material, 30
Alcohols, 123, 348
  branched-chain, 14
  free energy of hydration of, 12
  primary, 18
  secondary, 14
  straight-chain, 14
Alicyclic rings, fused, 25
Alkylamines, 26
n-Alkylamines, 23
n-Alkylamino acids, 23
5-Alkylcysteines, 23
n-Alkylguanidino acids, 23
Alumina, xv, 37, 215, 352
  acid, 352
  active, 352
  alkali in, 38, 39
  alkaline, 352
  degradations and isomerizations of steroids on, 38
  hydration of, 38
  hydrolysis of steroid 3-formates by, 39
  neutral, 352
  reaction with acetone, 38
  rearrangements of steroids on activated, 34
  regeneration of, 352
  separation of free α-ketols, 38
Aluminium alloy, expanded, 170
Aluminium silicate, 191
Amines, primary, free energy of hydration of, 12
Amino acids, 30, 60, 73, 124, 129, 201, 220, 229, 252
  acetylated, 4
  basic, 4
  dinitrophenyl, 49
Ammonia, 213

SUBJECT INDEX

Anion-exchange resin, 336
Anthracene, 16, 127, 328
Antibiotics, 213
Aromatic aldehydes, 382
Aromatic rings, 16
  substituents in, 24, 25
Artifact, 151, 153, 162
Atmospheric humidity, 233

Bases, 6, 31, 351
Beer's law, 234
Benzene, 17
Benzenesulphonamides, 286
Benzenesulphonyl chloride, 286
7, 8-Benzoquinoline, 117
6, 8-Benzoquinone, 328
Benzoylation, 359
Bismuthate, 112, 216, 271, 275, 291
Blue tetrazolium, 215, 224, 225, 229, 234, 265, 340, 357, 375
Bonds, double, xix, 37
Borohydride, 112, 291
Bromination, 368
Bromine, 382
n-Butanol, 31, 62
t-Butanol, 60, 66, 107, 124, 348
  solubility of, in water, 123
Butyl acetate, 67, 350

Caffeine, 213
Carbitols, 62, 350
Carbon tetrachloride, 349
Carborundum, 61
Carboxylic acids, 23
Carotenes, 42
Carotenoids, 328
D-Catechin, 25
CASSANDRA, 245
Cation-exchange resins, 336
Celite (kieselguhr), 4, 39, 145
  hydrophobic, 354
  ordinary, 353
  siliconed, 61, 63
Cellosolves, 62, 163, 350
Cellulose, 5, 30, 31, 131, 132, 162, 228, 354

Cellulose—*contd.*
  amorphous regions of, 3
  cotton linters, 354
  crystalline regions of, 3
  hydrolysis of, 230
  native, 3
  powder, 354
  powder columns, 146
  strong adsorption of steroids by, 137
  swelling of, 71, 73, 129
  water in, 3
Charcoal, 153
Chemical potential, 7, 11
Chloramphenicol, 213
  halogenated derivatives of, 213
Chloroform, 67, 349
Choline, 381
Chromatoblock, 187, 188
Chromatogram
  cloth, 3
  dried with warm air, 182
  Durchlauf method; "liquid" or "flowing", 38
  "mixed", 42
  mixed adsorption, 42
  overrun, 75, 79, 219, 263
  paper, xiv, 3, 70
    no-touch technique for preparing, 168
    transfer of material to, 138
    way of running, 173
  single length, 263
  streaking on, 29
  with water as stationary phase, 30
  two-dimensional, 212
Chromatograph
  tanks,
    double length, 174
    operation of, 170
    sealing of, 170
    spacing of, 171
Chromatography,
  adsorption, xiv, 36
  ascending, xiv
  cellulose, 3
  column, 144
  descending, xiv

Chromatography,
  gas–liquid, 117, 129
  multiple in one direction, 173
  negative evidence provided by, 269
  paper, 158
    preparative, 186
    quantitative methods based on, 170
    246
  partition, xiv, 3, 43, 191
  quantitative, 201
  reversed phase partition, xvii
  room, convector for, 176
  straight partition, xvii
  structural analysis and identification
    by, 256
  theory of, 1–32
Chromatopile technique, 137, 186
Chromic acid, 113, 206, 266, 272, 283,
    308, 309, 323, 344
  oxidation of adrenocortical side chain
    with, 364
  oxidations with, 363, 364
Chrysene, 127, 129, 300
Collidine, 276, 286, 351
Colour reagents,
  applied by dipping, 177
  hot solutions of, 176
  see "Scanning"
Column, xiv, 36, 148, 219, 221
  alumina, 344
  Celite, 312
  cellulose, 138, 147
  chromatographic, xv
  design and operation of, 148
  dimensions of, 148
  jacketed, 174
  methods, 250
  partition, 152
    with Celite or silica gel, 189
  preparation of adsorption, 144
  preparation of partition, 145
  rolled paper, 189
  swan neck, 148, 149
  testing the performance of, with dyes,
    150
Condensation
  tanks, on lids of, 69

Configuration
  inversions of, xix
  nor, homo, seco, bisnor, xix
Conditions,
  non-equilibrium, 96
Continuous extraction, 320
Conformation, 34
  flexibility of, 65
Conformational
  free rotation, 10
  inversion, bending, 10
  rigidity, 21
Contaminants, 213, 217, 262, 320, 326
Contamination, 223, 229
Co-ordination, 11, 17
Cotton linters, 354
Counter-current machine, Craig, 31, 99,
    131, 251, 335
Counter-current, theory, 4
  separating funnel concept of, 6
  systems, 127
Crystallization, 1
Cyclic structures, 16
Cyclohexane, 17, 120

D.D.T. derivatives, 45, 64
Decalin, 17, 101, 116, 117, 128, 164,
    167
  derivatives, 96
Degradations,
  Barbier–Wieland, 302
  classical, 301
  see "Microchemical reactions"
Dehydration, 283, 300, 368
Dehydrohalogenation, 286, 368
Detection, methods of, 176, 372
  contact fluoroscopy, 374
  on paper chromatograms, 372
  photography, 372
  see "Colour reagents"
Development, xiv
Diamines, 26
Dicyclohexylbenzene, 16, 25
Diethyl ether, 67
Diffusion, 3
  coefficient, 6
  constants of solutes, 6

Digitonin, 337
Di-isopropyl ether, 67, 332, 333, 335
Dimethylformamide, 350
m-Dinitrobenzene, 215, 228, 231, 233, 265, 271, 275, 356, 375
3, 5-Dinitrobenzoic acid, 299, 377
Dinitrophenyl derivatives, 286
Dinitrophenyl radical, 21
Dinitrophenylhydrazine, 377
Dinitrophenylhydrazones, 45, 63, 215, 217, 345, 361
1, 2-Diols, 291, 295
    complexes with borate, 295
cis/trans-Diols, 49
Dispersion forces, 24, 119
Displacement, 2, 249
    conditions, xiv
    effects, 30, 43, 66
Displacer, xiv, 191
Distribution, coefficient, xiv, 1, 15
    effect of concentration on, 2
    isotherms, liquid–liquid, 27
    linearity of, 28
    liquid–liquid, 3
Double bonds, xix, 37
    see "Groups"
Drugs, interference from, 312
Dyes, 180
    fat-soluble, 180
    used as markers, 151

Edging, 70, 71
Effluent, xiv, xv, 82
    continuous recording of, 15, 71
    profile of, 150
Electron, donation and acceptance, 10
Eluate, xv
    for infrared spectroscopy, 186
Eluent, xv
Elute, xv
Elution, 37, 185, 223, 248
    apparatus, 186
    chromatographic,
        by simple soaking, 183
        solvents to be used for, 183
    from paper, 182, 184

Elution—contd.
    gradient, xv, 41, 70, 153, 155, 158, 208, 221
    methods, 222
    of large zones, 179
    of steroid acetates, 186
    partial, of zones, 218
L-Epicatechin, 25
Equilibration, 60, 66
Equilibrium, 68
    between phases, 6
Esterification, 283, 291
Esters, 38, 309, 350
    hydrolysis of, 268
Estimation, specificity of, 318
[^{35}S] Ethanedithiol, 316
Ethanol, 31, 84, 348
Ethers, 349
    free energy of hydration, 12
    formation of, 367
Ethyl acetate, 67, 350
Ethyl methyl ketone, 351
Ethylene dichloride, 67, 349
Ethylene dithiol, 215
Ethylene glycol, 13, 31, 84, 163
[^{3}H] Ethylene glycol, 316
Ethylene ketals, 215, 326
    formation of, 363
    hydrolysis of, 360
Ethylenediaminetetra-acetic acid, 322, 354
Extraction, Soxhlet, 185
    addition of sodium hydroxide to plasma, 194
    complicated methods of, 195
    emulsion-free, 198
    neutral steroids from blood plasma, newer methods of, 195
    of bile acids, 196
    of conjugates from plasma, 197
    of oestrogens, 196
    of steroid conjugates, 196
    of sweat, faeces, and urine, 196, 198
Extracts
    by continuous ether extraction, 198
    concentration of fatty, 144
    of animal tissues, 192

Extracts—*contd.*
  of blood and plasma, 193
  of non-polar steroids, 192
  of post-prandial samples
    of whole blood, 194
  preparation of, 191
  typical methods for, 193

Fading of colours, 231
Fats, 63, 143
Fatty acids, 60
  dinitrophenylhydrazides, 21, 65
  hydroxamates, 22
Ferric chloride, 382
Flavones, flavonoids, etc., 24, 281
Fluorescein, 382
Fluorescence, 69, 340
  NaOH reaction, 224, 266
  quenching of, 179
  reagents, acid, 381
  reagents, alkali, 379
Fluorescence reactions, 297, 373
  with phosphoric acid, 299
Fluorimetric methods, 343
Fluorimetry,
  direct, on paper, 319
  in potassium t-butoxide, 321
  specificity of, 203
  using NaOH, 344
Fluorodinitrobenzene, 286
Folin-Ciocalteu reagent, 381
Forceps,
  for handling paper chromatograms, 168
Forces,
  capillary, xiv
  gravitational, xiv
Formamide, 17, 31, 60, 84, 162, 163, 350
Formic acid, 213
Fraction collector, 153
  gravity-operated, 157
  linear, 157
Frames, 171
  for strip chromatograms, 168, 169
  Tufnol, 171
  wood, 171

Front, xiv
  solvent, 4, 69
Furfural, 31
Furfuryl alcohol, 31

Glass, 61
Glassware, 355
  chromic acid adsorbed by, 355
  metallic impurities in, 350
  sources of danger in, 355
Glucose polymers, 26
β-Glucuronidase, 337
Glycerol, 13
Glycols, 17, 350
Girard hydrazones, 362
Girard's reagent, 199, 277, 337, 362
Group, 10, 44, 87, 88, 283
  acetoxyl, 369
  6β-acetoxyl, 369
  11α-acetoxyl, 89, 104
  16α-acetoxyl, 39
  17β-acetoxyl, 104
  21-acetoxyl, 89, 103
  acid, 209
  aldehyde, 270
  alkoxyl, 288, 290
  amino, 270, 286
  complex, 84, 88
  basic nitrogen, 289
  carbonyl, 9, 10, 283, 288
  carboxyl, 9, 10, 209, 270, 283, 289
  chromophore, 214
  complete elimination of, 300
  complex functional, 95, 290
  complex functions, 387
  conjugated sugar, 209
  diol, 95
  cis-diol, 94
  trans-diol, 94
  1,2-diol, 94, 122
  6β, 11β-diol, 119
  11β, 12α-diol, 121
  16, 17-diol, 122
  20, 21-diol, 294
  double bond, 111, 289
    characterization of, 287
    conjugated, 290, 388

SUBJECT INDEX   425

Group—*contd.*
9(11)-double bond, 101
epimeric acetoxyl, 209
epoxide, 286
ester, 209
ester sulphate, 289
α-fluoro-carboxyl, 289
9α-fluorine, 91, 92
halogen, 286, 388
hemiacetal, 98, 100, 294
hydrocarbon, 288
hydroxyl, 18, 37, 85, 96, 97, 119, 124, 209, 387
  and related groups, 283, 288
  axial, 9, 91, 345
  characterization of groups relating to, 285
  close, 94
  epimeric, 9, 25, 96
  equatorial, 9, 88, 91, 345
  general methods of characterizing, 284
  hindered, 91
  isolated, 96
  least hindered, 88
  meta, 25
  ortho, 25
  para, 25
  phenolic, 119, 286
  primary, 9, 10, 94
  secondary, 9, 10, 308
  steric hindrance of, 291
  tertiary, 92
2α-hydroxyl, 87, 98
Δ^5-3β-hydroxyl, 105, 109
5α-hydroxyl, 303
5α(H)-3α-hydroxyl, 104, 109
5α(H)-3β-hydroxyl, 101, 104
5β(H)-3α-hydroxyl, 104, 105, 109, 116, 130
6β-hydroxyl, 87, 89, 90, 92, 97, 100, 107, 108
7-hydroxyl, 91
7α-hydroxyl, 87, 98, 107
11α-hydroxyl, 87, 89, 90, 91, 98, 99, 103, 104, 107, 108

Group—*contd.*
11β-hydroxyl, 87, 89–91, 94, 97, 99, 100, 103, 107, 108, 130
12α-hydroxyl, 121
12β-hydroxyl, 122
14α-hydroxyl, 89, 92
15α-hydroxyl, 87, 89, 98, 107
15β-hydroxyl, 87, 89, 98, 107
16α-hydroxyl, 89, 121, 309, 364
16β-hydroxyl, 309, 364
17-hydroxyl, 280
17α-hydroxyl, 87, 89, 94, 97, 99, 108, 294
17β-hydroxyl, 104–106
18-hydroxyl, 94
19-hydroxyl, 87, 91, 98, 104, 107, 120
20-hydroxyl, 273, 305
20α-hydroxyl, 106
20β-hydroxyl, 106
21-hydroxyl, 87, 89, 90, 94, 97, 103
isotopically labelled, 214
α-ketol, 297, 388
ketone, 85, 96–123, 209, 289, 388
  terminal, 100
  αβ-unsaturated, 268
Δ^4-3-ketone, 105, 109, 116, 130
Δ^8-11-ketone, 276
3-ketone, 327
5α(H)-3-ketone, 122, 109, 130
5β(H)-3-ketone, 105, 109
6-ketone, 87, 89, 99
7-ketone, 303
11-ketone, 87, 89–91, 97, 99, 100, 103, 107, 108, 130
17-ketone, 94, 102, 105, 106, 108, 109, 327
20-ketone, 106
lactone, 209, 290
methyl, 12, 91, 93, 111, 128
  angular, 121
  at C-20, 122
2α-methyl, 273
13-methyl, 279
methylene, 12, 95, 111
  active, 297
  ΔR_{Mg} of, 21

Group—contd.
  Δ⁴-3-oxo-octahydronaphthalene, 297
  phenol, 209, 289
  phenolic 3-hydroxyl, 105
  primary amine, 286
  secondary amine, 270, 289
  C_{19}-skeleton, 108
  strong acid, 289
  tertiary amine, 270, 289
  unconjugated double bonds, 286

Heat,
  adsorption due to, 140
  application of, 231
  radiant, 233
  sources of, 175
Helix pomatia,
  succus entericus of, 337
n-Heptanol, 62
High-voltage electrophoresis, 329
Homologous series, 12, 19, 22, 23, 25
Hydrazones, formation of, 361
Hydrocarbon, 127, 347
  halogenated, 48, 349
  higher boiling, 67
  polycyclic, 281
  polynuclear, 304
  skeleton, 10, 281, 288, 300
  unsaturated, 290
Hydrochloric acid, 368, 369
Hydrogen bonds, 83, 121, 123, 124, 128, 131, 213, 278, 289
  steric limitations of, 17
Hydrolysis,
  of ethylene ketals, 360
  of ketonides, 361
  of labile steroid sulphates, 369
  with succus entericus of Helix pomatia, 370
Hydrolysis of esters, 359
  11β-acetoxyl group, 360
  using digestive juice of Helix pomatia, 360
  using perchloric acid, 360
  using potassium carbonate, 360
  using sodium carbonate, 360

Hydrolysis of esters—contd.
  using sodium hydroxide, 360
Hydroxylation 371

Ideal,
  conditions, 6, 31, 65
  behaviour, 32
  solution, 7
Impregnation of paper, 31, 140
  technique of, 31, 163, 164, 165
    modified, 141, 166
  variations in, 74
Impurities, 42, 264
  metallic, 4
Infrared, see "Spectroscopy"
Interaction, molecular, 11, 13, 14, 18
  additive treatment of, 15
  complex, 16
  dilution, 14
  energy of, 12, 17
  electronic, 115
  hydrocarbon–hydrocarbon, 14
  liquid–solute, 19
  non-specific, 127
  orientated, 21
  polar, 128
  random, 21
  short range, 31
  solvent–solute, 11, 115, 118, 125
  statistical treatment of, 16
  stereochemical effect and, 44
  steric, 115
  steroid–polysaccharide, 130
  types of, 10
  van der Waals, 64
  with bulk phase, 41
  with cellulose, 15
  with polar solvents, 17
Iodine, 379
p-Iodobenzenesulphonic acid, 206
  anhydride of, 252
p-Iodobenzenesulphonyl chloride ("pipsyl" chloride), 251
β-Ionones, 207
Isomerization, 40
Isonicotinoyl hydrazide, 378
Isopropyl ether, 30, 332, 333, 335

Ion exchange, xv, 4, 15
Isopropanol, 84, 348
Isotherm, adsorption, 2, 10

Kerosene, odourless, 64, 165, 327
Ketalation, 283, 316
Ketols, 38
Ketones, 11, 351
   free energy of hydration of, 12
Ketonides,
   formation of, 367
   hydrolysis of, 361
Kieselguhr (Celite), 4, 5, 353
Kober reagent, 277
   reaction, 314

Labelling,
   double isotopic, 217
Lactones, 38
Light, scattering of, 235
Ligroin, 66
Liquids, 17
   non-polar, 17
   short range order, 17, 118, 119
Logistics, of methods, 220
   of quantitative estimatations, 246
Lophenol, 288
Lutidine, 351

Magnesium hydroxide, 37
Magnesium silicate, 37, 39, 352
Manganese dioxide, 272, 274, 365
Marker substances, 206
   dyes, 338
Melting point, 42
Methanol, 17, 31, 60, 84
   aqueous, 18, 167
Method,
   semi-automatic, 223
Methyl orange, 180, 182
Methylation, 204, 314
Methylcyclohexane, 117, 128
Methylene chloride, 67, 349
Microchemical reactions, 358
Modifying reaction, 253, 326

Modifying step, 204, 209, 210, 211, 214, 216, 218
   combination of two, 210
Mole fraction, 27
Molecules,
   with large dipole moment, xvi
   large or asymmetrical, 17
Monomolecular layers, 41

Naphthalene, 328
Neotetrazolium, 225
Ninhydrin, 229
Nitromethane, 60, 62
p-Nitrophenylhydrazine, 377
p-Nitrophenylhydrazones, 215, 362
Nomenclature, (*see* structure and nomenclature of the steroids)
Non-ideal conditions, 70

Oils, 31
Oppenauer reactions, 364
Organic acids, 351
Osmium tetroxide, 371
Overloading, 68
Overrunning, 60, 66, 69, 219

Packer, Martin, 136, 146, 147
Paint-resins, 45
Paper, 43, 61, 159
   acetylated, 4, 63, 162, 304
   carboxyl groups in, 32
   chemical cleanliness of, 159
   cross washing of, 161
   curtain of, 159
   dirt from, 222
   elution from, 182
      failures of, 182
   esterified with stearatochromyl chloride, 162
   ether soluble matter in, 158
   glass, 186
   height of one theoretical plate on, 219
   hydration of, 73
   hydrophobic material in, 69

Paper—contd.
  impregnated with alumina, 162
  impregnation with silicic acid, 162
  impregnation with stationary phases of, 162
  longer strips of, 342
  oils in, 160
  pH of, 158
  preparation of, 158
  porosity of, 31
  rubberized, 4
  sheets, 354
    metallic impurities in, 354
  siliconed, 61, 162
  Soxhlet extraction, 160
  special sizes and shapes of, 159
  strong acids and bases on, 158
  thick sheets of, 66, 189, 219
  unwashed, 69
  washing procedures for, 160, 162
Paper chromatogram,
  curtain pattern of, 156
  methods of detection on, 372
Paraffins,
  high-boiling, 164
  long-chain, 64
Partition,
  ideal, 4
  coefficient, 4, 8, 9, 11, 15, 19, 27, 29, 44, 74, 83, 118
  systems, 7, 18
    liquid–liquid, xvi
    liquid–gas, xvi
Pentobarbitone, 213
Peracids, 286
Periodate, 216, 260, 271, 292
  oxidations with, 364
Periodate-Zimmermann reaction, 210, 231
Peroxides, 349
  in solvents, 38
  on silica gel, 40
  strongly adsorbed on alumina, 38
Petroleum,
  high boiling fractions of, 128
  light, 128
  low boiling fractions of, 128

Petroleum jelly, 162
Phase,
  bulk, 3
  composition, 124
  changes in, 116
  distribution of t-butanol between, 123
  gel, 3
  immiscibilities of two, 75
  inhomogeneities in, 6
  intermiscibility of, 122–124, 127
  methanol content of, 127
  mobile, xiv, xv, 101, 129
    velocity of, 173
    water content of, 124, 125
  moving, 1
  relative cross-sectional area of, 4
  reversed, 130, 131, 167
  stationary, xv, 1, 5, 44, 83, 129, 137
    application by spraying of, 165
    evaporation of, 68, 72
    maldistribution of, 191
    non-polar, 104
    velocity of, 6
    water content of, 26, 115
  surfaces of solid, 2
  viscous, 101
  water content of phases, 125
Phenanthrene, 16, 117, 127, 300, 328
Phenols, 119, 373
Phenolsulphonic acid, 382
Phenyl cellosolve, 163, 350
Phosphates, 4
Phosphomolybdic acid, 339, 340, 356, 378
  using n-propanol as vehicle, 299
Phosphoric acid, 381
Phosphorus oxychloride, 369
Porphyrins, 27
Pipettes, 133, 140
  micrometer syringe, 134
  self-filling, 134
Plasma, extracts of, 63
Polar, xvi
  groups, 14, 37

SUBJECT INDEX 429

Potassium t-butoxide, 381
Potassium hydroxide, 232, 356
  ethanolic, 273
Potassium permanganate, 371
Precipitation, 1
Propane-1,2-diol, 94
Propane-1,3-diol, 94
n-Propanol, 31, 64, 348
Propylene glycol, 18, 60, 66, 84, 115, 162, 163, 350
  evaporation of, 66
  paper impregnated with, 30
Proteins, 234
Purification of reagents, 347, 356
  alcohols, 348
  alumina, 352
  bases, 351
  cellosolves, 350
  cellulose, 354
  m-dinitrobenzene, 356
  esters, 350
  formamides, 350
  glassware, 355
  glycols, 350
  halogenated hydrocarbons, 349
  hydrocarbons, 347
  ketones, 351
  kieselguhr (Celite), 353
  magnesium silicate (Florisil), 352
  organic acids, 351
  paper sheets, 354
  phosphomolybdic acid, 356
  pyridine, 351
  silica gel, 353
  sodium and potassium hydroxide, 356
  sulphuric acid, 356
  tetralin, 348
  tetrazolium salts, 357
Purines, 252
Pyran ring, 25
Pyridine, 286, 291, 351

Quantitative, estimation, 201 *et seq.*
Quantity of substance, 2

R value, xvi
  relations between the different, 74
  sources of error in determining, 68
R_F value, xvi, 5, 65, 68, 74, 180, 263
  calculations with, 383
  derivation of, 7
  error in, 75
  high, 70
  variations of, 73
R_M value, xvi, 8, 9, 11, 19, 44, 74, 84, 204
  calculation with, 383
  derivation of, 7–9
  exception to, 116
  sources of error in, 386
ΔR_M, xvi, 77, 83
  approximation of, 79
  different types of, 84, 86
ΔR_{Mg} value, xvi, 85, 93–97, 99, 111, 112, 118, 122, 123, 204, 209, 210, 214, 281, 289, 386
  deficits, 121
  for different series of root substances, 89, 90, 91
  for more reproducible systems, 108, 109
  for substitution with 9α fluorine, 92
  of non-polar parts, 109
  selective changes in, 117
  variations of, 90, 115
ΔR_{Mr}, xvi, 85, 96, 102, 103, 104, 105, 106, 112, 116, 204, 210, 215, 218, 275, 276, 279, 281, 283, 284, 286, 290, 300, 304, 305
ΔR_{Ms}, xvii, 85, 102, 107, 108, 109, 112, 204, 210, 214, 267, 270, 284, 288, 289, 294, 304, 305
  large effects, 307
  measurement of, 307
R_S value, 74, 75, 77
R_T, 79
Radiation, 133
  hot-spots caused by, 68, 174
Radioactive isotopes, 247, 250
  assay of tritium-labelled acetoxyl groups, 321

Radioactive isotopes—*contd.*
  automatic scanners using, 254
  autoradiography using tritium, 382
  detection of, 382
Raoult's law, and deviations from ideality, 27
Reaction rates, 306, 310
  of dehydration, 302
  of esterification, 302
  measurement of, 301
  of oxidation, 302
  of saponification, 302
Reagents,
  application of, by dipping, 224
  aqueous, 228
  background with, 232
  bulk movement of, 178
  mechanical application of, 225
  mechanical applicator for, 227
  purification, *see* Purification of reagents
  radioactively labelled, 251
  suitable for scanning methods, 226
  treatment of paper strips with, 224
  volume of, taken up by the paper, 226
Recovery,
  internal control on, 253, 254, 322
Reduction,
  catalytic hydrogenation, 366
  extensive, 366
  of thioketals, 300
  reductive elimination with zinc, 366
  with metal hydrides, 365
    selective, 365
      selective reduction of 3-oxo groups, 366
  with zinc in acetic acid, 368
  Wolff-Kishner, 300
Reference substance, xvi
Resin, xv
Retention volume, 25, 68, 74
  true, 74, 81
Reversed phase, xvii, 62
Rubber, 5, 63, 162
  vulcanized powder of, 61
Run-off, xv

Running-in, 142
Running-up, 139, 140
  apparatus for, 168, 169
  purification by, 143
  technique, 141

Salicyloyl hydrazide, 378
Salts, 6, 31
Sand, 61
Saponifications, 291
Scanner, 236
Scanning, 82, 222, 374
  across the strip, 237
  automatic apparatus for, 240, 241, 246
  direct fluorimetry, 224
  electronic recording apparatus for, 240
  geometric and planimetric methods of, 238
  integrating recorders for, 243
  mechanical, 244
  oiling strips before, 235
  optical requirements for, 234
  variation of background in, 238
Semicarbazones, 362
Separation, 1, 60, 65
  and resolving power, 80
  crude, 39
  difficult, 60
  ion-exchange, 81
  preparative scale, 42
  sub-fractionation, 173
Silica gel, 5, 37, 39, 43, 60, 145, 192, 198, 353
  adsorbs water and some polar solvents, 40
  hydrated, 4
Silicic acid, 40, 61
Silicone, 63
Silver nitrate, alkaline, 374
Siphons, 154
  faults of, 154
Sodium bismuthate, 210, 262, 345
  oxidations with, 365
Sodium borohydride, 276
Sodium hydroxide, 340, 356

SUBJECT INDEX

Sodium nitroprusside, 382
Solubility, 11, 29
Solute,
  forming two species, 32
  acid, 32
  ionizing, 32
Solvent,
  acetic acid, 107
  acetic acid in water, xviii, 109
  acidic and basic, 36
  associated, 18
  at the edges of thick sheets, excess flow of, 189
  benzene, xvii
  bubbles of air in, 145
  chromatographic, xv
  decahydronaphthalene, xvii
  degassing of, 145
  for eluting and "running up" steroids 184, 186
  formamide, xvii
  front, 4
    unstable conditions at, 70
  hydrophobic, 4
  hydroxylic, 83
  lattice, 123
  light petroleum, xvii
  methanol in water, xvii
  miscibility of, 31
  mixtures, 40
  non-polar, 17
  on alumina and effect of peroxide impurity, 38
  polar, xvi, 18
    comparison of benzene and toluene xvi
    lattice formed by, 118
  propylene glycol, xvii
  removing the, 229
  sequence for alumina, 39
  sequence of, 38
  suitable for scanning methods, 226
  toluene, xvii
  water-like, 18, 83
  xylene, xvii
Solvent system, 10, 44, 83, 87, 89, 98, 171

Solvent system—*contd.*
  absolute methanol—light petroleum, 127
  acetic acid, 116
  acid or alkaline, 151
  aqueous acetone, 100
  atypical, 96, 98, 99, 107
  B/50, 87, 88–90, 92, 93, 97, 98, 103, 105, 106, 111
  based on esters, 300
  benzene-methanol-water, 85
  BL_1, 105, 106
  butane-1, 3-diol, 100
  s-butanol, 127
  t-butanol, 96, 98, 107
  chloroform–formamide ($CHCl_3/F$), 85, 87, 88
  collidine–water, 127
  D/85, 108, 109, 111
  D/A90, 108, 109, 111
  dioxan–methanol, 107
  dioxan–methanol–water, 10, 98, 99
  DT21/A70, 111
  DT21/85, 111
  DX21/85, 108
  DX21/A90, 108
  E_4, 97, 98
  E_2B, 97, 98, 123
  L/80, 89, 90
  L/85, 87, 88, 106, 111
  L/96, 111
  LB21/80, 87–90, 93, 103
  LD11/85, 105
  LDT111/85, 103
  light petroleum–toluene–methanol–water, 128
  LT11/70, 89, 90, 93, 103, 106
  methylcyclohexane–dioxan–methanol 99
  pyridine–water–hydrocarbon, 100
  tetralin–90% aqueous methanol, 129
  toluene-ethyl acetate–methanol–water, 125
  typical, 85, 98, 107
    capacity of, 66
    volatile, 60

Solvent–solute complexes, 31
Solvolysis, 331
  of steroid sulphates, 370
Specific enzymic reactions, 283
Specificity, 205, 207
  chemical, 203
Spectroscopic properties,
  infrared, 297
  ultraviolet, in sulphuric acid, 261, 297
Spectroscopy,
  in ethanolic alkali, 274
  infrared, 36, 207, 264, 278, 297
  ultraviolet, 297
  ultraviolet in ethanolic potassium hydroxide, 274, 299
  ultraviolet, in sulphuric acid, 261, 297, 299
Spots, double, 4
  annular, 139
Standard solution,
  evaporation of, 134
  in measuring cylinders, 134
  in volumetric flask, 134
  of steroids, 133
Starch, 4
Stearatochromyl chloride, 63
Steric,
  considerations, 95
  dilution, 13
  factors, xvi, 10, 11, 13, 15, 16, 25, 44, 119, 125
  hindrance, 24
Steroids,
  acetates of, 260, 340, 342
  acetonides, 291
  acid, 27, 64
  adrenal, 37, 46, 50, 51, 53–55
    less polar, 48
  adrenocortical, monoacetates of, xviii,
  adrenosterone, 266
  adsorbed on paper, 4
  aetiocholane-11, 17-dione, 130
  aetiocholane-17-one, 130
  aetiocholanolone, 108, 130, 216, 304, 336
    sulphate, 329

Steroids—*contd.*
  aldosterone, xviii, 50–52, 54, 66, 100, 108, 116, 207, 248, 254, 261, 291, 293, 294, 296, 317
    diacetate, 49, 99, 319
    [^{14}C] diacetate, 322
    esters of, 52
    estimation of, 202
    in blood, 205
    in urine, 317
    paper chromatography of, 323
    R_F and R_M values of, 320
    tritiated, 245
    urinary metabolites of, 52
  alkaloids, 64
    holarrhena, 56
    solanum, 56
    veratrum, 56
  allocortol, 341
  allopregnane-3α,20α-diol, 344
  allopregnane-3β, 20β-diol, 344
  allopregnane-3, 20-dione, 344
  allopregnanetriol, 340
  allotetrahydrocortisol, 218, 336, 340, 344, 345
    diacetate, 345
  amines, 45, 54
  ammonium salts of the sulphates of, 329
  analogues of, 344
  androstane, 129
  androstanediol, 317, 341
  androstanedione, 345
  androstan-3β-ol, 108
  androstanetriol, 341
  androstenedione, xviii, 89, 101, 108, 116, 130
  androstenetriol, 341
  androsterone, 35, 66, 304, 327, 336, 341
    sulphate, 329
  3β-benzoyloxycholest-5-en-7-one, 298
  bile acids, 34, 46, 48, 49, 53, 56, 58, 99, 122, 249
    conjugates, 53, 54, 56, 58
    ethyl esters of, 47
  ^{14}C-labelled, 248

SUBJECT INDEX

Steroids—*contd.*
  carboxylic acids, 53, 56
  cardenolides, 51, 54
  17-iso-cardenolides, 305
  cardiac aglycones, xviii, 51–53, 55, 299
  cardiac glycosides, xviii, 52, 53, 55, 62
  cardiac poisons, acetates, 52
  cholenic acid, 30, 249
  cholesterol, 35, 46, 63
    esters, 143
  cholest-4-en-3-one, 328
  cholic acid, 35
  configuration, xix
  conjugates, 32, 180, 214, 328, 330, 332
  corticosteroids, 208, 248
    labelled, 248
  corticosterone, 294, 298, 338
  cortisol, xvii, 50, 62, 106, 127, 254, 264, 296, 298, 311, 313, 336
    degradation of, 267
    in human blood, 311, 313
  cortisone, 62, 106, 296, 336
  cortol, xviii, 55, 56, 336, 340, 341
  β-cortol, 341
  cortolones, 55, 337, 340, 341
  β-cortolone, 341
  dehydrocorticosterone, 339
  dehydroepiandrosterone, 66, 295, 336, 341
    sulphate, 329, 334
  deoxycholic acid, 29
  deoxycorticosterone, 35, 89, 208, 341
  diacetates, 114
  4,6-dienones, 297
  digoxigenin (aglycone), 35
  digoxin (glycoside), 35
  20-dihydrocortisol, 340
  20-dihydrocortisone, 340
  17α,19-dihydroxy-11-deoxycorticosterone, 94
  diketones, 100
  Δ^4-3,6-diketones, 273
  1,2-diols, epimeric, 121
  2,3-diols, 122

Steroids—*contd.*
  5α, 6β-diols, 369
  6β, 11β-diols, 120
  16, 17-diols, 119, 122
  16α, 17α-diols, 371
  Δ^{14}-16α,17α-diols, 371
  diosgenin (sapogenin), 35
  11,17-dioxosteroids, 267
  double labelling of, 248, 252, 254
  5-ene-3β-ol structures, 294
  epiandrosterone sulphate, 334
  5-epimeric 3-oxo-, 43
  epi-oestriol, 121
  16-epi-oestriol, 119, 122
    polar, 55
  epoxides, 43
  equilin, 35
  ergosterol, 35, 64
  $\Delta^{8(14)}$-ergosterol, 288
  esters, 48, 50
    of non-polar, xviii, 45
  etianic acid, 308
  14-iso-etianic acid, 308
  extraction of, 192
  9α-fluoroallotetrahydrocortisol, 278, 345
  9α-fluorocortisol, 275, 298, 345
  9α-fluorocortisone, 29
  9α-fluoro-11-ketones, 276
  9α-fluorotetrahydrocortisol, 345
  glucuronosides, 43, 57–60, 329, 335
    5α/5β epimers of, 335
    less polar, 330
  glucuronosidates of, 378
  16, 17-glycols of, 343
  glycosides, 53
    scilla and bufo, 55
  hydrocarbon skeleton, 111
    some features of, 95
  hydrolysis by alumina of steroid 3-formates, 39
  hydrophobic, xviii
  11β-hydroxyaetiocholanolone, 35, 117, 128, 216, 217, 336, 340
    3-acetate of, 275
  11β-hydroxyandrostenedione, 206, 265, 341

Steroids—*contd.*
  11β-hydroxyandrosterone, 117, 128, 336, 340
  17-hydroxycorticosteroids, 207
  6β-hydroxycorticosterone, 119
  19-hydroxycorticosterone, 94
  6β-hydroxycortisol, 298
  17-hydroxy-11-deoxycorticosterone, 294
  6α-hydroxy-Δ^4-3-ketones, 273
  6β-hydroxy-Δ^4-3-ketones, 297, 369
  11β-hydroxyl group in the, 268, 276, 298
  18-hydroxyoestradiol, 94
  16α-hydroxyoestrone, 199
  18-hydroxyoestrone, 276, 279, 309
  2α-hydroxy-Δ^4-3-oxosteroids, 297
  5α-hydroxy-3-oxosteroids, 369
  7α-hydroxy-Δ^4-3-oxosteroids, 297
  11β-hydroxy-17-oxosteroids, 298
  17α-hydroxypregnanolone, 113, 340, 341, 342
  11β-hydroxypregnanetriol, 341
  16α-hydroxypregnanolone, 341
  11β-hydroxyprogesterone, 341
  17α-hydroxyprogesterone, 90, 250, 341
  3-hydroxysteroids, 99
  Δ^5-3β-hydroxysteroids, 92, 378
  5β(H)-3α-hydroxysteroids, 128
  7α-hydroxysteroids, 92
  15-hydroxysteroids, epimeric, 309
  19-hydroxysteroids, 94
  2α-hydroxytestosterone, 297
  11β-hydroxytestosterone, 341, 342, 369
  iodine complexes with, 372
  α-ketolic, 260
  3-ketones, 99, 100, 101, 116
  Δ^4-3-ketones, 343
  5α-3-ketones, 299
  5β-3-ketones, 299
  11-ketones, 298
  15-ketones, 309
  16-ketones, 309
  17-ketones, xvii, 47, 48, 100, 308
    derivatives of, formed with alkaline *m*-dinitrobenzene, 216

Steroids—*contd.*
  17-ketones—*contd.*
    glucuronosides, 43, 331
  20-ketones, 94, 100
    urinary, 37, 207
  labelled, 42
  2α-methyl, 298
  2α-methyladrenosterone, 273
  2α-methylcortisol, 271, 274
  2α-methyl-11β-hydroxyandrostenedione, 273
  microchemical reactions for, 358
  monoacetates of, 114
  monosubstituted, 101
  non-alcoholic ketones, 345
  non-polar, 45, 48, 66, 327
    sulphates of, 329
  19-nor-steroids, 179
  oestrogens, 34, 49, 52–55, 204, 208, 312, 315
    and pregnanediol, 37
    diazo dyes, 47
    in human urine, 314
    less polar, 47
    metabolites, 312, 314
    18-nor-oestradiol, 280
    18-nor-oestrone, 279
    oestradiol (17β), 35, 314
      methyl ether, 314
    oestriol, 35, 119
      methyl ether, 314
    oestrone, 314
      methyl ether, 314
  Δ^5-3β-ols, 295
  oxidation of side chains of, 292
  7-oxocholesterol, 328
  7-oxo-Δ^5-3β-hydroxysteroids, 378
  11-oxoaetiocholanolone, 336, 340
  11-oxoaetiocholanolone 3-acetate, 275
  11-oxopregnanetriol, 336, 341
  Δ^{16}-20-oxosteroids, 371
  11-oxygenated, xviii
  "pipsyl" derivatives, 251
  polar, xviii, 18, 129
    more polar, 60
    very non-polar, 64

SUBJECT INDEX 435

Steroids—*contd.*
  polar—*contd.*
    very polar, xviii
  polyketones, 268, 270
  prednisolone, 218, 344, 345
    monoacetate of, 345
  prednisone metabolites, 52
  pregnane, 129
  pregnanediol, 205, 327, 339
  pregnane-3α, 20α-diol, 344
  pregnanedione, 345
  pregnanetriol, 51, 216, 217, 336
  pregnane-3α, 16α, 20α,-triol, 341
  pregnanetriolone, 341
  progesterone, 35, 46, 47, 87, 89, 99, 100, 108, 130, 217, 248, 298, 299, 316–318, 344, 345
    in blood and urine, 316
  radioactive, 249
  sapogenins, 47, 49–53, 56, 84, 122,131
    11,12-dihydroxytigogenins, 121
    diols, 95
    11,12-ketols, 95
    rockogenin, 122
  saponins, 56, 58
  saturated 3,6-diketones, 297
  saturated 3-ketones, 294
  D–seco–16, 17–dialdehydes, 342
  seco–dicarboxylic acids, 102
  shape, 34
  side chain, 291
  solubilities of, 34
    low, 61
  stability of, 34
  standard solutions of, 133
  steric complexity of, and long range steric effects, 19
  sterols, xviii, 45, 46, 129, 328
    and their esters, 45
  stigmasterol, 35
  $\Delta^{8(14)}$-stigmastenol, 288
  streaking, those prone to, 67
  structure and nomenclature of, xviii
  substituents, 34
  sulphates, 32, 43, 57–60, 329, 331
  testosterone, 35, 89, 108, 248, 299, 324, 325, 341, 369

Steroids—*contd.*
  testosterone, in blood or urine, 324
  cis-testosterone, 369
  testosterone 3-ketal, 326
  tetrahydrocortisol, 218, 336, 340, 345
  tetrahydrocortisone, 218, 336, 340, 345
  triamcinolone, 56, 291
  tritiated, 248
  αβ-unsaturated ketone group in, 268, 372
  αβ-unsaturated lactones, 373
  rearrangements on alumina, 34
  urinary, 336, 338
    conjugates, 355
    polar, 343
Stopcocks, 148
Stoppers,
  ribbed plastic, 134
Streaking, 67, 131, 132
Streptovaricin, 213
Structure,
  chromatographic evidence of, 259
    criticism of, 260
    relevance of circumstantial evidence in, 260, 261, 280
  classical method of establishing the, 256, 280
  determination of,
    examples of, 264
    general procedure for, 262
    indistinguishable melting points, 258
    infrared spectra in, 257
    melting points, 257
    microchemical reactions in, 260
  proof of, 256
  statistical criteria of, 259
  substances of large molecular weight, 259
Structure and nomenclature of steroids,
  abbreviated nomenclature, xvii
  asymmetric substituents, xix
  double bonds, xix
  examples of, xix–xx
  trivial names used, xxii

Substituent,
  axial, 282
  effect of position and orientation of, 9
  equatorial, 282
  groups, complex, 67
  halogen, xix
  hydrocarbon, xix
  neighbouring, 88
  nomenclature, xix, xx, xxi
  nuclear, 308
Sucrose, 37
Sugars, 26, 60, 124, 129
Sulphatase, 337
Sulphur dioxide, 230
Sulphuric acid, 356, 361, 381
Supporting material, xv, 16
  relative amounts of, 137
Surface tension, 68, 70, 72
Suspension, 136
Swelling of cellulose by liquids, 31
System, 18, 85, 209
  acetic acid, 99, 116, 207, 294, 306, 327, 346
  acetic acid–t-butanol, 32
  acid, to neutral or basic solvent, 23
  adsorption, 36, 328
    displacement in, 42
  amines in gas-liquid, 22
  anhydrous, 289
  aqueous, 138
  aqueous methanol, 115
    addition of ammonia to, 214
  atypical, 83, 213
  based on dioxan–acetic acid, 288
  based on formic and acetic acids, 289
  butane-1,3-diol, 294
  n-butanol–ammonia, 32
  t-butanol, 97, 122, 207, 209, 288, 289, 306, 331
  capacity of, 31
  carbitol, 327
  chromatographic, xv
  collidine–water, 30
  dioxan–methanol, 107
  dioxan, 209

System—*contd.*
  effect of concentration on, 2
  formamide, 305, 320
  gas–liquid, 2, 290, 301, 328
  HCl–containing, 32
  heptane–70% acetic acid, 29
  heptane–methanol, 20
  impregnated paper, 31, 84
    capacity of, 66
  isopropyl ether, 333
  liquid–liquid, 2, 28
    counter-current, 42
  mechanism of adsorption, 41
  methanol, 305
  methyl cellosolve, 327
  miscible or one-phase, 3, 63
  nitromethane, 209
  odourless kerosene, 327
  partition, xvii, 2
  phenyl cellosolve, 327
  phenyl cellosolve–heptane, 73
  propylene glycol, 115
  propylene glycol–t-butanol, 294
  reversed phase, xvii, xviii, 61, 63, 165, 316, 327
    on untreated paper, 163
    tetralin—90% methanol, 21
  solvent, *see* solvent systems.
  straight, xvii
  typical, 83
  water content of, 116
  with intermiscibility, 30
  with very considerable intermiscibility, 30

"Tailing", 41, 125
  Tank, *see* Chromatograph tank
Taurine, 289
Temperature, 6, 66, 68, 123–125, 127
  control, 174
  effect of, 31, 92
  elevated, 31, 60, 175
  gradients, 68, 69
  lower, 73
  reduced, 31
  uneven distribution of, 175

Temperature coefficient of expansion of ethanol, 133
Terpenes, 39, 45
Tetraethylammonium hydroxide, 381
Tetralin, 17, 101, 117, 164, 165, 167, 347
Tetrazolium salts, 357
Theoretical plate, 6
Thioketals, 215
Thionyl chloride, 369
Thiosemicarbazones, 215, 217, 345, 362
p-Toluenesulphonic acid, 382
Topanol A (2,4-dimethyl-6-t-butylphenol), 334, 349
Transfer operations, 133
Trichloroacetic acid,
 fluorescence with, 216, 381
Triphenyltetrazolium chloride, 357, 374
Triterpenes, long range steric effects in, 19
Trough, xiv
 for running many separate sheets, 188, 189
Tubing,
 precision bore, for chromatographic columns, 148

Ultraviolet light, 178
Urine, 51

Van der Waals forces, 10
Vaseline (petroleum jelly), 64
Vicinal effects, 387
Vitamin D group, 45
Volume,
 interstitial, xv
 molar, xvi
 retention, xvii
Volumetric error, 133

V_r, 68, 74
 log V_r, 74

Water, 12, 17, 31, 62, 83, 84, 119, 123, 229
 as a stationary phase, 61
 bound, 3
 interaction energy of, 18
 lattice, 18
 stationary phases with low or zero content of, 30
Waterlogging, 69, 73, 142

Xylene, 17, 62, 101, 116, 117

Zeolites, xv
Zimmermann reaction, 207, 216, 224–226, 232, 234, 240, 266, 297, 324, 339, 345
 common faults in, 374
 lactone group in, 299
 11-oxygen function in, 298
Zimmermann reagent, 344, 348, 356
Zone, 7, 72, 224
 diffuseness and overlap of, 72
 elution of a large, 179
 fluorescent, 372
 irregularities of, 178
 location of, 178, 183, 250
 mutual contamination of two, 80–82
 partial elution of, 218
 real and apparent distribution of, 223
 sharpening of, 69
 specific activity in different regions of, 250
 spreading of, 249
 starting,
  shape of, 138
  quantitative estimation of, 138
 width of, 80, 81